Evolutionary Change
and Heterochrony

Evolutionary Change and Heterochrony

edited by
Kenneth J. McNamara
Department of Earth and Planetary Sciences,
Western Australian Museum

JOHN WILEY & SONS
Chichester · New York · Brisbane · Toronto · Singapore

ACV 6636

Published in 1995 by John Wiley & Sons Ltd,
Baffins Lane, Chichester,
West Sussex PO19 1UD, England

National   01243 779777
International (+44) 1243 779777

Other Wiley Editorial Offices

QH
395
,E96
1995

John Wiley & Sons, Inc., 605 Third Avenue,
New York, NY 10158-0012, USA

Jacaranda Wiley Ltd, 33 Park Road, Milton,
Queensland 4064, Australia

John Wiley & Sons (Canada) Ltd, 22 Worcester Road,
Rexdale, Ontario M9W 1L1, Canada

John Wiley & Sons (SEA) Pte Ltd, 37 Jalan Pemimpin #05-04,
Block B, Union Industrial Building, Singapore 2057

Library of Congress Cataloging-in-Publication Data

Evolutionary change and heterochrony / edited by Kenneth J. McNamara.
    p. cm.
    Includes bibliographical references and index.
    ISBN 0-471-95837-9
    1. Heterochrony (Biology)   I. McNamara, Ken.
QH395.E96 1995   95-6588
575–dc20   CIP

British Library Cataloguing in Publication Data

A catalogue record for this book is available from the British Library

ISBN 0-471-95837-9

Typeset in 10/12pt Times by Vision Typesetting, Manchester
Printed and bound in Great Britain by Biddles Ltd, Guildford and King's Lynn

This book is printed on acid-free paper responsibly manufactured from sustainable forestation,
for which at least two trees are planted for each one used for paper production.

CONTENTS

LIST OF CONTRIBUTORS

Raimund Feist, Institut des Sciences de l'Evolution, Université de Montpellier II, 34095 Montpellier Cedex 5, France

Giuseppe Fusco, Dipartimento di Biologia, Università Degli Studi di Padova, via Trieste, 75-35121, Padova, Italy

John L. Gittleman, Department of Zoology, University of Tennessee, Knoxville, Tennessee 37996-1410, USA

Kenneth E. Glander, Department of Biological Anthropology and Anatomy, Duke University and Duke University Primate Center, Durham, North Carolina 27706, USA

Brian K. Hall, Department of Biology, Dalhousie University, Halifax, Nova Scotia, Canada B3H 4J1

Dieter Korn, Geologisch-Paläontologisches Institut der Universität, Sigwartstrasse 10, D-7400 Tübingen, Germany

Bradley C. Livezey, Section of Birds, Carnegie Museum of Natural History, 4400 Forbes Avenue, Pittsburgh, Pennsylvania 15213-4080, USA

John A. Long, Department of Earth and Planetary Sciences, Western Australian Museum, Francis Street, Perth, WA 6000, Australia

Michael L. McKinney, Department of Geological Sciences, University of Tennessee, Knoxville, Tennessee 37996-1410, USA

Kenneth J. McNamara, Department of Earth and Planetary Sciences, Western Australian Museum, Francis Street, Perth, WA 6000, Australia

David M. Meyers, Department of Biological Anthropology and Anatomy, Duke University, Durham, North Carolina 27706, USA

Alessandro Minelli, Dipartimento di Biologia, Università Degli Studi di Padova, via Trieste, 75-35121, Padova, Italy

Tsutomu Miyake, Department of Biology, Dalhousie University, Halifax, Nova Scotia, Canada B3H 4J1

Volker Mosbrugger, Geologisch-Paläontologisches Institut der Universität, Sigwartstrasse 10, D-7400 Tübingen, Germany

Matthew J. Ravosa, Department of Cell and Molecular Biology, Northwestern University Medical School, 303 East Chicago Avenue, Chicago, Illinois 60611-3008, USA

Rainer Schoch, Geologisch-Paläontologisches Institut der Universität, Sigwartstrasse 10, D-7400 Tübingen, Germany

Moya M. Smith, Division of Anatomy and Cell Biology, UMDS, Guy's Hospital, London Bridge, SE1 9RT, UK

Gregory A. Wray, Department of Ecology and Evolution, State University of New York at Stony Brook, Stony Brook, New York 11794, USA

PREFACE

Kenneth J. McNamara

The study of evolution has, for more than half a century, been supported by the two mighty pillars of genetics and natural selection. Thus, for instance, interpretations of "Darwin's" Galápagos finches invariably discuss the adaptive significance of the genetic mutations that produced birds with bills of different size and shapes, so enabling them to feed from a range of food sources. Largely neglected from such evolutionary studies has been an appreciation of just how a genetic mutation actually produces the variation in bill shape that will suffer selection. In other words exactly how does the phenotype evolve? What intrinsic factors operate to provide a wide variety of bill shapes on which natural selection can work?

The last decade has seen a renaissance in the study of evolutionary developmental biology stressing, in particular, the importance of comparative analyses of organisms' ontogenetic histories. For so long evolutionary relationships between organisms have been focused simply on adult phenotypes: thus the adults of one species of finch are seen as evolving from the adults of another species. This resurgence of interest in evolutionary developmental biology is revealing that to understand how phenotypes evolve we must look at the complete ontogenetic history of organisms and at their evolutionary relationships. It is becoming clear from recent studies that relative changes to the rates and timing of growth during ontogeny play a significant role in phenotypic evolution.

Seen in this light, the evolution of, say, finches with larger, stouter bills, is a consequence either of greater relative growth rates, or extensions of ancestral allometries, perhaps by a delay in the onset of maturation. Only after such intrinsically induced variation has been effected can natural selection operate. Articulated as *heterochrony* such changes are the embodiment of the relationship between ontogeny and phylogeny; for it is the effect of alterations to the timing and rate of ontogenetic development that is one of the cornerstones of evolution.

Following two symposia held in 1986, one in San Antonio, USA, the other in Dijon, France, two publications of contributed papers ensued (McKinney, 1988;

David *et al.*, 1989) that provided "state-of-the art" summaries of investigations into the role of heterochrony in evolution. A more general synthesis of heterochrony was published by McKinney and McNamara (1991) in which it was argued that rather than being some sort of quaint evolutionary phenomenon worthy of, at most, a couple of paragraphs in textbooks on evolution, heterochrony is a major, but much underrated, component of evolutionary theory. Bridging the gap between genetics and natural selection, the placement of heterochrony within a coherent, structured framework as a means of understanding how ontogenies evolve, is contributing toward a more cohesive theory of evolution.

With the burgeoning amount of research being undertaken in this field it was appropriate that the Fourth Congress of the European Society for Evolutionary Biology, held in Montpellier, France in August 1993, thought fit to include a symposium on this topic. In collaboration with my colleague Dr Raimund Feist of the University of Montpellier I invited a number of biologists, palaeontologists and anthropologists to deliver overviews of current activity in heterochrony in a broad range of disciplines. Some of these papers form the basis for the chapters in this book, supplemented by contributions from other workers who were unable to attend the meeting. Both the symposium and, perhaps even more so, the chapters in this book, highlight the all-pervasive influence of heterochrony in evolution, encompassing molecular biology, cell biology, palaeontology, macroevolution, speciation, ecology (in particular life history strategies), sexual selection and behaviour. The contributions range from broad reviews to more detailed case studies. Part One looks at the more theoretical aspects of heterochrony as the principal factor in ontogenetic evolution. Part Two contains chapters illustrating how phenotypic evolution in many groups of animals and plants is a function of the evolution of organisms' ontogenies. Part Three focuses on the interplay between heterochrony and ecological factors, demonstrating the complex interactions between organisms' ontogenies and the environment in which they live.

Hall and Miyake set the scene in Chapter 1, discussing time as it affects the developing organism, and in particular how embryos measure time in relation to cell cycles. While studies of heterochrony have, not surprisingly, focused on morphological change, McKinney and Gittleman (Chapter 2) argue that heterochrony is implicit in both life history strategies, where growth rates and longevity feature prominently, and in aspects of organisms' behaviour. Minelli and Fusco (Chapter 3) examine the role of heterochrony in the evolution of meristic structures, using terrestrial arthropods as their model. They explore, in particular, heterochronic patterns in segment origin and differentiation within arthropods. Part One of this book concludes with my examination (Chapter 4) of the effect of heterochrony on sexual dimorphism. The evolution of sex is almost universally looked at from the extrinsic perspective of sexual selection, with barely a glimpse at the intrinsic factors operating to engender sexual dimorphism.

Part Two presents a number of case studies that illustrate how heterochrony operates in morphological evolution. The effect of heterochrony in plant evolution has received relatively less attention than it has in animal evolution. Mosbrugger (Chapter 5) demonstrates that heterochrony provides an excellent theoretical concept to describe many crucial steps in the evolution of land plants. He also discusses some of the reasons why heterochrony is particularly common in land plants. A much unexplored area of heterochronic research is in its effects on changes to the morphology of vertebrates' skulls. Schoch (Chapter 6) puts this to rights by examining the impact of heterochronic changes to skull characters in amphibians, based in part on fossil material, but also examining heterochrony in the evolution of anurans and urodeles. Smith (Chapter 7) looks at how heterochrony may have influenced the tissue diversity of enamel through vertebrate evolution. In particular she looks at recent studies of the molecular controls operating at all stages of tooth development to demonstrate how a heterochronic mechanism can be proposed for the evolution of dental tissues. In Chapter 8 I examine, with my colleague John Long, dinosaur evolution in the context of heterochrony. With the wealth of new ontogenetic data forthcoming in the last few years, it is possible to argue that most evolutionary trends that have been documented in dinosaurs were heterochrony driven. In the last chapter in Part Two, Livezey (Chapter 9) demonstrates in a detailed review the role of paedomorphosis in avian flightlessness (De Beer, 1956). However many features of avian anatomy owe more to peramorphic processes.

To open the last section of the book, Part Three, which looks at the interplay between heterochrony and ecology, Wray (Chapter 10) reviews the evidence for the crucial role of heterochrony in the early developmental histories of echinoids, and its ecological consequence, in particular how it has affected their life history strategies. In his detailed examination of the phenomenon of eye reduction in an extinct group of arthropods, the trilobites, Feist (Chapter 11) shows not only how this was under heterochronic control, but how it was related to periods of global eustatic deepening. Similarly, Korn (Chapter 12) uses Palaeozoic ammonoids to show how environmental perturbations affected not only which heterochronic morphotypes evolved, but also the rates at which they evolved. In the final chapter Ravosa, Meyers and Glander (Chapter 13) look in detail at ecogeographic size variation in one group of Malagasy lemurs, and show the importance of heterochrony in its generation, in particular how ecological factors such as resource seasonality have induced such changes.

In this book chapter authors have essentially followed the modern nomenclature of heterochrony, as articulated by Alberch *et al.* (1979). If a descendant undergoes less growth during ontogeny than its ancestor this is known as *paedomorphosis*. If it undergoes more growth it is known as *peramorphosis*. Each of these states can be achieved in three basic ways, simply by varying the timing of onset, or offset, or rate, of development. If development is stopped at an earlier growth stage in the descendant than in the ancestor, for example by earlier onset of sexual maturity, ancestral juvenile features will be retained by the descendant adult—this is known as *progenesis*. If the onset of development of a particular structure is delayed in a descendant, it will develop less than in the ancestor—this is known as *postdisplacement*. The third process that produces paedomorphosis is *neoteny*, whereby the actual rate of growth is reduced.

The three ways that peramorphosis can be achieved are just the opposite to paedomorphosis. Development can start earlier and so get a head start—this is known as *predisplacement*; or the rate can be increased—this is *acceleration*; or development can be extended, perhaps by a delay in the onset of sexual maturity—this is *hypermorphosis*. In such cases ancestral allometries will be extended, producing descendant adults that are morphologically quite different from the ancestors.

The stimulus for producing this book arose from the Organising Committee of the Fourth Congress of the European Society for Evolutionary Biology who asked me to co-organise the Symposium: "Heterochrony in Evolution". In particular I would like to thank the Chairman, François Catzeflis, not only for inviting me, but also for asking me to give a Plenary Lecture, so allowing me to try and convince 800 evolutionary biologists of the importance of heterochrony in evolution. I would particularly like to thank my co-organiser, Raimund Feist, for all his support and encouragement, and, with Monique Feist, for providing me with such kind hospitality in Montpellier. I would like to express my thanks to Iain Stevenson at Wiley for his endless enthusiasm for all things concerning evolution. I am also grateful to Tiffany Robinson, Jenny Bevan, Nicky Christopher and Louise Metz for their help with the production of this book. Last, but certainly by no means least, I wish to thank all my colleagues for their excellent contributions and for helping to show that to understand evolution fully we must look at the way the entire ontogeny evolves.

Perth, Australia
November, 1994

REFERENCES

Alberch, P., Gould, S.J., Oster, G.F. and Wake, D.B., 1979, Size and shape in ontogeny and phylogeny, *Paleobiology*, **5**: 296–317.
David, B., Dommergues, J.-L., Chaline, J. and Laurin, B., 1989, *Ontogenèse et évolution*, Geobios, Mémoire spécial No. 12.
De Beer, G., 1956, The evolution of ratites, *Bull. Br. Mus. Nat. Hist. Zool.*, **4**: 59–70.
McKinney, M.L., 1988 (ed.), *Heterochrony in evolution: a multidisciplinary approach*, Plenum Press, New York.
McKinney, M.L. and McNamara, K.J., 1991, *Heterochrony: the evolution of ontogeny*, Plenum Press, New York.

Part One
HETEROCHRONY IN EVOLUTION

Chapter 1

HOW DO EMBRYOS MEASURE TIME?

Brian K. Hall and Tsutomu Miyake

> *Time enters as an essential element into our definition of organism.*
> E.S. Russell (1930, p. 171)

INTRODUCTION

Time is not an easy concept. Entire books have been written about it—time in history, cosmic, absolute, or relative time, time as destroyer, time and motion. A thousand page *Encyclopaedia of Time* and a 450 page bibliographic guide to 6000 books and articles on time in 100 sub-disciplines have just been published (Macey, 1994a,b). Time is surely on the increase. The eighth edition of the *Concise Oxford English Dictionary* provides 19 definitions of time as a noun and five of time as a transitive verb. The concern of this volume is with time in ontogeny and phylogeny, or heterochrony.

Heterochrony (Gk *heteros*, other, different; *chronos*, time) is a change in the relative time of appearance of a character in a descendant relative to its time of appearance in an ancestor. Heterochrony is therefore a phylogenetic concept and not an ontogenetic process, although its mechanistic explanations lie in ontogeny.

Change in relative timing can occur any time during ontogeny, i.e., anytime between fertilization and maturity. A truism—not always true (see Chapter 10, herein)—is that changes initiated early in ontogeny (i.e., in development), such as different times of origin of cells, tissues or organs, have greater effects than changes initiated later in ontogeny.

Ontogeny can be related to life history; see below and Chapter 2. The critical historian of development and heredity, E.S. Russell, recognized and articulated this 65 years ago.

> The living thing at any one moment of its history must be regarded as merely a phase of a life-cycle. It is the whole cycle that is the life of the individual, and this cycle is indissolubly linked with previous life-cycles—those of its ancestors right back to the dawn of time. This is what we mean by the continuity of life. (Russell, 1930, p. 171)

Evolutionary Change and Heterochrony. Edited by Kenneth J. McNamara © 1995 The Editor and Contributors.
Published in 1995 by John Wiley & Sons Ltd.

Only rarely has timing and/or rate of development of the same character or process in an ancestor been determined. Nevertheless, it is fair to say that researchers can determine relative timing of events in an organism and relate them to those seen in unrelated organisms. Such comparisons will not, of course, reveal heterochrony. Occasionally comparisons can be made in a phylogenetic context. Such comparisons can reveal heterochrony.

To identify a heterochronic event requires knowledge of the timing of that event in both descendant and ancestor. Without such knowledge, heterochrony—a phylogenetic concept—cannot be assigned. Sadly, only in few situations has heterochrony, so defined, been identified. Dwarfing of elephants on islands where the state of skeletal fusion at specific sizes provides a metric is one good example (Roth, 1992). Evolution of larval morphology during the post-Palaeozoic radiation of echinoids, where larval adaptation can be related to precise embryonic stages, is a second—see Chapter 10. The evolution of Cambrian trilobites through peramorphosis and paedomorphosis where size provides a metric for timing is a third—see Chapter 11 and McNamara (1986). It is also possible to employ the "comparative method", a new, rigorous way to analyse ontogeny in a phylogenetic context (Chapter 2). This relies on living species and avoids problems inherent in determining chronology in fossils. For other examples see McKinney (1988), and McKinney and McNamara (1991).

To search for an ontogenetic basis for heterochrony is to assume that we can time the appearance of a character, or determine the rate of a developmental process, in relation to timing to one or other of the following:

1. to timing and/or rate of development of *other features in the same organism*—this for practical reasons; and/or
2. to timing and/or rate of development of *the same character or process in an ancestor*, verified using a robust phylogeny—this to establish the change as heterochrony, and to document its ontogenetic basis.

Where heterochrony affects the entire organism—as in progenesis, or a shift in timing of metamorphosis, or birth—it is known as *global* or *systemic* heterochrony (see Chapters 4, 6, 12 and 13). When only a part of an organism is altered, we speak of *local* or *specific* heterochrony (see Chapters 3–5, 7–9, 11 and 12).

Establishing relationships to other features in the same organism (type 1 above) is accomplished quite readily for those model organisms in common use in developmental biology—*Xenopus*, embryonic chicks and mice, *Drosophila*, *Caenorhabdites elegans*, the zebrafish (*Danio* (*Brachydanio*) *rerio*). Tables and stages of normal development are available: the tables for *Xenopus*, the zebrafish and chick reproduced in Stern and Holland (1993), for birds in Starck (1993), for the mouse in Theiler (1972), and for amphibians in Rugh (1965) and Burggren and Just (1992). Type 1 is less easily achievable for stages of ontogeny initiated after hatching or birth (for which, by-and-large, staging tables do not exist), or for *any* stages of non-laboratory animals for which complete, or even partial, ontogenetic series are rarely available, some exceptions being for the squid,

Loligo pealii, cichlid fish *Oreochromis mossambicus*, and lesser spotted dogfish, *Scyliorhinus canicula* (Arnold, 1965; Anken *et al.*, 1993; Ballard *et al.*, 1993). Just how realistic is our assignment of stages and timing of events in ontogeny, especially those taking place during embryonic development? Does our ageing and staging of embryos accurately reflect how embryos measure time? Can embryos measure time? Do embryos measure time? If they do, how do they? These are the questions we attempt to answer in this opening chapter.

HOW DO WE MEASURE EMBRYONIC TIME?

A shift in relative timing may advance or slow an event during ontogeny. Advancement (acceleration), slowing-down (retardation), using terminology from Gould (1977) but concepts that are much older (see e.g., Garstang, 1922; de Beer, 1958). Gould defined acceleration and retardation respectively, as:

> A speeding up of development in ontogeny (*relative to any criterion of standardization*)
> . . .
> A slowing down of development in ontogeny (*relative to any criterion of standardization*) . . . (Gould, 1977, pp. 479, 486, our italics)

"*Relative to any criterion of standardization*" interests us here.

Changes in timing are assessed as time of onset, time of offset (termination) or rate of development (Alberch *et al.*, 1979). We have to be careful with the latter—rate changes do not necessarily lead to differences in timing unless duration is also altered.

Previous studies (Alberch *et al.*, 1979; Hall, 1984, 1990, 1992) have addressed the assumptions underlying definitions of stable reference points used in comparing timing between different organisms, or between different characters in the same individual. Such assumptions are critical if we are to turn definitions into working criteria.

The customary "criteria of standardization" fall into four categories—maturity, size, age and morphology.

1. Measures of maturity—hatching, birth, attainment of sexual maturity, metamorphosis, i.e., major developmental events.
2. Measures of growth—size at maturity, size at time x, growth as a rate variable.
3. Measures of chronological age—either absolute age (days, months or years of age) or relative age (percentage of time through ontogeny, and/or percentage of final size). Relative age is affected by temperature, body size and metabolic rate.
4. Attainment of a particular morphology or morphological stage.

There is no objective way of selecting which criterion or combination of criteria to use. Convenience is the arbiter of choice, although heterochrony must always be relative timing of developmental events.

These criteria are not necessarily correlated with one another. They are certainly not under the same control during ontogeny. Size and sometimes morphology can be influenced by temperature, nutrition, litter size, etc. Catch-up growth, regulation (the ability to compensate for deviation from normal development), or indeterminate growth—when final size is not fixed—can and do influence final size; see Cock (1966). However, maturity—to some extent—and chronological age—to a large extent—are independent of factors influencing size; see Bernardo (1993) for a discussion of the factors (especially size and age) that influence onset of maturity, and Sattler (1992) and Chapter 4 for discussions of size and age. Maturity and age are criteria of standardization established in ancestors—the result of generations of selection. Size and morphology are more direct ontogenetic criteria, subject to the vicissitudes of life history, including ecological factors that affect heterochrony—see Hall (1992) and the chapters in Part Three of this volume.

Heterochrony *is* often associated with extreme change in size, especially reduction (resulting from progenesis). Small organisms from three groups in which heterochrony has played a role in evolutionary change are *Thorius*, the smallest salamander (1.3 cm snout–vent (S–V) length; Hanken, 1993—the largest living salamander is one of the Asiatic giant salamanders, *Andrias davidianus* at 1.52 m in S–V length), *Idiocranium russeli*, a very small caecilian (1.1 cm S–V length; Wake, 1986—the largest caecilian is *Caecilia thompsoni*, S–V length of 1.52 m), and small elephants on islands (*Elephas falconeri* from Sicily at 130 kg compared with *Loxodonta africana*, the modern African elephant, weighing-in at up to 10 tonnes; Roth, 1992).

HOW DO EMBRYOS MEASURE TIME?

This heading assumes that embryos measure time and not some other parameter of maturity such as size. Debate over whether time or size should be the metric of choice in analyses of development, growth, allometry or heterochrony has raged for decades—see German and Meyers (1989) for a discussion and for access to the literature.

Size and time

A classic demonstration of size determination was provided by Hans Driesch who separated four- or eight-cell stage echinoid embryos (blastulae) into separate cells (blastomeres) using vigorous agitation (Driesch, 1892). The resulting single cells each produced complete—but small—embryos and larvae. The chronology of their development was no different from that of their intact siblings. Size not time is being controlled in this, the first documented case of regulative development; see MacLean and Hall (1987) for further examples.

Similarly, normal large tadpoles developed on time when Conrad Wadding-ton—the father of canalization, genetic assimilation and epigenetics—added additional material to frog gastrulae (Waddington, 1938). These experiments, spanning such divergent organisms, are examples of what Sinervo (1993) has called "allometric engineering". Egg size is correlated with such life history traits as timing of metamorphosis, size at metamorphosis, and developmental rate (McLaren, 1965; Raff, 1992).

A quite different situation occurs in mammalian development, where mid-gastrula stage mouse embryos are normal both in size and in timing of development following removal of one-half or one-quarter of the cells at the two- or four-cell stage respectively (Tarkowski, 1959; Power and Tam, 1993). Destruction of as much as 80% of a seven-day embryo has minimal effects on the size, timing, or normality of mid-gestation embryos, and does not influence time of birth (Snow and Tam, 1979). Increased cell size in tetraploid mice produces mid-gestation stage embryos that are 85% of normal size but have half the normal number of cells (Henery *et al.*, 1992). Cell size, not cell number, is the metric used by these embryos to measure time.

Catch-up phenomena in pre-implantation embryos have been ascribed by Ishikawa *et al.* (1992) to an "ovulation clock" that times development in relation to whether mating is delayed until the morning in animals which ovulated the night before. These authors, and others cited by them, see embryos timing development by this ovulation clock rather than by the "fertilization clock", where development is timed beginning at fertilization; see also Satoh (1982) for the view that a single "fertilization clock" is unlikely to time embryonic development.

Starck (1993) has recently synthesized studies on the evolution of avian ontogenies, building on the classic studies by Ricklefs (1967, 1987 and references therein). In dividing avian ontogenies into three periods—embryonic (fertilization to hatching), postnatal (hatching to sexual maturity), and adult (sexual maturity to death)—he emphasized that, as in mammals, biological time is dependent on body size. Consequently, physiological rates rather than chronological time can be used as a time standard in comparative studies. Place such studies in a phylogenetic context and you have, at least for extant organisms, a basis for detecting heterochrony.

Starck notes the difficulties associated with obtaining a single metric to measure time over the three periods of ontogeny. Duration of the three periods varies enormously, both within and between periods (Table 1.1, column 2). For practical reasons, Starck advocates morphological stages, growth and physiological rates, as metrics for the embryonic, postnatal and adult periods respectively (Table 1.1, column 3). Clearly, such metrics can only be applied to extant (and readily accessible) species and will not aid in determination of heterochrony involving extinct or fossil organisms.

The embryonic period
The advantage of normal stages for defining the embryonic period is that local

Table 1.1 The periods of avian ontogeny, their duration, the metrics suggested for their quantification, and estimated mass-specific energy metabolism for each period.

Period	Duration	Metric	Metabolized energy (kJ^{s}g)
Embryonic	Fertilization to hatching 11–80 days	Incubation time morphological stages	2 ± 0.8
Postnatal	Hatching to sexual maturity 20 days to years	Growth	20–40
Adult	Sexual maturity to death[a] 8–120 years	Physiological rates such as mass-specific energy metabolism	2400–4300

From data in Starck (1993).

[a] Estimating longevity in birds is notoriously difficult and many records are anecdotal. Accurate estimates from recovery of banded wild birds in N. America yield upper limits of 21 years 9 months for the white-winged dove, *Zenaida asiatica*, 20 years 11 months for the common grackle, *Quiscalus quiscula*, and 17 years 5 months for Clark's nutcracker, *Nucifraga columbiana* (*J. Field. Ornithol.* [1983], **54**: 123–137, 287–294).

heterochrony can be recognized when an organ system deviates from the sequence of normal stages. The Russian, I.I. Schmalhausen, and the Englishman, Julian Huxley, recognized this 60 years ago in their growth quotients and growth coefficients—the ratio of the growth of an organ to the rest of the body at time *x*, or the growth of the same organ at different times. As Huxley concluded:

> We cannot discover the true growth-coefficient of an organ during its early stage without precise information as to the time-relations of development. (Huxley, 1972, p. 143)

Examples of the utility of this approach are the uncovering of variability in timing of appearance of skeletal elements and disassociation between cartilage and bone formation in skull development in the Oriental fire-bellied toad, *Bombina orientalis*, and of heterochronic appearance of neomorphic skull elements at the end of ontogeny in anuran amphibians (Hanken and Hall, 1984, 1988; Trueb, 1985; and see Chapter 6).

Starck emphasizes that the embryonic periods of all the 14 or so avian species for which stages are available can be compared using just 42 standard developmental stages. Stages 1 to 33, representing three-quarters of the embryonic period, are surprisingly constant between species. Stage 39 is most variable, both in duration, and in time to hatching; the time between stage 39 and hatching is prolonged in precocial species (four to nine days) but short in altricial species (two days in the budgerigar and rock dove).

Starck concluded that the embryonic period ends when energy metabolized/g body weight is around 2 kJ/g, but emphasizes the extreme variability in such a physiological measure for timing transitions between ontogenetic periods.

The postnatal period
This period, which can last from 20 days to several years, is timed by growth. Ricklefs (1967) suggested relating growth to time required to increase body weight from 10 to 50% of final size, but growth independent of final size can also be used. In the postnatal period the physiological unit, energy metabolism/g body weight lies between 20 and 40 kJ/g (Table 1.1, column 4).

Very closely timed measurement of growth can reveal surprising patterns that could influence use of growth as the metric during this period. Lampl *et al.* (1992) observed unexpected saltatory growth during early human postnatal growth. They concluded that between 90 and 95% of development up to 21 months of age is not accompanied by growth, i.e., this much development occurs during periods of stasis.

The adult period
A number of physiological rates can be used to "time" the adult period of ontogeny. Energy metabolism/g is one, falling between 2400 and 4300 kJ/g (Table 1.1, column 4). These values are 100 fold higher than in the postnatal period, which in turn is 10–20 fold higher than in the embryonic period. This unit of metabolic rate is obviously useful for comparisons between periods, both within and between species, but is not necessarily useful for comparing different species within one ontogenetic period—variation is just too high. Heart rate, respiration, or basal metabolic rate have also been used as physiological measures for this adult period.

Concerning heterochrony and these metrics for measuring time in birds Starck concludes that:

> the constancy of time patterns in birds implies that heterochrony is not a mechanism of changing ontogenies . . . in contrast to other vertebrate groups (e.g., amphibians) where heterochrony . . . plays an important role in the evolution of ontogenies. (Starck, 1993, p. 292)

Physiological time

One measure of physiological time—energy metabolized/g body weight—has been noted and used to compare adult ontogenetic phases. For a detailed discussion see Calder (1984). These and other measures of physiological time are potential metrics precisely because not all organisms live on the same physiological time scale; but bear in mind that metabolic rate is not necessarily correlated with development rate. Small organisms live on a shorter time scale, i.e., "time" passes more quickly for them, than larger organisms, for whom "time" passes more slowly. Data for heartbeat and wingbeat shown in Table 1.2 illustrate this phenomenon. The range is enormous—20 to 700 heartbeats/minute and < 1 to over 2200 wingbeats/second. However, some elements of physiological time are much less variable—1 breath/4 heartbeats or 25×10^6 heartbeats/lifetime. Of course, mechanical or hydrodynamic constraints may confound such correlations of metabolism with time.

Obviously, physiological time must be normalized to be useful in measuring the elapse of time. Normalization may either be absolute, as to body mass, or relative to some other physiological measure, as discussed above. Given the effects of temperature on physiological rate, physiological measures as a metric to measure time must be used with caution.

Reiss (1989) advocated mass-specific metabolism as the physiological unit of choice for physiological and developmental time. He listed seven criteria that must be met for this metric to be meaningful. Three involved independence from morphology, size and temperature. A fourth was dependence on only one preceding causal event, this to avoid double counting. The fifth involved assurance that homologous events at similar developmental stages were being compared in closely related organisms. The sixth was a requirement for a positive correlation between the metric and chronological time, the seventh that the metric be measurable.

Using calories/g dry weight as the physiological metric, Reiss found a range of values from 6800 to 61 000 across selected species of fish, turtles, snakes and birds. By relating this physiological time unit to morphological development, Reiss arrived at a comparison of relative duration of development; see his Figure 3. That congeneric species of albatrosses and terns have closely similar values for physiological time was used to argue for the reality of this metric.

Table 1.2   Heartbeat (beats/minute) and wingbeat (beats/second) for some representative small and large animals.

Heartbeat		Wingbeat	
Shrew	700	Midges	2220
Mice	500	*Chironomus*	650
Daphnia	450	Mosquito	580
Rabbit	200	Honeybee	230
Cat	125	Hummingbird	100
Moths	140	Sparrow	15
Dog	80	Stork	3
Humans	72	Condor	<1
Crayfish	60		
Elephants	40		
Whales	20		
Clams	22		

Source: Prosser and Brown (1961).

Developmental time

Morphological stages

Placing embryos or postnatal organisms into morphological stages as a reliable means of measuring time has already been discussed, although without alerting the reader to the nuances of this deceptively simple method.

While considerable information may be obtained from fresh specimens, more—and different—information can be obtained from fixed specimens.

Other features may not be revealed unless specimens are examined with scanning electron microscopy. This is especially true for small invertebrates where surface characters are so critical to systematics; see Tyler (1988). Yet other features may only be evident in serial sections of particular stages, as in visualization and timing of the onset of the cellular condensations (the membranous skeleton) preceding formation of cartilaginous and bony elements of the vertebrate skeleton (Hall and Miyake, 1992).

An example of the increasing resolution that can be obtained when "staging" embryos using several levels of analysis is shown in Figures 1.1 and 1.2, which represent facial development in C57BL/6 mouse embryos. The three diagrammatic representations and the scanning electron micrograph are all from embryos designated by Theiler (1972) as stage 21. Closer analysis of frontonasal development and of initiation of whisker primordia reveals development of morphology that would allow identification of sub-stages. Such increased precision of staging is important if we are to understand the subtle changes in timing that accompany (elicit?) alterations in developmental processes, which in turn are the mechanistic basis of heterochrony.

Since the ages of mouse embryos vary both within and between litters, gestational age is not a reliable metric for "ageing" mouse embryos (Miyake *et al.*, 1993). The problem can be seen if the relationship between gestational age and morphological stage is compared (Figure 1.3). A given gestational age (shown in days) may contain embryos of more than one morphological stage (shown as Theiler stages). Embryos "aged" within the first half of the eleventh day could be at any of the six Theiler stages between 14 and 20 (Figure 1.3). One stage does not equal one day. Later in development, for example, during the last half of the twelfth day, most embryos are at one morphological stage, viz. Theiler stage 21 (Figure 1.3). There is more likely to be one stage/day at these later gestational ages; intra- and inter-litter variations decrease toward the end of gestation (Miyake and Hall, unpublished observations). Is this indicative of catch-up phenomena as described above or a normal consequence of the relationship between stages and time?

For these inbred mice the relationship of number of stages to gestational age falls into two clusters with a sharp transition between Theiler stages 19 and 20 (Figure 1.3). Such time shifts could not have been seen in embryos "aged" on gestational age alone, or without "staging" individual embryos using morphological features. Such data can be used to calculate the duration of individual stages

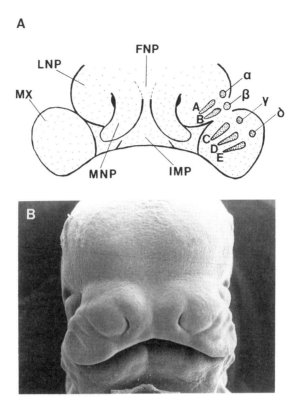

Figure 1.1   **A**. Schematic diagram of the frontonasal region and whisker primordia in C57BL/6 mice at Theiler (1972) stage 21. A–E, horizontal whisker ridges; α–δ, rudiments of mystacial whiskers; FNP, frontonasal process; IMP, intermaxillary process; LNP, lateral nasal prominence; MNP, medial nasal prominence; MX, maxilla. **B**. Scanning electron micrograph, Theiler stage early 21. The frontonasal processes have fused with the medial nasal prominence. The median nasal prominences expand laterally, creating the intermaxillary process. × 34.3.

and—in studies of heterochrony—to determine timing of the transition in the relationship between time and stage, and changes in the duration of stages. We are using this approach to compare inbred strains of mice, organs in individuals of a single strain, and phylogenetically-related organisms (Miyake and Hall, unpublished observations).

Cell cycles
In a major review including much previously inaccessible Russian work, Dettlaff, *et al.* (1987) grappled with the problem of using relative characteristics of developmental duration as a measure of developmental time. They specifically addressed the duration of developmental periods and processes as possible metrics for developmental time. By taking temperature into account they were

A

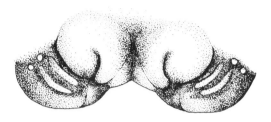

B

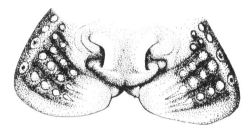

Figure 1.2. **A**. Facial region of mouse embryo of early stage 21, corresponding to Figure 1.1B. Two mystacial rudiments (γ and δ in Fig. 1.1A) and two rows of whisker primordia (C and D in Fig. 1.1A) are present on the maxilla. **B**. Facial region of mouse embryo of mid stage 21. The muzzle is almost completely formed above the intermaxillary process. Whisker rudiments have formed on horizontal whisker ridges—A and B on each LNP and C–E on the maxillae. Four rudiments of mystacial whiskers have developed, two on each LNP (α and β) and two on each maxilla (γ and δ). A concavity on mystacial rudiments γ and δ indicates formation of papillae. Abbreviations as in Figure 1.1A.

able to develop what they consider a "biological time parameter". Duration of one cell cycle (t_o)—determined during the synchronized (early) phase of cleavage divisions—was chosen as the unit of biological time. This approach is not new—it goes back to studies on amphibians by N. J. Berrill 60 years ago. Berrill used the interval between the first and second cleavage furrows as the unit of developmental time; see discussion in Dettlaff *et al.* (1987).

The metric, defined by Dettlaff and colleagues as the relative duration of the mitotic phase of the first cleavage division, ranges between 36 and 800 min in seven species of fishes and is 34 and 100 min in two amphibians, *Xenopus laevis* and *Ambystoma mexicanum*. Temperature differences explain some of this variation. The relative duration of various early developmental processes as measured by t_o is summarized in Table 1.3; t_o is remarkably stable.

Other workers such as Satoh (1982) have also emphasized DNA replication and cell cycles as the mechanism timing cell differentiation. Dettlaff and

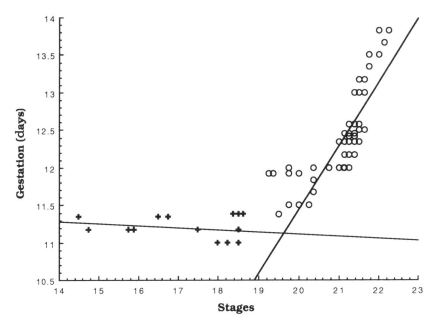

Figure 1.3   The relationship between gestational age (days) and morphological stage (Theiler (1972) stages) in inbred C57BL/6 mice. Before 11.5 days of gestation there is more than one stage/day—six stages within the first half of the eleventh day. After 11.5 days of gestation there is typically only one stage/day. The transition in the relationships between number of stages and gestational age occurs between stages 19 and 20. ○, $Y = -5.559 + 0.85X$; $R^2 = 0.751$. +, $Y = 11.667 - 0.028X$; $R^2 = 0.06$. Data from Miyake and Hall (unpublished observations).

colleagues discuss evidence relating cycles of DNA replication (t_o) to onset of RNA synthesis and to selective activation of enzyme synthesis. Holliday (1991) took this to the genetic level by proposing methylation of repeated sequences of DNA as the molecular basis of such a clock.) Both these biochemical changes are intimately related to control of developmental processes. What we now need is an extension of these analyses to other groups of organisms and an application of this metric to cases of suspected heterochrony, or to cases of proven heterochrony where other metrics have been used. Although we are unaware of any such analyses, the study by Rothe *et al.* (1992) on evolution of rapid development in *Drosophila* comes close. (For a general discussion of factors that affect rate and time of development in *Drosophila* see Karr and Mittenthal, 1992.) Rothe and colleagues demonstrated that duration of the cell cycle is coordinated with gene size, a coordination that constrains evolution of rapid development and that would presumably also constrain evolution through heterochrony, at least during early embryonic development when cell division is rapid and rampant.

A series of genes that regulate timing of the switch from larva to adult in the

Table 1.3 Biological time (t_o) in minutes for selected phases of embryonic development in selected fishes and amphibians.

Period	Fishes	Amphibians
Fertilization to first cleavage furrow	3	3
Fertilization to fall in mitotic index	12–14	
First cleavage furrow to fall in mitotic index	14–20	10–26
Fall in mitotic index to gastrulation	6–16	12–27
Fertilization to onset of RNA synthesis	10–18	
Fertilization to onset of gastrulation	18–29	
Fertilization to 10 pairs of somites	50–61	
Fertilization to end of epiboly	35–66	
Gastrulation to slit-like blastopore	18–19	12–25
Slit-like blastopore to fusion of neural folds	7[a]	12–21
Gastrulation to ten pairs of somites	28–34	
Gastrulation to end of epiboly	17–37	
Fertilization to fusion of neural folds	45[a]	47–67

Based on data in Dettlaff *et al.* (1987) for seven species of fishes and eight species of amphibians. See text for details.
[a] Based on single species, *Acipenser güldenstädtii.*

roundworm *Caenorhabdites elegans* has been identified (Ambros, 1989). These *lin* (cell lineage abnormal) genes control timing by regulating the number of cell divisions in stem cells. Such "heterochronic genes" are just beginning to be discovered in other organisms, although Davidson (1991) was able to organize all embryos into three types, two of which he argues to be heterochronic derivatives of the first, and genes regulating circadian clocks in *Drosophila* and *Neurospora* were identified 20 years ago; see Page (1994) for an overview. Very recently, Swalla *et al.* (1994) identified a gene in the ascidian *Molgula citrina*, in which tissues and organs found in adults of other species develop precociously in the larva. A muscle actin gene, expressed in adults of other species but in larvae of *M. citrina*, is a potential candidate "heterochrony gene", although whether this gene regulates the time shift or is a consequence of the time shift, is unclear. *Hoxd*-13, a member of a growing family of *Hox* genes involved in patterning and morphogenesis, has been shown to act heterochronically during limb development in the mouse (Dollé *et al.*, 1993). *Hoxd*-13 retards limb development by as much as four days, resulting in formation of neotenic limbs, a classic heterochrony.

Critical developmental events
Developmental biologists have identified the major developmental events responsible for transition from stage to stage during ontogeny; see MacLean and Hall (1987) and Hall (1992) for examples. These transitions may be between morphological stages, between stages of cell differentiation (the transition from determined to differentiating cell types), between hierarchical inductive events (the sequence of epithelial–mesenchymal interactions in development of the

teeth), or between developmental processes (from differentiation to growth). Satoh (1982) marshalled evidence that the clock timing initiation of cell differentiation is not the same as that timing morphogenesis, the former being set by DNA replication and cell cycles (see previous section), the latter by cytoplasmic components. *Critical periods* are often emphasized in developmental studies, especially in teratological. The transitions just enumerated are *critical events* (this distinction is discussed by Hall, 1985).

Using studies on embryonic development in *Xenopus laevis*, Cooke and Smith (1990) argued that embryos measure time elapsing between fertilization and major inductive events such as induction of the mesoderm. The stages that we measure when we assess time since fertilization, are, according to their view, a true projection of how embryos measure time. Such a metric operating in individual cells (*"a cell-autonomous timing mechanism"*) does not measure number of cell divisions, but rather the state of commitment of the cell at a given time during ontogeny. It is not a clock but an oscillator.

Servetnick and Grainger (1991) obtained evidence for an autonomous developmental timer in their studies on lens induction in *Xenopus laevis*. Ability to form lens from ectoderm is timed by the gradual loss of neural competence and by the appearance and subsequent rapid loss of competence to form lens. This is an autonomous timing mechanism independent of lens induction. Cooke and Smith (1990) documented that actin genes are turned on at mid-gastrulation, irrespective of when mid-gastrulation occurs. They conclude:

> that all the cells of the embryo monitor the time elapsed since fertilization, in the form of a schedule that lists the activities that cells of the embryo might be directed to undertake and says when these should be undertaken. (Cooke and Smith, 1990, p. 891)

The "schedule" is not of independent events but a sequence of events that must proceed in a particular order. Competence of regions of the early embryo to form more structures than actually formed (what in the older literature was called prospective potency) is tied intimately to such early embryonic timing mechanisms. The hierarchical integration of inductive events (the epigenetic cascade) is a sequence of causes and effects that can be quantified and whose disruption can lead to heterochrony. Examples for inductive interactions in organogenesis within vertebrates are discussed by Hall (1983) in the context of integration between timing of cell migration and of epigenetic interactions. While we can assess the occurrence of heterochrony by measuring arbitrary stages, phases, or rates of ontogeny, understanding the mechanistic basis of heterochronic change will require understanding autonomous timing mechanisms used by embryos to measure time. Clocks lead to descriptions, oscillators lead to mechanisms.

CONCLUSIONS

Heterochrony is a phylogenetic concept with a mechanistic basis in ontogeny. Although we can readily identify a basis for shifts in *timing* during ontogeny—shift in onset, rate, duration, or offset, either as acceleration or retardation—the mechanistic basis for *heterochrony* has been only rarely demonstrated. Change in relative timing can occur at any time or stage during ontogeny (life history) but earlier changes normally have greater effect than later changes. We discuss stable reference points used to standardize timing shifts—maturity, size, age and morphology. Whether embryos measure size or time is discussed in some detail. Included in this discussion is the use of physiological rates, chronological time or morphological stages as measures of "time".

The balance of the chapter concentrates on "developmental time". We discuss and advocate the use of morphological stages to enable investigators to measure the passage of time during organismal ontogeny. Embryos measure time in relation to cell cycles and causal sequences of critical developmental events. While other metric of time are clocks invented for our convenience, these biological parameters are the cell-autonomous oscillators used by embryos to measure time.

ACKNOWLEDGMENTS

We thank NSERC of Canada, NIH of USA and the Killam Trust of Dalhousie University for financial support, and Alan Pinder for thoughtful comments on the chapter. James Hanken provided access to the data on the largest caecilians and salamanders, and Ian McLaren access to the longevity data for North American birds in Table 1.1.

REFERENCES

Alberch, P., Gould, S.J., Oster, G.F. and Wake, D.B., 1979, Size and shape in ontogeny and phylogeny, *Paleobiology*, 5: 296–317.

Ambros, V., 1989, A hierarchy of regulatory genes controls a larva-to-adult developmental switch in *C. elegans, Cell*, **57**: 49–57.

Anken, R.H., Kappell, Th., Slenzka, K. and Rahman, H., 1993, The early morphogenetic development of the cichlid fish, *Oreochromis mossambicus* (Perciformes, Teleostei), *Zool. Anz.*, **231**: 1–10.

Arnold, J.M., 1965, Normal embryonic stages of the squid, *Loligo pealii, Biol. Bull.*, **128**: 24–32.

Ballard, W.W., Melinger, J. and Lechenault, H., 1993, A series of normal stages for development of *Scyliorhinus canicula*, the Lesser Spotted Dogfish (Chondrichthyes: Scyliorhinidae), *J. Exp. Zool.*, **267**: 318–336.

Bernardo, J., 1993, Determinants of maturation in animals, *TREE*, **8**: 166–173.

Burggren, W.W. and Just, J.J., 1992, Developmental changes in physiological systems. In M.E. Feder and W.W. Burggren (eds), *Environmental physiology of the amphibians*, University of Chicago Press, Chicago: 467–530.

Calder, W.A. III, 1984, *Size, function, and life history*. Harvard University Press, Cambridge.

Cock, A.G., 1966, Genetical aspects of metrical growth and form in animals, *Quart. Rev. Biol.*, **41**: 131–190.

Cooke, J. and Smith, J.C., 1990, Measurement of developmental time by cells of early embryos, *Cell*, **60**: 891–894.

Davidson, E.H., 1991, Spatial mechanisms of gene regulation in metazoan embryos, *Development*, **113**: 1–26.

de Beer, G.R., 1958, *Embryos and ancestors*, Oxford University Press, Oxford.

Dettlaff, T.A., Ignatieva, G.M. and Vassetzky, S.G., 1987, The problem of time in developmental biology: its study by the use of relative characteristics of development duration, *Sov. Sci., Rev. F. Physiol. Gen. Biol.*, **1**: 1–88.

Dollé, P., Dierich, A., LeMeur, M., Schimmang, T., Schuhbaur, B., Chambon, P. and Duboule, D., 1993, Disruption of the Hoxd-13 gene induces localized heterochrony leading to mice with neotenic limbs, *Cell*, **75**: 431–441.

Driesch, H., 1892, Entwicklungsmechanische Studien. I. Der Werth der beiden ersten Furchungszellen in der Echinodermentwicklung. Experimentelle Erzeugen von Theil-und Doppelbildung, *Z. wissenschaft. Zool.*, **53**: 160–178, 183–184.

Garstang, W., 1922, The theory of recapitulation. A critical restatement of the biogenetic law, *J. Linn. Soc. (Lond.)*, **35**: 81–101.

German, R.Z. and Meyers, L.L., 1989, The role of time and size in ontogenetic allometry: I. Review, *Growth Differ., Aging*, **53**: 101–106.

Gould, S.J., 1977, *Ontogeny and Phylogeny*, Harvard University Press, Cambridge, Ma.

Hall, B.K., 1983, Epigenetic control in development and evolution. In B.C. Goodwin., N. Holder and C.C. Wylie (eds), *Development and evolution*, The Sixth Symposium of the British Society for Developmental Biology, Cambridge University Press, Cambridge: 353–379.

Hall, B.K., 1984, Developmental processes underlying heterochrony as an evolutionary mechanism, *Can. J. Zool.*, **62**: 1–17.

Hall, B.K., 1985, Critical periods during development as assessed by thallium-induced inhibition of growth of embryonic chick tibiae in vitro, *Teratology*, **31**: 353–361.

Hall, B.K., 1990, Heterochronic change in vertebrate development, *Seminars in Devel. Biol.*, **1**: 237–43.

Hall, B.K., 1992, *Evolutionary developmental biology*, Chapman and Hall, London.

Hall, B.K., and Miyake, T., 1992, The membranous skeleton: the role of cell condensations in vertebrate skeletogenesis, *Anat. Embryol.*, **186**: 107–124.

Hanken, J., 1993, Adaptation of bone growth to miniaturization of body size. In B.K. Hall (ed.), *Bone Volume 7: Bone Growth B*, CRC Press, Boca Raton, Florida: 79–104.

Hanken, J. and Hall, B.K., 1984, Variation and timing of the cranial ossification sequence of the Oriental fire-bellied toad, *Bombina orientalis* (Amphibia, Discoglossidae), *J. Morphol.*, **182**: 245–255.

Hanken, J. and Hall, B.K., 1988, Skull development during anuran metamorphosis. II. Role of thyroid hormone in osteogenesis, *Anat. Embryol.*, **178**: 219–227.

Henery, C.C., Bard, J.B.L. and Kaufman, M.H., 1992, Tetraploidy in mice, embryonic cell number, and the grain of the developmental map, *Devel. Biol.*, **152**: 233–241.

Holliday, R., 1991, Quantitative genetic variation and developmental clocks, *J. Theor. Biol.*, **151**: 351–358.

Huxley, J.S., 1972, *Problems of relative growth*, 2nd Ed. Dover Publications, New York.

Ishikawa, H., Omoe, K. and Endo, A., 1992, Growth and differentiation schedule of mouse embryos obtained from delayed matings, *Teratology*, **45**: 655–659.

Karr, T.L. and Mittenthal, J.E., 1992, Adaptive mechanisms that accelerate embryonic development in *Drosophila*. In J. Mittenthal and A. Baskin (eds), *Principles of organization in organisms*, SFI Studies in the Sciences of Complexity, Proceedings Volume XIII, Addison-Wesley, Menlo Park, CA: 95–108.

Lampl, M., Veldhuis, J.D. and Johnson, M.L., 1992, Saltation and stasis: a model of human growth, *Science*, **258**: 801–803.

Macey, S.L. (ed.), 1994a, *Encyclopaedia of time*, Garland Publishing, New York and London.

Macey, S.L. (ed.), 1994b, *Time: a bibliographic guide*, Garland Publishing, New York and London.

MacLean, N. and Hall, B.K., 1987, *Cell commitment and differentiation*, Cambridge University Press, Cambridge.

McKinney, M.L. (ed.), 1988, *Heterochrony in evolution: a multidisciplinary approach*, Volume 7 in Topics in Geobiology, Plenum Press, New York.

McKinney, M.L. and McNamara, K.J., 1991, *Heterochrony: the evolution of ontogeny*, Plenum Press, New York.

McLaren, I.A., 1965, Temperature and frogs eggs. A reconsideration of metabolic control, *J. Gen. Physiol.*, **48**: 1071–1079.

McNamara, K.J., 1986, The role of heterochrony in the evolution of Cambrian trilobites, *Biol. Rev.*, **61**: 121–156.

Miyake, T., Cameron, A.C. and Hall, B.K., 1993, Detailed timing of onset of the mandibular skeleton in inbred C57BL/6 mice, *Molec. Biol. Cell*, **4**: 145a.

Page, T.L., 1994, Time is the essence: molecular analysis of the biological clock, *Science*, **263**: 1570–1572.

Power, M.-A. and Tam, P.P.L., 1993, Onset of gastrulation, morphogenesis and somitogenesis in mouse embryos displaying compensatory growth, *Anat. Embryol.*, **187**: 493–504.

Prosser, C.L. and Brown, F.A. Jr, 1961, *Comparative animal physiology*, 2nd Ed, W.B. Saunders, Philadelphia.

Raff, R.A., 1992, Direct-developing sea urchins and the evolutionary reorganization of early development, *BioEssays*, **14**: 211–218.

Reiss, J.O., 1989, The meaning of developmental time: a metric for comparative embryology, *Amer. Nat.*, **134**: 170–189.

Ricklefs, R.E., 1967, A graphical method of fitting equations to growth curves, *Ecology*, **48**: 978–980.

Ricklefs, R.E., 1987, Comparative analysis of avian embryonic growth, *J. Exp. Zool.*, *Suppl.* **1**: 309–323.

Roth, V.L., 1992, Inferences from allometry and fossils: dwarfing of elephants on islands. In D. Futuyma and J. Antonovics (eds), *Oxford Surveys in Evolutionary Biology* **8**, Oxford University Press, Oxford: 259–288.

Rothe, M., Pehl, M., Taubert, H. and Jäkle, H., 1992, Loss of gene function through rapid mitotic cycles in the *Drosophila* embryo, *Nature*, **359**: 156–159.

Rugh, R., 1965, *Experimental embryology. Techniques and procedures*, Burgess Publishing Co., Minneapolis, Minn.

Russell, E.S., 1930, *The interpretation of development and heredity. A study in biological method*, Clarendon Press, Oxford.

Satoh, N., 1982, Timing mechanisms in early embryonic development, *Differentiation*, **22**: 156–163.

Sattler, R., 1992, Process morphology: structural dynamics in development and evolution, *Can. J. Bot.* **70**: 708–714.

Servetnick, M. and Grainger, R.M., 1991, Changes in neural and lens competence in *Xenopus* ectoderm: evidence for an autonomous developmental timer, *Development*, **112**: 177–188.

Sinervo, B., 1993, The effect of offspring size on physiology and life history, *BioScience*, **43**: 210–218.

Snow, M.H.L. and Tam, P.P.L., 1979, Is compensatory growth a complicating factor in mouse teratology?, *Nature*, **279**: 555–557.

Starck, J.M., 1993, Evolution of avian ontogenies, *Current. Ornithol.*, **10**: 275–366.

Stern, C.D. and Holland, P.W.H., 1993, *Essential developmental biology. A practical approach*, Oxford University Press, Oxford.

Swalla, B.J., White, M.E., Zhou, J. and Jeffery, W.R., 1994, Heterochronic expression of an adult muscle actin gene during Ascidian larval development, *Devel. Genet.*, **15**: 51–63.
Tarkowski, A.K., 1959, Experiments on the development of isolated blastomeres of mouse eggs, *Nature*, **184**: 1286–1287.
Theiler, K., 1972, *The House Mouse. Development and normal stages from fertilization to 4 weeks of age*, Springer-Verlag, Berlin.
Trueb, L., 1985, A summary of osteocranial development in anurans with notes on the sequence of cranial ossification in *Rhinophrynus dorsalis* (Anura, Pipoidea, Rhinophrynidae), *S. Afr. J. Sci.*, **81**: 181–185.
Tyler, S., 1988, The role of function in determination of homology and convergence— examples from invertebrate adhesive organs, *Fortsch. Zool.*, **36**: 331–347.
Waddington, C.H., 1938, Regulation of amphibian gastrulae with added ectoderm, *J. Exp. Zool.*, **15**: 377–381.
Wake, M.H., 1986, The morphology of *Idiocranium russeli* (Amphibia: Gymnophiona), with comments on miniaturization through heterochrony, *J. Morphol.*, **189**: 1–16.

Chapter 2

ONTOGENY AND PHYLOGENY: TINKERING WITH COVARIATION IN LIFE HISTORY, MORPHOLOGY AND BEHAVIOUR

Michael L. McKinney and John L. Gittleman

Which came first, the chicken or the egg? Neither; the embryo gave rise to both.

INTRODUCTION

The above answer to evolution's best-known riddle is a guaranteed flop at cocktail parties. But it illustrates how evolutionists have consistently asked the wrong questions over much of the last century, thereby ensuring unsatisfactory answers. By viewing evolution as a branching tree of adults or genes, theorists have omitted what selection really acts upon: ontogeny. Ontogenies evolve, not genes or adults. Mutated genes are passed on only to the extent that they promote survival of ontogenies; adulthood is only a fraction of ontogeny.

This deficiency has led to renewed interest in ontogeny, including a new discipline of "evolutionary developmental biology" (Hall, 1992). In part, this will require detailed quantitative genetic models and empirical experimental data on how covariation responds to selection (e.g., Cowley and Atchley, 1992) in many different taxa. This will help unite developmental with population biology, to see how epigenetic rules translate into microevolutionary change. But even this synthesis is incomplete in two ways. Firstly, developmental biology has focused on morphological ontogeny whereas ontogeny (and phylogeny) also include changes in behaviour and life history traits. Quantitative genetic models should ultimately include these. Secondly, microevolutionary studies of this kind must be supplemented by macroevolutionary information from palaeontology, phylogenetic analyses and other areas of evolutionary biology that examine

Evolutionary Change and Heterochrony. Edited by Kenneth J. McNamara © 1995 The Editor and Contributors. Published in 1995 by John Wiley & Sons Ltd.

large-scale, long-term patterns. Only these will reveal the full picture of how ontogenies evolve to produce phylogenies.

This chapter attempts to build a conceptual framework, with initial data, on how to synthesize all of this into a truly complete view of how all aspects of ontogeny evolve to produce phylogenetic change. Heterochrony plays an important role in this.

HETEROCHRONY AND EVOLUTIONARY DEVELOPMENTAL BIOLOGY

The concept of heterochrony has its origins in patterns of morphological covariation. It originated from obvious similarities between juveniles of one species and adults of another. This occurs because most evolution is from minor modifications of highly integrated ontogenies. The phylogenetic covariation patterns of closely related species are thus usually only slight modifications of ontogenetic covariation patterns (reviews in Gould, 1977; McKinney and McNamara, 1991). These patterns of covariation are usually analysed using bivariate and multivariate allometry, to translate ontogenetic allometric patterns into phylogenetic ones.

But this traditional concept of heterochrony is an incomplete view of evolution, in two important ways. Until these deficiencies are corrected, heterochrony will continue to be dismissed by many evolutionists as a common, but not necessarily interesting, mechanism of allometric change (e.g., Raff and Kaufman, 1983; Thomson, 1988). The first, and most widely noted deficiency is that the study of heterochrony has largely been a descriptive taxonomy of patterns, with little understanding of the underlying processes (discussions in McKinney and McNamara, 1991 and Hall, 1992). This deficiency is slowly being addressed, with empirical advances in developmental biology and quantitative developmental genetics, and growing interest in these fields by evolutionary biologists (e.g., McKinney and McNamara, 1991; Hall, 1992).

This chapter addresses the second deficiency in the traditional concept of heterochrony: its focus on morphology. Even if the first deficiency is corrected, by improved understanding of genetic, cellular and tissue changes during development, heterochrony will provide an incomplete view of evolution as long as it is focused mainly on morphology. This is because ontogenetic modification in evolution is not limited to morphology. Two other types of traits also show suites of covariation during ontogeny: life history traits and behavioural traits. Both of these areas already have rich traditions of study that developed separately from heterochronic studies. These studies show that, like morphology, ontogenetic covariation in life history and behavioural traits originate from time and rate changes in development and almost always translate into patterns of phylogenetic covariation. Ontogenetic constraints on evolutionary "tinkering" via heterochrony are thus not limited to morphology. In this chapter, developmental constraints are defined as "biases on the production of variant phenotypes or limitations on phenotypic variability caused by developmental system" (Maynard Smith *et al.*, 1985).

We will attempt to outline how heterochrony can be combined with studies of life history and behaviour to provide a much more complete view of how evolution occurs. None of these three areas, heterochrony, life history, and behaviour, can claim theoretical precedence over the others because they are complementary. Indeed, the terminology and concepts of heterochrony can be directly applied to the evolution of life history and behaviour. In fact, most of the evolution of ontogeny studied by these fields is heterochrony, as it involves change in rate or timing of life history or behavioural development. Conversely, study of morphological evolution alone omits environmental selection on life history and behaviour that must also occur during evolution. Evolutionary trade-offs required by ontogenetic constraints involve covariation of all aspects of development: morphological, behavioural, and life history.

We will also note the applicability of these ideas to the increasingly popular "comparative method" of life history analysis (Harvey and Pagel, 1991). Many who use this seem unaware, or at least uninterested, that the nested patterns of phylogenetic constraint they analyse are actually patterns of modified ontogenies. "Phylogenetic constraint" is produced by the constraints of inherited ontogenetic covariations in morphology, life history and behaviour.

LIFE HISTORY AND HETEROCHRONY

Both life history and heterochrony have a long history in the literature, and there has been a steady increase in interest of both fields over the last 20 years. Yet despite an enormous degree of subject overlap, such as evolutionary change in maturation and many other key ontogenetic events, workers in life history and heterochrony have almost studiously ignored the research in the other field. For example, two recent books summarizing the "state of the art" in life history evolution (Roff, 1992; Stearns, 1992) have no mention of the term "heterochrony" anywhere in either book! We suspect that there are two main reasons why life history workers have avoided heterochrony: (1) its past focus on morphology, and (2) its confusing jargon. With the application of heterochronic ideas beyond morphology, and simplification of heterochronic jargon (McKinney and McNamara, 1991), life history workers will hopefully become more attracted to heterochronic ideas.

In contrast, diffusion of life history ideas to workers in heterochrony has already begun. Gould (1977) is often credited with the first systematic attempt to integrate heterochrony with life history concepts, such as "*r*-selecting" environments promoting progenesis (review in Hanken and Wake, 1993). A number of workers (e.g., McKinney, 1984, 1986; Allmon, 1994) have attempted to test and refine this integration using fossils, but methodological impediments in the analysis of both life history and heterochrony have slowed it. The main life history impediment is that the extremely simple r–K dichotomy does not work for population-level processes; instead we see an emerging picture of complicated trade-offs among growth rate, body size, age of maturation, senescence, and other life history events (Roff, 1992; Stearns, 1992; Charnov, 1993). The main obstacle in most heterochronic analyses so far has been the lack of ontogenetic age data (McKinney and McNamara, 1991). By using size as a proxy for age, these have essentially invalidated any heterochronic diagnoses because different rate and timing processes can produce the same size–shape patterns (Godfrey and Sutherland, in press).

Nevertheless, such obstacles can be overcome in a variety of ways to provide a much more complete view of how ontogeny evolves. Before demonstrating this, we first review the (often confusing) concept of life history, and how it is used in phylogenetic analyses.

WHAT IS LIFE HISTORY?

Few definitions are available of "life history traits", even in books devoted to the subject (see Roff, 1992; Stearns, 1992; Charnov, 1993), yet everyone seems to know what they are. This is probably because any trait that has a bearing on reproductive output in terms of an age-specific value is included as a catch-all "life history". For example, Stearns (1992) lists the following as principal life history traits: size at birth; growth pattern; age at maturity; size at maturity; number, size, and sex ratio of offspring; age- and size-specific reproductive investments; age- and size-specific mortality schedules; length of life. Thus, Stearns defines life histories as a catalogue of traits rather than by conceptual or functional definition. The situation seemingly becomes even more diffuse when considering the number of ways to study life histories. A multitude of approaches and techniques is used to evaluate what factors effect life history variation across taxa. Generally, the primary types of explanation lie with demography, quantitative genetics and reaction norms, trade-offs, and lineage-specific (comparative) effects. Given the n (life history traits) times n (approaches) scale of problems in life history evolution it is hardly surprising that general patterns are marked more by exceptions than general rules; as one study may show, for example, that litter size is compensated by birth weight another study shows just the opposite (see Stearns, 1992). Life history research is thus in a state of appraisal, with critical summary statements undoubtedly provoking a new generation of theory and empirical data. Perhaps the time is right for bringing in other perspectives such as "ontogeny". The following is a brief summary of an aspect of life history study, one that offers much potential overlap with ontogeny or heterochrony: comparative (lineage-specific) trends.

Comparative study: variables, methods and results

All comparative (cross-taxonomic) studies of life history traits underscore two factors: phylogeny and allometry. Although neither is necessarily causal, both factors are important for examining comparative patterns. Before describing empirical results it is necessary to carefully define "phylogeny" in the context of comparative life history studies. Essentially all comparative analyses that have searched for correlates of life history variation with taxonomic rank (i.e., species, genera, families, etc.) and/or with phylogenetic topology (i.e., branch lengths or symmetry) have found them (Harvey and Clutton-Brock, 1985; Gittleman 1986a,b, 1993; Read and Harvey, 1989; Miles and Dunham, 1992). For example, across the terrestrial Carnivora, birth weight is autocorrelated (i.e., significant correlation, Cliff and Ord, 1981) with phylogenetic distance at $r = 0.72$ (0.001 level of significance); furthermore, as expected from trait similarity typically observed among more closely related taxa, this correlation mainly results from patterns of species within genera and genera within families (Gittleman, 1993, 1994; see Figure 2.1). This pattern need not be the case, however. Other life history traits, such as age at sexual maturity, actually reveal little correlation among species within genera but much correlation at higher levels (Figure 2.1). This raises two important points about phylogenetic relations in comparative biology in general and life histories studies in particular. First, phylogenetic patterns may surface in different ways with different traits. Selection varies in intensity and rate, thus revealing differential trait evolution through cladogenesis or speciation (Gittleman and Kot, 1990; Gittleman *et al.*, in press). Second, on a more methodological note, empirical studies across taxa must evaluate with each trait how phylogeny is, if at all, correlated with trait variation. Simply assuming that phylogeny is important for understanding trait evolution does not make it so (Gittleman and Luh, 1992, 1994; Gittleman *et al.*, in press).

A phylogenetic approach conveys a larger message for testing hypotheses within an evolutionary context. Although phylogenetic relations are today considered mundane, only 5–10 years ago to state that a trait was constrained or correlated with phylogeny was in the least new and at most blasphemy to a strict adaptationist perspective (e.g., Gould and Lewontin, 1979). In the study of life history evolution in particular, phylogenetic correlates were unexpected because of the rich tradition of ecological theory predicting that reproductive rate and output would match environmental demand. The construct of r- and K-selection, suggesting that life history traits will allow an appropriate increase in population numbers to environmental selection, is a prime example of why historical factors would be minimized. A phylogenetic perspective on life history evolution has (1) significantly improved comparative methodology by accounting for hierarchical structure and thus statistical independence among taxa, and (2) incorporated questions of evolutionary origin within tests of adaptation (see Harvey and Pagel, 1991; Gittleman and Luh, 1992, 1994). The empirical observation that

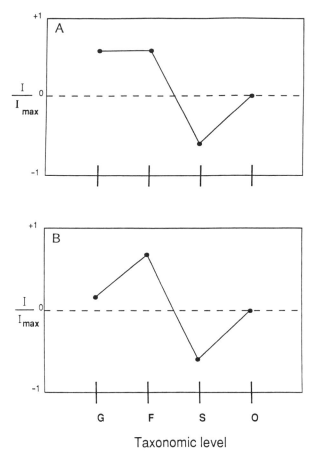

Taxonomic level

Figure 2.1 Phylogenetic correlates across the terrestrial carnivores of two life history variables, birth weight (A) and age at sexual maturity (B). I/I_{max} is a simple autocorrelation that indicates degree of similarity. Correlations demonstrate how phylogenetic correlation varies with taxonomic rank for each life history trait. Taxonomic rank is denoted by G for genus, F for family, S for superfamily, and O for order.

phylogeny and life history traits are closely linked is the touchstone between life histories and heterochrony, as we will discuss later. Last, developments in comparative methodology greatly increase the accuracy with which phylogeny can be estimated and incorporated into life history studies (Harvey and Pagel, 1991; Gittleman and Luh, 1992, 1994; Purvis *et al.*, 1994).

Allometry is the other variable that has received much treatment in comparative study. Size variables including body weight or brain weight are significantly correlated with all life history traits. Such correlations were used as evidence for indicating allometric "constraint" of life history patterns and, as consequence, considered a causal explanation (e.g., Harvey and Clutton-Brock, 1985;

Gittleman, 1986a). More recently, however, size relations are viewed as inevitable consequences which convey little in the way of actually explaining variation in life histories (Harvey *et al.*, 1989; Gittleman, 1994): a mouse is not going to give birth to an elephant-sized neonate, even though size must be subtracted prior to systematic comparisons across taxa. Thus, allometry in life history studies has correctly returned to the original use of allometric equations (see Schmidt-Nielson, 1984), namely a basis for comparisons in order to reveal deviations from a general pattern across diverse taxa (Harvey *et al.*, 1989; Gittleman, 1993).

In addition to phylogeny and allometry, factors influencing comparative trends of life histories fall into three primary categories: physiology (metabolism), ecology, and mortality schedules. Support for the first two variables is scant. Metabolism, specifically basal metabolic rate, is hypothesized to explain life histories because reproductive rate should be optimal. The primary limitation to optimizing reproduction is metabolic capacity (McNab, 1983; 1986; Calder, 1984; Thompson and Nicoll, 1986; Gordon, 1989). Comparative studies of absolute life history traits and basal metabolic rate (i.e., without accounting for body size allometry) do show significant correlations. However, when body size effects are removed from each trait, as necessary, metabolism is not related to life histories in birds (Trevelyan *et al.*, 1990), carnivores (Gittleman, 1993), or eutherian mammals (Harvey *et al.*, 1991).

Failure to find ecological correlates of life history differences across taxa is more surprising. The only robust pattern is that primate species living in tropical rain forests have lower potential reproductive rates than species residing in savannahs or secondary growth forests (Ross, 1987). This pattern is explained in terms of *r–K* selection, with tropical forest species being more *K*-selected due to relative environmental stability. Harvey *et al.* (1989) suggest that few comparative findings between ecology and life histories relate to two factors. First, because ecology and body size are so closely related: if body size effects are removed in allometric analysis of life histories (as described above) then essentially ecological effects are also removed. Second, phylogenetic effects of life histories that are independent of size are among major taxonomic groups (i.e., at ordinal levels). When these effects are statistically accounted for, much of the ecological variation among species is eliminated. At least two other reasons may be given for why ecology thus far reveals negative effects in comparative life history studies (Gittleman, 1993): poor resolution in the ecological data, with most comparative studies relying on general species categorization (e.g., mammals are classified by diet in terms of carnivore, omnivore, frugivore, etc.); and, life histories may adjust to a shorter time scale (i.e., populational differences) that is not picked up by cross taxonomic comparisons. Certainly, all of these reasons emphasize the need for future studies to measure ecologies in relation to life history patterns and simultaneously consider life history evolution at various hierarchical levels (genetic, population, species, etc.).

New findings, coupled with a theoretical basis, give renewed promise for comparative studies of life histories with one other variable: mortality rate. Charnov (1991, 1993) presented a model that predicts a negative relationship between average adult mortality and the age at maturity; that is, species with high mortality rates will be compensated with more rapid reproduction. The model assumes that (a) populations are stable, (b) mortality rates stop decreasing prior to maturity, (c) growth in body size (before maturity) and reproductive output (after maturity) are proportional $W^{0.75}$ where W equals body mass, and (d) adult mortality rates are selected by environmental factors which result in changes in age at maturity in order to maximize fitness.

Three independent analyses support the model. First, across 21 families from 10 orders of placental mammals, temporal life history variables (e.g., gestation length, age at weaning) are negatively correlated and number of offspring is positively correlated with age-specific mortality (Harvey *et al.*, 1989; Promislow and Harvey, 1990). Second, using a statistical method that integrates phylogenetic information into comparative tests, identical correlational patterns are found across terrestrial carnivores (Gittleman, 1993, 1994; see example in Figure 2.2). Third, also using new comparative statistics across a more extensive mammalian data set, the precise values of slope are consistent with those predicted by the Charnov model between age at maturity and average adult mortality rate

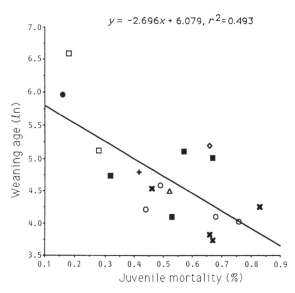

Figure 2.2 Weaning age and juvenile mortality in carnivores. After removing phylogenetic effects using an autoregressive model (see Gittleman and Kot, 1990), weaning age (ln) is significantly correlated with juvenile mortality across six families and two species (giant and red pandas) of Carnivora. ○, Canidae; □, Ursidae; △, *Ailurus*; ◇, *Ailuropoda*; +, Procyonidae; ×, Mustelidae; ●, Hyaenidae; ■, Felidae.

(Berrigan *et al.*, 1993). Undoubtedly, the next phase of testing this model is to determine causal relationships between these life histories and examine what precise forms of mortality (e.g., floods; food deprivation; meteorites) are driving the observed patterns.

HETEROCHRONY IN LIFE HISTORY EVOLUTION

We now show that it is surprisingly easy to relate heterochrony to life history and comparative studies, to produce a more complete way to visualize how ontogeny evolves. Figure 2.3 shows the traditional classification scheme of heterochrony, with three types of ontogenetic changes: onset, offset and rate. Peramorphosis or "overdevelopment" (e.g., greater size) is attained via early onset, late offset, or increased rate. Paedomorphosis or "underdevelopment" is attained with the opposite changes.

The changes in Figure 2.3 can apply to tissues in local growth fields (e.g., hypertrophy of a limb). Selection can directly produce such "dissociated" heterochronies to cause allometric shape changes. This is undoubtedly very common in evolution; it can be analysed with various quantitative genetic models of epigenesis (e.g., Slatkin, 1987; Cowley and Atchley, 1992). These can be very powerful in quantifying the constraining effects of development when applied to experimental data on genetic covariance among various morphological traits (e.g., Zelditch, 1988).

Heterochrony and body size

Life history theory has generally treated such morphological (and also behavioural) change as being funnelled through life history traits. That is, local allometric shape alterations will be selected to the extent that they affect life histories (Stearns, 1992). Whether or not this is always true is debatable (see below). But it does point out that a good way to relate heterochrony directly to life history is through body size because this is a commonly analysed life history trait that is also commonly discussed in the heterochrony literature. Thus, the changes in Figure 2.3 can just as well apply to body size as to the size of specific organs. "Global" heterochrony refers to a whole-organism change, such as delayed onset of maturation that leads to larger body size, whereas dissociated heterochronies affect specific organ sizes (McKinney and McNamara, 1991).

The next step is that of modelling body size evolution via selection of developmental trajectories. This has been done using a variety of quantitative genetics models that incorporate selection and phenotypic covariance patterns (e.g., Kirkpatrick and Lofsvold, 1992). Such models incorporate two substantial improvements over the rather simplistic traditional views of heterochrony depicted in Figure 2.3. These improvements are: (1) a populational view of

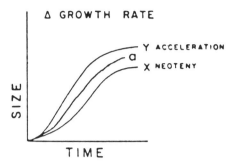

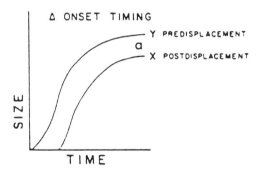

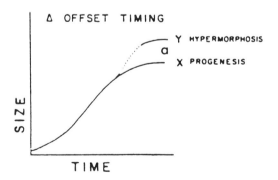

Figure 2.3   Six major types of heterochrony in terms of size vs age (time) plots; a represents ontogenetic trajectory of ancestor. Acceleration = increased rate; neoteny = retarded rate; predisplacement = early onset; postdisplacement = late onset; hypermorphosis = late offset; progenesis = early offset. Modified from McKinney and McNamara (1991).

developmental trajectories, and (2) a multiphasic view of developmental trajectories. (Both of these also apply to local organ and tissue trajectories.) The population view acknowledges that both size and growth vary among individuals in a species, depending on both genetic and environmental factors (Bernardo, 1993 for review). Thus, if we infer that species A evolved from species B through one of the heterochronic changes in Figure 2.3, we really mean that the species' *average* growth trajectory has been altered in one of the ways shown in Figure 2.3.

The multiphasic view acknowledges that growth usually follows some sort of sigmoidal pattern with lag, exponential, and asymptotic phases. When combined with the population view, individuals in a species vary according to how these phases differ. Some individuals have a more delayed maturation (asymptotic phase), others have greater growth rate during the exponential phase, and so on. Another key aspect of the multiphasic view is that selection can act directly on a single phase, up to a point, without affecting the other phases. One of the major findings of comparative studies is that, for example, weaning age, maturation age, early natal growth rate, lifespan, and many other variables that affect adult body size can be "adjusted" by selection to alter growth curve shapes in complex ways. For example, the red panda has a much longer gestation length and slower growth rate than related species of comparable body size, but age of sexual maturity is generally similar to those same related species (Gittleman, 1994). Similarly, humans show the heterochronic process of "bimaturism" wherein the sexes mature at different ages (Shea, 1988; see McNamara—Chapter 4 herein). Yet despite their earlier maturation (progenetic pattern), human females tend to live longer than males.

The traditional heterochronic changes of Figure 2.3 are thus oversimplifications in depicting "pure" heterochronic changes. For example, it implies that when growth rate change (such as acceleration) occurs, age of maturation will stay constant. This *can* occur, but it may not. In reality, both rate and offset of growth can change simultaneously. For example, a delay in age of maturation will often "accelerate" growth (beyond that of the ancestor) during the exponential phase because the individual stays longer in that multiplicative phase. Figure 2.4 illustrates this, and also how this pattern is made even more complicated when two offset changes occur (delay of the lag phase, such as delay of prenatal growth, and a similar delay of the exponential phase, such as delay of maturation).

Such *sequential heterochrony*, i.e., similar changes in temporal onsets or offsets in each growth phase, are of great interest for two reasons. One is that humans are largely the product of this (McKinney and McNamara, 1991). Second is that sequential heterochrony in a species often represents fairly extreme adaptations, indicative of unusual ecological and/or phylogenetic circumstances. Gittleman (1994), for example, shows how the giant panda delays gestation and maturation but grows at extremely slow rates (Figure 2.5).

The crucial point is that body size growth patterns are subject to complex evolutionary alterations. Body size is determined by growth trajectories that are multiphasic, with onset, rate and offset of each phase: (1) varying among

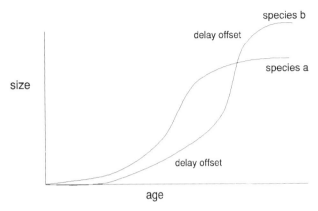

Figure 2.4 Sequential hypermorphosis in species b is seen as delay in offset of the lag growth phase and delay in offset of the exponential growth phase. Note changes in slope (growth rate) that also occur.

individuals intraspecifically, and (2) being subject to natural selection. The growth trajectories of even related species may therefore differ in complex ways, due to selection on specific phases in the trajectory (e.g., red vs giant pandas).

This growth curve complexity has largely been the undoing of the core theories of both life history and heterochrony. In life history, the classic *r–K* theory is clearly an inaccurate way of relating history traits to each other and to environmental selection at the population and species level (Roff, 1992; Stearns, 1992). A species with early maturation (global growth progenesis) does not always show, for example, early death or early age of weaning. It is thus not true that disturbed environments ("*r*-selection") will consistently select for small size, early maturation, early birth, early death and all the other classic life history covariant traits. As Stearns (1992) details, this is because there are often many complex trade-offs being made among these traits. Selection may act on body size at a specific stage, growth rate at certain stage, timing of certain events, or all of the above (Stearns, 1992; Bernardo, 1993). One of the main contributions of the comparative method has been to document just how much phylogenetic (i.e., ontogenetic) flexibility there can be in life history traits, allowing selection ("ecology") to alter life history traits in fairly complex ways. But of course some ontogenetic constraints do exist so that life history traits are not completely free to vary from one another. Age of maturation and growth rate, for example, are not entirely uncorrelated so that change in one of these may limit the degree of change in the other (Bernardo, 1993). At coarser scales these general trait constraints are visible; *r–K* is too simple for populations and species, but we discuss below how *r–K* patterns are seen when comparing higher taxa.

In heterochrony, the complexity of body size growth trajectories has undermined many of the naive notions of how rate and timing changes in ontogeny produce evolution. Multiphasic growth trajectories mean that *each phase* can be altered

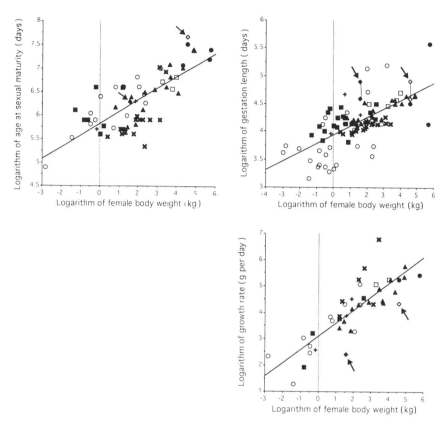

Figure 2.5   Life history traits plotted against female body weight for major carnivore groups and the red panda (*Ailurus*), the giant panda (*Ailuropoda*). Arrows identify the pandas. Modified from Gittleman (1994). ×, Canidae; ●, Ursidae; ◆, *Ailurus*; ◇, *Ailuropoda*; +, Procyonidae; ○, Mustelidae; ■, Viverridae; □, Hyaenidae; ▲, Felidae.

by (the same or different) rate or timing changes (e.g., Figure 2.4). The lag phase (e.g., during embryogenesis) can be extended by late offset, which may (or may not) also alter the growth rate of that phase. Later phases may or may not be affected in the same way, or at all. Taken to an extreme, the application of heterochronic terms thus becomes cumbersome. A descendant may be described as hypermorphic and (simultaneously) accelerated during part of its body size growth, but progenetic, neotenic, or unaffected for other growth phases. This becomes even more cumbersome with mosaics of dissociated heterochronies of local growth fields, such as changes in limb growth.

   This does not mean that heterochronic terms and theory should be discarded. Rather, it points out, as with r–K, that its utility as a descriptor of how ontogenetic covariation translates to phylogenetic covariation is a question of scale and context: heterochrony, like r–K theory, is most useful when applied to the coarse covariation patterns where it was first observed. In the case of heterochrony, this was in comparing general suites of similarities between juveniles and adults of related species. Such suites are very common in evolution because of the constraints of developmental covariation on selection. Thus, "juvenilization" (paedomorphosis) and "overdevelopment" (peramorphosis) will likely always be valuable descriptors of, and more importantly, provide clues about, evolutionary processes. Indeed, we next discuss, how to unite heterochrony with body size and other life history traits into a general scheme of describing ontogenetic and phylogenetic covariation through evolutionary time.

Life history traits and heterochrony

Despite their traditional separation, heterochrony and life history are relatively easy to unify conceptually. Figure 2.6 (top) shows one of the common ways of classifying heterochronic change: a plot of size vs ontogenetic age shows whether the descendant is accelerated, has delayed maturation, and so on (*sensu* Figure 2.3). Size can refer to body size, or any organ size or dimension (e.g., limb length) and thus can include components of shape (Chapter 2 in McKinney and McNamara, 1991 for discussion). Even complex growth trajectories can be compared to see if certain phases are offset, accelerated (have higher slope), or show change in onset. Figure 2.6 (top) shows the simple case where each of three phases (lag, exponential and asymptotic) is accelerated, with no change in offset or onset of each phase.

The bivariate plot of Figure 2.6 unifies the traditional heterochronic focus on morphological (size, shape) change with life history if body size is the y axis. Body size, and its many determinants, are major concerns of life history studies. Such determinants include not only growth rate but also temporal life history events, e.g., age of birth, weaning, maturation or death.

Figure 2.6 (bottom) shows a multivariate plot which unifies heterochrony with temporal, as well as the other major set of life history traits, size-related (i.e., size-constrained) traits such as birth weight and litter size. This can even include population abundance, which is size-related (Calder, 1984). Life history traits on this axis can be "mapped" as they ontogenetically (and phylogenetically) change in concert with age and size. This shows the developmental trajectory of an individual in terms of size, size-related traits, and age. Ontogeny becomes phylogeny when the average trajectories of different species are compared. The size and age axes put this ontogenetic–phylogenetic comparison in terms of classical heterochrony. All three axes put the comparison in terms of classical life history concerns: size, size-related traits, and temporal life history traits (events). We believe that this interconnected view is a more accurate one of the trade-offs that define ontogenetic (i.e., phylogenetic) constraint. Such trade-offs involve not just morphology (size and shape), as emphasized in the heterochronic literature, nor just size nor life history traits and events. Instead, natural selection acts simultaneously on all of these as a single suite of covariant traits.

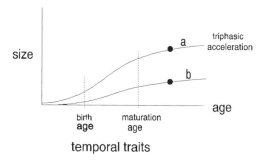

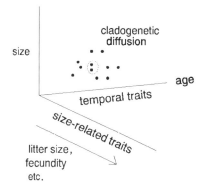

Figure 2.6   Top: Acceleration of growth rate of each of three phases in species a, with no change in offset time for each phase relative to ancestral species b. Bottom: Hypothetical projection of adult morphs from top graph onto an evolutionary plot that demonstrates how the ontogenetic changes translate into phylogenetic changes (cladogenesis). Points show adult morphs of related species whose location in "phylogenetic space" here is determined by changes in their ontogenetic trajectory *sensu* top plot. Circled points represent ancestral morphs from which the other adult morphs were derived (by altering ancestral morph growth trajectory). This plot also adds a third axis, size-related traits such as litter size, that covary with ontogenetic size change. ● = adult morph.

The strength of covariance among these traits in Figure 2.6 will vary, as will the strength of selection on each trait. Indeed, there is a large literature where it is unclear whether evolution occurred from selection solely on morphology, size, or some other life history traits, or some combination of these (e.g., McKinney, 1984; Allmon, 1994). Much of the most promising future work will focus on teasing these apart, especially with experimental and statistical studies using quantitative genetics (e.g., Hall, 1992) and "allometric engineering" (e.g., Sinervo, 1993). However, we hasten to add that macroevolutionary studies, of fossil and phylogenetic patterns, can also provide data on selection and ontogenetic change. Wei's (1994) excellent work on planktic foraminifera exemplifies how the fossil record can be used to correlate palaeoenvironmental changes with changes in size and shape (see also Korn, Chapter 12, herein). Similarly, we show next that phylogenetic analysis is also very useful in determining the role of selection on ontogenetic change. One of its advantages is that, by incorporating ontogenetic age data on living species, phylogenetic analysis can test specific null hypotheses about the timing of life history events and rates of growth in cladogenetic patterns.

A TEST: TEMPORAL VS SIZE-RELATED CONSTRAINTS

As shown above, life history traits are correlated with phylogenetic history due to ontogenetic constraints. We might therefore ask if rate events (i.e., temporal life histories) show more variable patterns with phylogeny than life histories that are constrained by size (i.e., birth weight; litter weight)? A simple descriptive analysis involves analysing the relationship between taxonomic ranks (a surrogate of phylogenetic distance) and various life history variables. In Table 2.1 we see that, across mammalian carnivores, temporal variables such as gestation length and weaning age have lower autocorrelation coefficients than size variables with phylogeny. This makes intuitive sense. Size variables are genetically and geometrically constrained with individual female characteristics (heritability; birth canal). By contrast, temporal variables are likely to be more variable and perhaps to relate more closely with environmental and/or behavioural factors. For example, if food availability declines then a female mammal will have to terminate lactation for energetic reasons (Gittleman and Thompson, 1988). This type of causal

Table 2.1 Comparative relations of phylogeny and life history traits in carnivores. All autocorrelations (*r*) are significant at the 0.001 level; however, temporal life history variables, such as gestation length or weaning age, reveal lower correlations than variables mediated by size such as birth weight or female brain weight (see Gittleman, 1993, for further details of empirical results).

Life history	Sample size	Autocorrelation (*r*)
Female body weight	113	0.799
Female brain weight	113	0.866
Gestation length	97	0.566
Birth weight	68	0.717
Weaning age	71	0.437
Longevity	55	0.411
Age at sexual maturity	65	0.579
Age eyes open	70	0.507
Growth rate—individual	43	0.560
Growth rate—litter	43	0.534
Litter weight	66	0.685

relation is therefore more amenable to heterochronic changes than size variables (see also Figure 2.6). The point is that in searching for life history differences along phyletic lines (i.e., observed heterochrony), those life history traits which exemplify rate change may be more instructive than size-related life histories.

Testing models of heterochrony

In the following, we present an example for testing patterns of phyletic differences in size and temporal change using comparative data on the morphology, life histories and phylogeny of carnivores. All comparisons are based on information from Gittleman (1986a,b, 1993) except for divergence times which are based on estimated first occurrences in the fossil record as given in Wayne *et al.* (1991). We should note that many of the results are not significant. We present them in the spirit of developing an empirical methodology for applying and testing models of heterochrony to life history data.

Form of phyletic change

To test the form of life histories in relation to a model of heterochrony we first remove the effect of body size on each life history trait. Residuals are calculated for each carnivore species so that spurious effects among lineages would not result from size-based allometry alone. Two types of analyses are then performed: species residual values are used as a dependent variable (Table 2.2) or average

Table 2.2 Correlations (r) of estimated first occurrence in the fossil record (mya) with bivariate intercepts and slopes of life history traits and body weight across the terrestrial Carnivora. Sample size equals 8 for all family comparisons except where otherwise noted in parentheses. Correlations of slope are based on reduced major axis estimates of slope for each life history trait plotted on body weight. Levels of significance are as follows: ** $= p < 0.01$; * $= p < 0.08$. See text for discussion and data sources.

Variable	Estimate of first occurence in fossil record (mya)		
	Family	Species (n)	Family slopes
Body weight	−0.31	0.12 (40)	—
Brain weight	0.40	0.20 (40)	−0.29
Gestation length	−0.27	0.28 (38)	−0.09
Birth weight	0.14	−0.12 (30)	0.50
Weaning age	−0.20	0.11 (21)	−0.90**
Longevity	0.40	−0.16 (22)	−0.52
Age at maturity	−0.05	0.25 (23)	−0.69 (7)*
Eyes open	0.20	0.20 (27)	0.64*
Growth rate	0.05	0.28 (39)	−0.52 (6)

family intercepts are used. These measures of life history traits are then correlated with the average age of taxonomic origin; for taxonomic families in the carnivore data set most analyses are restricted to a sample of 8 whereas species samples are as large as 40. As shown in Table 2.2, no correlations are significant between observed (average family) intercepts or relative species values and phyletic time of origin. This shows how the form of a life history variable can be tested against a (null) model of heterochrony using the comparative method.

Rate of phyletic change

The rate of life history evolution can similarly be tested in terms of heterochrony by considering estimates of slope for each life history variable against phyletic change. From the carnivore data set, values of slope were estimated across taxonomic families for each life history on body size using a reduced major axis model (see Harvey and Pagel, 1991). In general, temporal life history slopes are negatively correlated with time of phylogenetic origin such that older taxa (e.g., Ursidae; Mustelidae) have shallower slopes than recent taxa (e.g., Hyaenidae; Herpestidae); an illustration of this pattern is shown in Figure 2.7 using weaning age as a dependent variable. Such observed differences in slope show a consistent pattern in how phylogenetic covariation is systematically related to ontogenetic

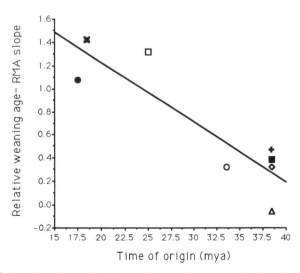

Figure 2.7   Reduced major axis slopes of weaning age vs body size are higher in carnivores that originate later in evolutionary time. There is thus a systematic tendency for phylogenetically later clades to delay weaning age relative to growth rate. These patterns may have become "embedded" in their clade ontogenies as illustrated in Figure 2.6 (bottom). ○, Canidae; □, Procyonidae; △, Ursidae; ◇, Mustelidae; +, Viverridae; ×, Herpestidae; ●, Hyaenidae; ■, Felidae.

covariation: younger branches in the carnivore clade have later weaning ages, at smaller sizes.

EVOLUTIONARY DIFFUSION OF MODIFIED ONTOGENIES

Figure 2.6 (bottom) can be visualized as a diffusional process, where sorting of modified ontogenies produces new species through time. This diffusion is often seen as a result of intrinsic (ontogenetic) innovation which is constrained by a combination of intrinsic limits and environmental sorting of available ontogenies (e.g., Schindel, 1990). All relevant variables in Figure 2.6 (bottom), such as age, life history (temporal and size-related), and size, are subject to environmental selection. The end point (adult species mean location in morphospace) is thus determined by how selection sorts these variables at the aggregate (population) level. Such movement in morphospace in the fossil record has traditionally been measured in terms of adult morphology (e.g., Schindel, 1990; Foote, 1993).

A variety of trends can occur in the diffusional process, representing non-random movement of the "particles" (species loci) in Figures 2.6 and 2.8. McShea (in press) discusses rigorous tests for trend determination (see also McKinney, 1990). We see here that such trends represent selective forces acting on size, shape and morphology throughout ontogeny in some way that affects many species collectively. The well-known trend toward body size increase is often explained as increasing variance from the diffusional process (McKinney, 1990; McShea, in press). In Figures 2.6 and 2.8, this would be depicted as movement of "particles" upward, in the direction of the vertical axis.

Evolution is limited to ontogenetic "tinkering" on size and life history

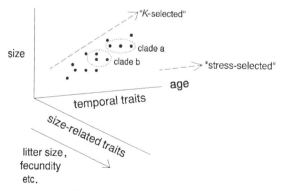

Figure 2.8 Cladogenetic diffusion of species occurs by changes in ontogenetic trajectories. Species in more closely related clades tend to occur in closer proximity in this depiction of morphospace and "life history" space. The general pattern of *r–K*, while not proven for populations, is visible in most coarse-scale, clade comparisons such as this. See text for discussion. ● = adult morph.

variables, so the phylogenetic diffusion process causes autocorrelation of traits with phylogenetic proximity. As shown in Figure 2.8, clades are formed when cladogenetic diffusion in morpho- and life history space produces groups of descendant species. The strength of morpho- and life history space trait autocorrelation may vary among taxonomic levels compared, depending on the trait (e.g., Figure 2.1). This reflects differences in how the individual ontogenetic traits were modified to produce the nested phylogenetic pattern: traits strongly shared (autocorrelated) by higher taxa often represent developmentally "embedded" traits that occur earlier on ontogeny and are thus relatively more constrained than traits appearing later in ontogeny. Conversely, these ontogenetically late appearing traits also tend to appear later in phylogeny, and thereby shared mainly by lower taxa (e.g., species within genera). While exceptions exist to this "neo-von Baerian" view that most evolutionary change occurs later in ontogeny (Wray, 1992), many evolutionists now subscribe to it as a general, statistical tendency (e.g., Levinton, 1988; McKinney and McNamara, 1991). Mabee's (1993) extensive study of centrarchid fishes indicates that terminal alterations to ontogeny account for about 75% of evolutionary character state change (although she notes that artifactual biases affect this estimate).

Given the hierarchical nature of clade branching, we might expect that the patterns of change (radiation and decline) in morphospace discussed above would be seen at many phylogenetic levels, from genera to families and orders. Burlando (1993) discusses such self-similarity in the fractal patterns found in the taxonomic and phylogenetic patterns of fossil and living taxa. He analyses the well-known "hollow curves" of taxonomy in which most genera contain only one species, most families contain only one genus, and so on. His inference is that this arises because radiations consist of many isolated lineages and a few phyletic clumps, with the latter consisting of smaller isolated lineages and phyletic clumps, and so on. At each scale, some taxa proliferate into many parts of morphospace to produce clumping, while others produce only a few or no branches. The dynamic of intrinsic constraint and extrinsic selection of ontogenies occurs at many spatial and temporal scales. This conforms to the hierarchical nature of ontogeny and phylogeny.

Figure 2.8 also conforms to the findings of Stearns (1992) that classic r–K patterns occur at higher taxonomic levels, despite being largely absent at the population and species levels. Large size, late maturation, slower growth, and other "K-selected" traits co-occur as a suite when large-bodied phyla, classes, and orders are compared with smaller-bodied taxa (e.g., comparison of mammals with insects, or artiodactyls with rodents). Stearns (1992) attributes this to life history decisions "made long ago", that have become embedded in the ontogeny of higher taxa. Population abundance, for example, will vary for any given size, but there is a general tendency for smaller sizes to have greater abundances. Similar patterns are observed for other size-related traits, as well as temporal traits. Such patterns are by phylogenetic branching (diffusion) outward from an area of ontogenetic space initially occupied by the ancestors of the higher taxon.

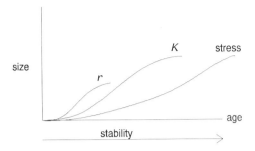

Figure 2.9   Stable, favourable environments select for late maturation and slow growth; stable but stressful environments favour very late maturation and very slow growth.

The location of space initially occupied determined the general region of space where expansion could occur.

Figure 2.8 also shows how "stress-selected" life histories may be viewed. Keddy (1994) has reviewed the works of Grime and Southwood who have promoted stress selection as a supplement to the r–K continuum. A basic aspect of stress selection (Figure 2.9), is that stable, but stressful, environments favour both very late maturation (hypermorphosis) and very slow growth (neoteny) to produce relatively large individuals. When applied to Figure 2.8, this implies that species with very slow growth for their size (rightmost points) may be stress-adapted. Pandas, noted above, have such slow growth, which may reflect the poor nutritional value of bamboo, a usually stable but relatively low-quality food.

HETEROCHRONY AND BEHAVIOUR

Thus far we have considered morphological (size, shape) covariation with life history. But ethological covariation is analogous in many ways. Figure 2.10 illustrates how the ontogenetic trajectory of behaviour includes both temporal and size-related traits. Temporal traits would include onset and offset times of certain behaviours, such as the onset of various motor skills or verbalization. This axis would also include critical periods. The size-related axis would include behaviours directly related to body size. As brain size scales with body size, learned behaviour and behavioural complexity are vectors on this axis.

Figure 2.10 attempts to show how modification of behavioural ontogenies can lead to cladogenetic diffusion in a manner analogous to morphology. Evolution thus involves not only diffusion into morphospace, but also diffusion into "ethospace", by a similar process of ontogenetic modification. Such ethological aspects have received much less study than morphospace, despite the critical role of behaviour in evolution. Indeed, behaviour evolves faster than morphology in

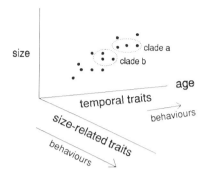

Figure 2.10   Same plot as Figure 2.8 but with the addition of ontogenetic changes in behaviours that translate into phylogenetic diffusion. Temporal behavioural traits include onset and offset times of behaviours, including critical periods. Size-related behavioural traits include those that covary with body size via brain–body allometry.

many groups (Gittleman *et al.*, 1995). One would expect that learned behaviours would be much less constrained in their movement in Figure 2.10. Fortunately, there is growing interest in studying the phylogeny of behaviour, although only a small minority of such studies focus on ontogenetic change (e.g., Burghardt and Gittleman, 1990).

Behavioural allometry

The evolution of many, perhaps most, behaviours occurs from heterochrony. Some of these are evident in phylogenetic analyses of living species (Irwin, 1988; Gittleman *et al.*, 1995). In other cases, breeding experiments can produce specific heterochronic changes (review in Price, 1984). One of the best studies is behavioural juvenilization in domesticated dogs (Wayne, 1986). Cairns (1976) selected for a developmental lag in the onset of aggressive behaviour in a strain of mice.

When plotted as a function of age, most behaviours show the same triphasic patterns seen with organ or body growth, with lag, exponential, and asymptotic phases (e.g., Parker and Gibson, 1990; McKinney and McNamara, 1991). This permits the use of *behavioural allometry* as an analytical tool. The basic bivariate allometric equation, $y = bx^k$, is the solution to a differential equation that reduces to:

$$k = (dy/y)/(dx/x)$$

where k is the allometric coefficient (review in Batschelet, 1979).

Huxley's allometric equation thus eliminates time in the growth rates being compared. The relative growth rates of behavioural phenotypes can therefore be compared in a manner analogous to the growth rates of morphological phenotypes. Constant $k =$ simple allometry with $k > 1$ indicating positive allometry; changing $k =$ complex allometry. Multivariate extensions of this (Shea, 1985) are also analogous. We now discuss two ways to apply behavioural allometry: (1) comparison of behavioural growth to body size growth, and (2) behaviour–behaviour comparison.

McKitrick (1993) and Hanken and Wake (1993) discuss many examples of how behaviour can covary with body size and life history evolution. In many cases, these covariation patterns are relatively simple extrapolations that apparently occur from the direct correlation of brain size with body size. Hanken and Wake (1993), for example, describe fascinating examples where body size miniaturisation results in behavioural simplification, such as in ants and copepods. The apparent cause is an allometric reduction in brain volume.

It is also possible to compare the multiplicative patterns of two behavioural ontogenetic changes (or more than two if multivariate analyses are done). Figure 2.11 shows a bivariate example of positive allometry, whereby the descendant, species a, has triphasic acceleration of behaviour y and no change in x.

Why do behaviours follow a power law?

The genesis of allometric patterns in morphology (size-shape) is cell multiplication (Blackstone, 1987). It is relatively easy to develop models that relate cell multiplication to tissue growth (Katz, 1980). The genesis of behavioural allometry has received very little attention, but must obviously be ultimately related to multiplication of neuron growth and the multiplicative patterns of

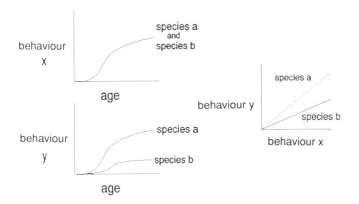

Figure 2.11   Bivariate example of behavioural allometry during the ontogeny of two behavioural traits, x and y. When compared directly, normalizing for time, species a is allometrically accelerated for trait y; the cause of the allometric ("shape" change in "ethospace") is triphasic ontogenetic acceleration of trait y in species a. This assumes species a is the descendant; if species b is the descendant then this shows allometric neoteny of species b.

connectance. Such patterns are manifested in power laws that can be fruitfully analysed with a growing number of fractal-based methods. Hastings and Sugihara (1993) review such methods, including evidence that fractal patterns of neurons arise from the dynamics of pattern formation during development. Diffusion-limited aggregation seems especially important. The power law pattern of behavioural phenotypes thus seems to reflect underlying patterns of information storage and motor cell behaviour that are determined by the underlying neural patterns. Even complex learned behaviours may therefore exhibit this pattern. An example would be the exponential aspect of the well-known "learning curve".

Such allometric analyses may be powerful descriptive and analytical tools for evolution of behaviours through ontogenetic modification. The complex development of many behaviours implies that complex allometry may be found. Human infants, for example, initially lag behind other primates in sensorimotor development, but later far exceed them due to sequential hypermorphosis. Further exploration of behavioural allometry is to be found in McKinney (in press).

ACKNOWLEDGMENTS

M.L.M. gratefully acknowledges support from NSF grant EAR-9316417.

REFERENCES

Allmon, W.D., 1994, Patterns and processes of heterochrony in lower Tertiary turritelline gastropods, U.S. Gulf and Atlantic Coastal Plains, *J. Paleont.*, **68**: 80–95.
Batschelet, E., 1979, *Introduction to mathematics for life scientists*, Springer-Verlag, Berlin.
Bernardo, J., 1993, Determinants of maturation in animals, *Trends Ecol. Evol.*, **8**: 166–173.
Berrigan, D., Purvis, A., Harvey, P.H. and Charnov, E.L., 1993, Phylogenetic contrasts and the evolution of mammals life histories, *Evol. Ecol.*, **7**:270–278.
Blackstone, N.W., 1987, Allometry and relative growth: patterns and process in evolutionary studies, *Syst. Zool.*, **36**: 76–78.
Burghardt, G.M. and Gittleman, J.L., 1990, Comparative behavior and phylogenetic analyses: new wine, old bottles. In M. Bekoff and D. Jamieson (eds), *Interpretation and explanation in the study of animal behavior*, Westview Press, Boulder: 192–223.
Burlando, B., 1993, The fractal geometry of evolution, *J. Theor. Biol.*, **163**: 161–173.
Cairns, R.B., 1976, The ontogeny and phylogeny of social interactions. In M. Hahn and E. Simmel (eds), *Communicative behavior and evolution*, Academic Press, New York: 112–143.
Calder, W., 1984, *Size, function and life history*, Harvard University Press, Cambridge.
Charnov, E.L., 1991, Evolution of life history variation among female mammals, *Proc. Nat. Acad. Sci.*, **88**: 1134–1137.
Charnov, E.L., 1993, *Life history invariants*, Oxford University Press, Oxford.
Cliff, A.D. and Ord, J.K., 1981, *Spatial processes: models and applications*, Pion, London.
Cowley, D.E. and Atchley, W.R., 1992, Quantitative genetic models for development, epigenetic selection, and phenotypic evolution, *Evolution*, **46**: 495–518.
Foote, M., 1993, Discordance and concordance between morphological and taxonomic diversity, *Paleobiology*, **19**: 185–204.
Gittleman, J.L., 1986a, Carnivore life history patterns: allometric, phylogenetic, and ecological associations, *Am. Nat.*, **127**: 744–771.

Gittleman, J.L., 1986b, Carnivore brain size, behavioral ecology and phylogeny, *J. Mamm.*, **67**: 23–36.

Gittleman, J.L., 1993, Carnivore life histories: a re-analysis in the light of new models. In N. Dunstone and M.L. Gorman (eds), *Mammals as predators*, Clarendon Press, Oxford: 65–86.

Gittleman, J.L., 1994, Are the pandas successful specialists or evolutionary failures? *BioScience*, **44**: 456–464.

Gittleman, J.L. and Kot, M., 1990, Adaptation: statistics and a null model for estimating phylogenetic effects, *Syst. Zool.*, **39**: 227–241.

Gittleman, J.L. and Luh, H.-K., 1992, On comparing comparative methods, *Ann. Rev. Ecol. Syst.*, **23**: 385–404.

Gittleman, J.L. and Luh, H.K., 1994, Phylogeny, evolutionary models, and comparative methods: a simulation study. In P. Eggleton and D. Vane-Wright (eds), *Phylogenetics and ecology*, Academic Press, London: 103–122.

Gittleman, J.L. and Thompson, S.D., 1988, Energy allocation in mammalian reproduction, *Am. Zool.*, **28**: 863–875.

Gittleman, J.L., Anderson, C.G., Luh, H.-K. and Kot, M., In press, Comparing behavioral versus morphological evolution. In E.P. Martins (ed.), *Phylogenies and the comparative method in animal behavior*, Oxford University Press, Oxford.

Godfrey, L.R. and Sutherland, M., 1995, Flawed inference: why size-based tests of heterochronic process do not work, *J. Theor. Biol.*

Gordon, I.J., 1989, The interspecific allometry of reproduction: do larger species invest relatively less in their offspring? *Funct. Ecol.*, **3**: 285–288.

Gould, S.J., 1977, *Ontogeny and phylogeny*, Harvard University Press, Cambridge.

Gould, S.J. and Lewontin, R.C., 1979, The spandrels of San Marco and the Panglossian paradigm: a critique of the adaptationist programme, *Proc. R. Soc. Lond.*, Series B, **205**: 581–598.

Hall, B.K., 1992, *Evolutionary developmental biology*, Chapman and Hall, London.

Hanken, J. and Wake, D.B., 1993, Miniaturization of body size: organismal consequences and evolutionary significance, *Ann. Rev. Ecol. Syst.*, **24**: 501–520.

Harvey, P.H. and Clutton-Brock, T.H., 1985, Life history variation in primates, *Evolution*, **39**: 559–581.

Harvey, P.H. and Pagel, M.D., 1991, *The comparative method in evolutionary biology*, Oxford University Press, Oxford.

Harvey, P.H., Read, A.F. and Promislow, D.E.L., 1989, Life history variation in placental mammals: unifying the data with theory, *Oxford Surveys in Evol. Biol.*, **6**: 13–31.

Harvey, P.H., Pagel, M.D. and Rees, J.A., 1991, Mammalian metabolism and life histories, *Am. Nat.*, **137**: 556–566.

Hastings, H.M. and Sugihara, G., 1993, *Fractals: a user's guide for the natural sciences*, Oxford University Press, Oxford.

Irwin, R.E., 1988, The evolutionary importance of behavioural development: the ontogeny and phylogeny of bird song, *Anim. Behav.*, **36**: 814–824.

Katz, M.J., 1980, Allometry formula: a cellular model, *Growth*, **44**: 89–96.

Keddy, P.A., 1994, Applications of the Hertzsprung-Russell star chart to ecology, *Trends Ecol. Evol.*, **9**: 231–235.

Kirkpatrick, M. and Lofsvold, D., 1992, Measuring selection and constraint in the evolution of growth, *Evolution*, **46**: 954–971.

Levinton, J.S., 1988, *Genetics, paleontology, and macroevolution*, Cambridge University Press, Cambridge.

Mabee, P.M., 1993, Phylogenetic interpretation of ontogenetic change: sorting out the actual and artefactual in an empirical case study of centrarchid fishes. *Zool. J. Linn. Soc.*, **107**: 175–291.

Maynard Smith, J., Burian, R., Kauffman, S., Alberch, P., Campbell, J., Goodwin, B.,

Lande, R., Raup, D. and Wolpert, L., 1985, Developmental constraints and evolution, *Quart. Rev. Biol.*, **60**: 265–287.

McKinney, M.L., 1984, Allometry and heterochrony in an Eocene echinoid lineage: morphological change as a byproduct of size selection, *Paleobiology*, **10**: 407–419.

McKinney, M.L., 1986, Ecological causation of heterochrony: a test and implications for evolutionary theory, *Paleobiology*, **12**: 282–289.

McKinney, M.L., 1990, Trends in body size evolution. In K.J. McNamara (ed.), *Evolutionary trends*, Belhaven Press, London and University of Arizona Press, Tucson: 75–118.

McKinney, M.L., In press, Biological evolution and cognitive development. In J. Langer and C. Killen (eds), *Piaget, evolution, and development*, Erlbaum, Hillsdale, NJ.

McKinney, M.L. and McNamara, K.J., 1991, *Heterochrony: the evolution of ontogeny*, Plenum Press, New York.

McKitrick, M.C., 1993, Phylogenetic constraint in evolutionary theory: has it any explanatory power? *Ann. Rev. Ecol. Syst.*, **24**: 89–118.

McNab, B.K., 1983, Energetics, body size, and the limits to endothermy, *J. Zool.*, **199**: 1–29.

McNab, B.K., 1986, Ecological and behavioral consequences of adaptation to various food resources. In J.F. Eisenberg and D.G. Kleiman (eds), *Advances in the study of mammalian behavior*, American Society of Mammalogists, Lawrence, Kansas: 664–697.

McShea, D.W., In press, Investigating mechanisms of large-scale evolutionary trends, *Evolution*.

Miles, D.B. and Dunham, A.E., 1992, Comparative analyses of phylogenetic effects in the life-history patterns of Iguanid reptiles, *Am. Nat.*, **139**: 848–869.

Parker, S.T. and Gibson, K.R., 1990, *"Language" and intelligence in monkeys and apes*, Cambridge University Press, Cambridge.

Price, E.O., 1984, Behavioral aspects of animal domestication, *Quart. Rev. Biol.*, **59**: 1–32.

Promislow, D.E.L. and Harvey, P.H., 1990, Living fast and dying young: a comparative analysis of life-history variation among mammals, *J. Zool.*, **220**: 417–438.

Purvis, A., Gittleman, J.L. and Luh, H.-K., 1994, Truth or consequences: effects of phylogenetic accuracy on two comparative methods, *J. Theor. Biol.*, **167**: 293–300.

Raff, R.A. and Kaufman, T.C., 1983, *Embryos, genes, and evolution*, Macmillan Co., New York.

Read, A.F. and Harvey, P.H., 1989, Life history differences among the eutherian radiations, *J. Zool.*, **219**: 329–353.

Roff, D.A., 1992, *The evolution of life histories: theory and analysis*, Chapman and Hall, New York.

Ross, C.R., 1987, The intrinsic rate of natural increase and reproductive effort in primates, *J. Zool.*, **214**: 199–220.

Schindel, D.E., 1990, Unoccupied morphospace and the coiled geometry of gastropods. In R. Ross and W.D. Allmon (eds), *Causes of evolution: a paleontological perspective*, University of Chicago Press, Chicago: 270–304.

Schmidt-Nielson, K., 1984, *Scaling: why is animal size so important?* Cambridge University Press, New York.

Shea, B.T., 1985, Bivariate and multivariate growth allometry: statistical and biological considerations, *J. Zool.*, **20**: 367–390.

Shea, B.T., 1988, Heterochrony in primates. In M.L. McKinney (ed.), *Heterochrony in evolution: a multidisciplinary approach*, Plenum Press, New York: 237–268.

Sinervo, B., 1993, The effect of offspring size on physiology and life history, *Bioscience*, **43**: 210–218.

Slatkin, M., 1987, Quantitative genetics of heterochrony, *Evolution*, **41**: 799–811.

Stearns, S.C., 1992, *The evolution of life histories*, Oxford University Press, New York.

Thompson, S.D. and Nicoll, M.E., 1986, Basal metabolic rate and the energetics of reproduction in therian mammals, *Nature*, **321**: 690–693.

Thomson, K.S., 1988, *Morphogenesis and evolution*, Oxford University Press, Oxford.

Trevelyan, R., Harvey, P.H. and Pagel, M.D., 1990, Metabolic rates and life histories in birds, *Funct. Ecol.*, **4**: 135–141.

Wayne, R.K., 1986, Cranial morphology of domestic and wild canids: the influence of development on morphology. *Evolution*, **40**: 243–261.

Wayne, R.K., Van Valkenburgh, B. and O'Brien, S.J., 1991, Molecular distance and divergence time in carnivores and primates, *Mol. Biol. Evol.*, **8**: 297–319.

Wei, K., 1994, Allometric heterochrony in the Pliocene-Pleistocene planktic foraminiferal clade *Globoconella*, *Paleobiology*, **20**: 66–84.

Wray, G.A., 1992, Rates of evolution in developmental processes, *Am. Zool.*, **32**: 123–134.

Zelditch, M.L., 1988, Ontogenetic variation in patterns of phenotypic integration in the laboratory rat, *Evolution*, **46**: 28–41.

Chapter 3

BODY SEGMENTATION AND SEGMENT DIFFERENTIATION: THE SCOPE FOR HETEROCHRONIC CHANGE

Alessandro Minelli and Giuseppe Fusco

INTRODUCTION

This chapter reviews heterochronic patterns in segment origin and differentiation within arthropods. Some of our results will also reasonably apply to other groups of segmented animals, but should be viewed with a fair degree of caution. Most developmental processes that all segmented animals have in common, as revealed by the recent advances in molecular genetics of development, have little to do with segmentation proper. In particular, annelids and arthropods have seemingly achieved segmentation independently, rather than inheriting this condition from a common segmented ancestor (Minelli and Bortoletto, 1988). This concept is also supported by increasing evidence from molecular biology, suggesting that these two phyla are not sister groups (Field *et al.*, 1988; Eernisse *et al.*, 1992). On the other hand, transient segmental patterning may occur in animals without any morphological evidence of segmentation in their post-embryonic stages. Wood and Edgar (1994), for instance, point to the fact that although the first larval stage of the nematode *Caenorhabditis elegans* is morphologically unsegmented, it nevertheless has a partially metameric organization, in that it possesses blast cells with similar developmental potential, which are distributed at regular intervals along the antero-posterior axis of the worm.

In arthropods, the layout of body architecture is the combined outcome of several processes (reviewed in Bates and Martinez Arias, 1993), such as (a) establishing anteroposterior and dorsoventral polarities; (b) activating a few foci of earlier and/or more active differentiation, e.g., at body ends, or at the prospective mesothorax in insects; (c) articulating the developing body into

Evolutionary Change and Heterochrony. Edited by Kenneth J. McNamara © 1995 The Editor and Contributors. Published in 1995 by John Wiley & Sons Ltd.

cell-lineage units (compartments), segments, and regions (tagmata); (d) patterning
those units through specific cues mostly coded for by homeotic genes; (e)
moving towards more advanced (larval, adult) patterns, based on those laid
down in earlier developmental stages. Any one of these processes is a potential
field for the action of heterochrony.

SEMANTIC ISSUES

Which kind(s) of body units?

When dealing with heterochronic changes involving the origin and/or the fate of
a series of body units, we need to approach a semantic issue first. There are
several, rather than one, possible meanings for the term "body unit". Units
defined under different perspectives, such as anatomical vs ontogenetic, do not
necessarily overlap but, in the current practice, they are often mixed together,
inadvertently, in many "comparative" treatments.

At least three different kinds of body units need be distinguished here: those of
descriptive morphology, those of functional biology and those of morphogenetics.
For instance, we can speak of segments both within descriptive morphology and
morphogenetics; sometimes, the segments of descriptive morphology are
synonymous with the segments defined in morphogenetics, but it does not
necessarily need be so. One of the most unexpected findings of *Drosophila*
developmental genetics has been the repatterning of the body in definitive
segmental units following an earlier articulation in provisional segments, or
parasegments, which are well individualized morphogenetic units without any
obvious morphological counterpart (Martinez Arias and Lawrence, 1985).
Again, to describe a *Drosophila* wing in terms of compartments (cell-genealogical,
i.e., morphogenetic units) equates with the recognition of two halves, each of
them being a part of a different compartment, which do not correspond to any
meaningful anatomical or functional unit (Figure 3.1).

Within the framework of morphogenetics, we can point to the widespread
occurrence of repatterning events, best exemplified by peracarid crustaceans,
where a strictly patterned cell lineage only provides for "segmented" raw
materials, out of which the definitive, truly segmental structures, such as the
anlagen of appendages, evolve, with no respect for the borders of cell clones
(Dohle and Scholz, 1988).

How to homologize body units

As soon as we restrict our attention to either descriptive morphology, or to
morphogenetics, we need to delve deeper, in order to define the units of discourse
for further comparisons. That means that we must identify suitable homology
criteria in order to split "gross" homology into individual components with

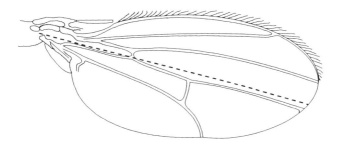

Figure 3.1 A straight line divides the wing of *Drosophila* into anterior and posterior compartments. These cell-lineage units do not correspond to any anatomical or functional units.

recognizably distinct (genetic or epigenetic) informational background (Minelli, in press; Minelli and Peruffo, 1991).

In terms of descriptive morphology, in many arthropods there is a mismatch between dorsal and ventral aspects, with regard to the number of segments or tagmata (body regions), or the borders between them. Let us give some examples of dorsoventral mismatch in segmentation first.

In Diplopoda, the presence of two pairs of legs, but only one tergal or pleurotergal plate, in each typical trunk segment, is only the most conspicuous expression of a still incompletely understood mismatch between the dorsal and ventral aspect of the body. In the platydesmid millipede *Brachycybe nodulosa*, the number of dorsolateral plates (pleurotergites) is not closely correlated with the number of leg-pairs. Individuals with 18 pleurotergites can have any number of leg-pairs between 25 and 29; alternatively, specimens with 28 leg-pairs can have either 17, 18, 19 or 20 pleurotergites (Enghoff *et al.*, 1994: 131–3).

In Pauropoda, the usual nine leg-bearing segments, clearly recognizable on the ventral aspect of the animal, are matched dorsally by five tergal plates only, roughly corresponding to legs I, II + III, IV + V, VI + VII, and VIII + IX, respectively (Figure 3.2).

In Symphyla, on the other hand, the tergal plates are often more numerous (15 to 24) than the "true" trunk segments (14), or the pairs of legs (12).

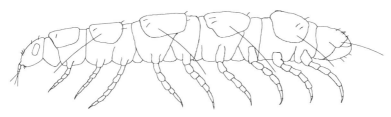

Figure 3.2 Pauropods exemplify the widespread mismatch between dorsal and ventral body segmentation.

In scutigeromorph centipedes, only eight dorsal plates cover the 15 leg-bearing segments of the adult.

Moving from dorsoventral mismatch in segments to dorsoventral mismatch in tagmosis, this last condition is the rule in most groups of malacostracan crustaceans. Just think of the different development of the carapace in the different groups.

How to compare developmental stages

When comparing the ontogenetic stages of different, although closely related, arthropods, we have to decide which stage of species B, if any, is more correctly and meaningfully comparable with a given stage of species A. Simply referring to the ordinal number of the stage, i.e. to the number of moults the animal has undergone, is not a safe procedure, because of the different number of stages that A and B may possibly go through before attaining maturity, or during their whole life. It would be hardly better than "attempting to homologize the 37th vertebra in one species of snakes with the 37th vertebra of a second species", a hypothetical attitude rightly ridiculed by Bock (1989, p. 343). Sometimes, in addition, we are also faced with intraspecific variation in the number of post-embryonic stages. Grandjean (1938, 1951) introduced the concept of *stase*, to indicate a temporal segment spanning between two moults, provided that both of them are accompanied by discontinuous changes in one or more character(s); any number of moults may happen during one stase, if these have no sensible consequence in morphology. The number of stases is small and fixed (but for subsequent reduction) within a given (large) phyletic line. Grandjean's stase may appear as an attractive unifying concept of arthropod developmental biology but, to date, it has found little useful application outside mites, in spite of some efforts to analyse in similar terms the development of spiders and a few other groups (e.g., Vachon, 1953; André, 1989). Heterochrony is possibly one major difficulty in recognizing stases: stase-specific traits are, almost by definition, those we can trace across species at corresponding developmental levels.

In a study of heterochrony in lithobiomorph centipedes (Minelli *et al.*, in press) we have recently suggested a pragmatic approach to the homologization of developmental stages in those arthropods where post-embryonic development is punctuated by a large and variable number of moults not accompanied by major structural changes: we have re-scaled the post-embryonic stadia of these centipedes on a common scale, within three fixed points that identify two contiguous segments, the first spanning between the first larval and the first postlarval stadium, the second between the first postlarval and the first mature stadium (Figure 3.3). A comparable approach was taken by Weisz (1946, 1947) in his study of the development of *Artemia*.

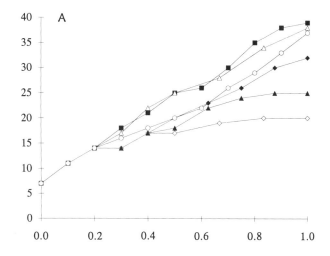

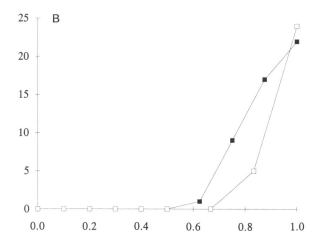

Figure 3.3 Heterochrony in the post-embryonic development of some *Lithobius* species (Chilopoda). The ontogenetic trajectories for individual characters have been plotted onto a measure of "normalized post-embryonic development", with first larval stage and first mature stage at the two ends of the scale (0.0 and 1.0, respectively); the mid-point of the *x* axis corresponds to the first postlarval (but still immature) stage. Individual points along the trajectories refer to the individual post-embryonic stages. A, Antennal segments. ■, *L. forficatus*; ◆, *L. melanops*; ◇, *L. crassipes*; ▲, *L. microps*; △, *L. calcaratus*; ○, *L. tenebrosus*. B, Setae of female first genital segment. ■, *L. microps*; □, *L. calcaratus*. (After Minelli *et al.*, in press.)

HETEROCHRONIC PATTERNS

Heterochrony in embryonic stages

In dipterans, such as *Drosophila* and the house fly, delaying cellularization until a late blastoderm stage speeds up the distribution of cellular markers, thus allowing synchronous segmentation of the whole body, but in other insects the whole embryonic development is accomplished in a cellularized context. In these insects, segments differentiate in anteroposterior sequence, during hours or days. This is, clearly, an instance of heterochrony at the blastoderm stage. The relative timing of segment formation and gastrulation allows two main kinds of insect embryos to be distinguished: those with short-germ ("Kurzkeime", such as dragonflies, or grasshoppers) and those with long-germ ("Langkeime", such as flies); see Sander (1976) for a review.

It is interesting to compare, in these different kinds of insect embryos, the expression patterns of those genes which are responsible for establishing the segmental body architecture. One of these genes is *wingless* (*wg*). In the short-germ embryo of the beetle *Tribolium*, its expression pattern is strictly comparable with that in the long-germ *Drosophila* embryo. The similarity of *wg* expression pattern in these two insects suggests that *wg* serves similar functions in both, as do the expression patterns of the *Tribolium* homologues of other *Drosophila* genes involved in segmentation (gap, pair-rule, segment polarity and homeotic genes). However, as pointed by Nagy and Carroll (1994, pp. 462–3):

> The expression patterns of the *Tribolium* segmentation genes homologues appear in an anterior-posterior progression, rather than simultaneously as in *Drosophila*, consistent with the sequential formation of segments in *Tribolium*. Thus, it could be argued that the difference between long- and short-germ segmentation could be achieved by heterochrony, or a simple change in timing of gene expression. However, there are differences in the early blastoderm expression patterns of the segmentation gene homologues that suggest that the molecular mechanisms operating to create the pattern in the posterior blastoderm of *Tribolium* are more divergent.

Similar comparisons are also possible within one insect order. Patel *et al.* (1994) have investigated the expression patterns of the segmentation (pair-rule) gene *even-skipped* in three beetles: the short-germ *Tribolium*, the intermediate-germ *Dermestes* and the long-germ *Callosobruchus*. The expression pattern of this gene consists of an array of eight transverse stripes, thus giving rise to a two-segment periodic pattern. To quote Patel *et al.* (1994, pp. 431–4):

> At the cellular blastoderm stage of intermediate germ *Dermestes*, even-skipped protein is detected in all cells posterior to 35–40% egg length. In long-germ *Callosobruchus*, this expression boundary is located at 25–30% egg length. In both of these beetles, as in *Tribolium*, eight primary *even-skipped* stripes form sequentially in an anterior-posterior progression. . . . A conspicuous difference between these three beetles is, however, seen in the temporal relationship between morphological

development and the molecular progression of segmentation. In *Tribolium*, the first primary *even-skipped* stripe appears as cell condensation begins, and a second primary stripe resolves at the onset of gastrulation. In *Dermestes*, two primary *even-skipped* stripes have formed at the time cells condense towards the ventral side, two more stripes have resolved by the onset of gastrulation, and the remaining four primary stripes appear during caudal extension. In *Callosobruchus*, the first three primary *even-skipped* stripes are formed by the time cells begin to condense on the ventral side, and three additional primary stripes form by the time gastrulation begins . . . Germ-type designations do not necessarily correlate with a particular mechanism of pattern formation, but . . . at least within the Coleoptera, do accurately predict the temporal aspects of segmentation . . . We postulate that the transitions between short-, intermediate- and long-germ development within the Coleoptera is simply the result of heterochrony. The germ-type differences between *Drosophila* and *Schistocerca*, on the other hand, are more likely to have resulted from changes in the mechanisms that generate anterior-posterior patterns.

In spite of the recent progress in insect developmental genetics, no authentic "heterochrony gene" seems to have been characterized in this group. A few such genes, however, are known in the nematode *Caenorhabditis elegans* (Ambros and Moss, 1994), where these genes affect larval, rather than embryonic, development, and even in plants (Freeling *et al.*, 1992).

Heterochrony in segment building (post-embryonic)

Different timing in the production of body segments has often been advocated as a reason for suggesting (or negating) homology between body regions of more or less closely related animals. In annelids, anterior or larval vs posterior or postlarval segments have been generally distinguished (e.g., Siewing, 1963; Korn, 1982). At variance with the postlarval ones, which are (more or less distinctly) sequentially formed, the larval segments form synchronously, as those of the forehead of several arthropods, i.e., the naupliar segments of crustaceans, from the anostracan brine shrimp *Artemia salina* (Benesch, 1969) to the malacostracan crayfish *Cherax destructor* (Scholz, 1992); early and more or less synchronously formed "naupliar segments" are also recognized in the centipede *Lithobius* (Hertzel, 1984).

The temporal schedule of post-embryonic segment formation is often quite conservative, but a few groups do exhibit an astonishing amount of diversity. For example, all millipedes (Diplopoda) are anamorphotic, but their developmental schedule follows one or the other of the three schemes, which have been recently redefined by Enghoff *et al.* (1994) as *euanamorphosis, hemianamorphosis* and *teloanamorphosis*. In euanamorphosis (Julida, Colobognatha) the number of segments steadily increases at every moult, without any obvious end-point, and only terminates with the death of the animal. In hemianamorphosis (Polyxenida, Glomerida and other groups) only the first few moults are accompanied by an addition of new segments; as soon as the final segment number is achieved, the animal continues to moult without any further addition of segments. Finally, in teloanamorphosis (Chordeumatida, Polydesmida) the addition of new segments ends at a fixed stadium, after which no further moult occurs. In most of those millipedes where sternites do not fuse to dorsal and lateral plates to form complete "segmental rings", there is no fixed and consistent rule as for moult-to-moult segment increase and very little correlation between the moult-to-moult progression in the number of tergites vs sternites. In pill millipedes (Glomerida), for instance, "the gradient for the differentiation of the ventral side (leg-pairs and sternites) and the dorsal side (tergites and pleurites) must be independent to a certain degree" (Enghoff *et al.*, 1994, p. 124). A choice group in which to look for heterochronic patterns of segmentation is the Chordeumatida. In all chordeumatids development is teloanamorphotic. Within each species, the number of stadia is fixed, as is the schedule of stadium-after-stadium increase in segment number and, thus, the number of pleurotergites (equivalent to trunk "segments") in the adult. According to species, this number is 25, 27, 28, 29, 30, or 31, most commonly 29. Sometimes there is a difference between the sexes, the males having 25 rather than the 27 pleurotergites possessed by the females, or 27 vs 29, or 29 vs 31. As Enghoff *et al.* (1994, p. 137) point out, "It would be natural to think that species with 25 pleurotergites were animals maturing in stadium VII, the anamorphosis having been abbreviated by two stadia. However, *Opisthocheiron canayerensis* and *Chamaesoma brolemanni*, studied by Geoffroy (1984) and David (1989), respectively, do not mature until stadium VIII [at which stadium species with 29 pleurotergites in the adult have 27 pleurotergites], but add one pleurotergite less than normal in stadia VI and VII" (Figure 3.4). Outgroup comparisons clearly demonstrate the derived condition of the developmental schedules in these two species, which thus evolved through progenesis and neoteny together.

Most millipedes hatch as larvae with three pairs of fully developed legs, but there are exceptions. A few species anticipate the full differentiation of one or more pairs of appendages. Four pairs of legs in the first stadium are perhaps the rule in Colobognatha, but this is a (quantitatively) minor deviation from the rule, if compared with *Pachyiulus flavipes* (Julida), which hatches with no less than 27 leg-pairs (Dirsh, 1937), or with *Dolistenus savii* (Colobognatha), described by Silvestri (1949) as hatching with 41 pairs of legs!

Still, within diplopods, the pill millipedes *Glomeris* and *Trachysphaera* begin their post-embryonic development with the same number (three pairs) of fully-formed legs and develop anamorphically until they attain the full number of segments (10 tergites in both genera). However, they follow different progression rules as for segment number after each larval moult: the stadium-after-stadium progression in the number of fully-formed leg-pairs is 3, 8, 10, 13, 15, 17 in *Glomeris* and 3, 6, 8, 11, 14, 16, 17 in *Trachysphaera*. Many more examples of this kind of heterochrony can be gleaned from the extensive data on millipede development published by Enghoff *et al.* (1994).

Similar differences in stage progression are found within centipedes (Chilopoda). *Lithobius* and *Scutigera* both have 15 pairs of legs at the end of the initial

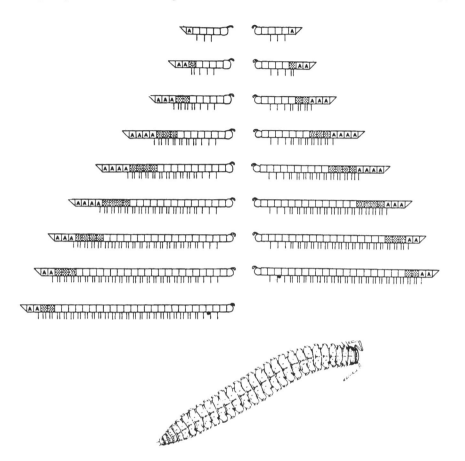

Figure 3.4 In most chordeumatic millipedes (whose habitus is exemplified by that of *Nanogona polydesmoides*), the post-embryonic development runs through the nine stages sketched in the left sequence. In *Opistocheiron canayerensis* and *Chamaesoma brolemanni* the developmental schedule (right sequences) is modified through progenesis and neoteny.

anamorphotic progression, but they start at different levels and go on through different schedules: in *Lithobius*, the number of leg-pairs in the individual larval stadia, beginning with the first larva, is 7, 8, 8, 10, 12, 15; in *Scutigera*, it is 4, 5, 7, 9, 11, 13, 15.

Heterochrony in segmental specialization (differentiation)

The differentiation of the individual metameric structures does not proceed at a uniform pace. In arthropods, the structures of ectodermal origin (including those of the central nervous system) often precede the differentiation of the mesodermal structures of the corresponding segments. For instance, the histological observations of Itow (1986) have demonstrated that first instar larvae of the horseshoe crab (which still have incomplete sets of appendages), and even advanced embryos of the same animal, already have the same number of ganglia as the adult. Itow correctly interprets this finding as meaning that these early stadia already have the same number of segments as the adults, although they lack several pairs of appendages.

Shortage of space prevents us from dealing in detail with heterochronic patterns involving the post-embryonic differentiation of the individual joints of arthropod appendages. Such patterns are very variable within Crustacea, as best exemplified by the very detailed data presented in a recent monograph on copepod evolution (Huys and Boxshall, 1991), mainly based on the study of segmentation of antennules and other appendages, with numerous examples of heterochrony. The same patterns we see in crustaceans are also common in other groups; for instance, in the evolution of the antennae of Heteroptera, Zrzavý (1992) identifies two opposite trends in the evolution of segmental numbers, i.e., segment fragmentation and anarthrogenesis, the latter including both fusion (peramorphosis) and non-division (paedomorphosis).

In several groups of crustaceans heterochronic patterns are clearly detectable within a genus, thus showing both the lability of seemingly basic developmental patterns and the enormous scope for heterochrony in the evolution of their body plan. A lot of useful information on these arthropods has been summarized by Schram (1986).

Menshenina (1990) has documented the progression in the development of thoracopods through several post-embryonic stages (calyptopis III through furcilia VI) in four species of *Euphausia* (krill, Euphausiacea): *E. frigida, E. triacantha, E. crystallophias* and *E. superba*. In the first species, appendage differentiation is gradual and an anteroposterior gradient in appendage differentiation is clearly visible; the other species show increasingly anticipated and tendentially synchronous differentiation of the appendages; this trend reaches its extreme in *E. superba*, where calyptopis III larvae already have fairly developed thoracopods II–VI (all of them at degree 3 of differentiation, scaled 0 to 6; for comparison, all of them are at degree 0 in *E. frigida*, whereas *E. triacantha* has a very small thoracopod II and *E. crystallophias* has small or very small thoracopods II–V, but still lacks thoracopod VI). At the following stage (furcilia I), thoracopods II to VI in *E. superba* are all at degree 4 and all of them attain the final level 6 at furcilia II stage, when *E. frigida* larvae still lack the last two pairs of thoracopods completely.

Heterochronic shifts in the position of the boundary between tagmata

Two main types of tagmosis can be distinguished within copepods, the *podoplean* and the *gymnoplean* condition. In podoplean copepods (all orders other than

Platycopoida and Calanoida), the articulation between prosome and urosome lies between the fourth and the fifth pedigerous somites, whereas in gymnoplean copepods the definitive articulation is between the fifth and the sixth postcephalosomic somites. This last condition, however, is only achieved at the third copepodid stage, following a second copepodoid stage with a functional flexure between the fourth and fifth postcephalosomic somites, i.e., in the same condition as the podopleans, where this is also the definitive position of the boundary. According to Huys and Boxshall (1991, pp. 318–19), the fixation of the tagma boundary one moult earlier in development may indicate a heterochronic origin for the podoplean tagmosis.

Intercalary and progradous growth and differentiation

Anteroposterior polarity is a property we find, time and again, at all major levels of organization, from the cell to the segment to the whole body. Indeed, the major significance of the intersegmental boundary is probably just that: the periodical resetting of the value of anteroposterior positional information, reinstating it to (more or less) the same value it has at the corresponding anterior border of the preceding segment (e.g., Lawrence, 1966; Held, 1992).

A nearly universal rule, for animals in general and arthropods in particular, is the progressive development and differentiation of body parts in an anteroposterior and proximo-distal progression. Molecular genetics of development has recently provided a physical explanation for this correspondence between space and time, i.e., between the position of a given structure along the main body axis and the development time, at which the same structure is specified. This is the well-orchestrated activity of the HOM/Hox homoeotic genes, the pivotal elements of a threefold correspondence: to their linear arrangement along the chromosomes correspond both the regular timing of their sequential expression and the regular anteroposterior sequence of the body parts they specify. This evolutionarily conserved co-linearity is, perhaps, one of the fundamental features of the control these genes exercise over the body architecture of all metazoans (Slack *et al.*, 1993).

There are, however, a few interesting exceptions to this general correspondence between anterior and earlier vs posterior and later, with animals exhibiting *intercalary* differentiation of mid-body segments following the earlier differentiation of the hindmost ones. In larval stomatopods, pleopods (i.e., the appendages of the pleon, the posterior tagma) may develop, according to species, either before or after the pereiopods (i.e., the appendages of the pereion, the intermediate tagma). For instance, in *Squilla* the pseudozoea I of pleopods are already differentiated but pereiopods are not, whereas the opposite happens in antizoea I of *Lysiosquilla*.

In other instances, the delayed differentiation of the appendages of intermediate segments becomes a permanent feature of the adult. That happens to both pairs of maxillae of *Artemia*, which never become functional, in spite of the increasing number of postmaxillary appendages which progressively attain their full development, starting with stage 6. Comparisons with other anostracans are quite interesting. In *Branchinecta*, all appendages differentiate, inclusive of maxillae; moreover, the wave of progressive differentiation regularly moves from the head in a caudal direction. However, developing functional maxillae does not seem to be easy: although formed at an early stage (stage 3, i.e. first postnaupliar stage), until stage 12 the only functional appendages are still the naupliar ones (antennae, antennules and mandibles), i.e., the premaxillary ones; then, at stage 12, there is a burst of differentiation, involving, in one stroke, both pairs of maxillae as well as two following pairs of trunk appendages; in later stages, the differentiation wave goes on much more slowly, simply adding one more differentiated pair of appendages at each stage.

An even more interesting example is *Derocheilocaris* (Mystacocarida). Many heterochronic patterns can be seen when comparing the post-embryonic developmental schedules of three species, such as *D. remanei, D. typicus* and *D. katesae*. First, one of these species (*D. remanei*) has a free naupliar stage, which does not appear in the other ones. Second, in *D. remanei* the number of metanaupliar stages (9) is larger than in the other species (6). Third, and more interesting in this context, the different pairs of postnaupliar appendages appear and/or acquire functional state according to species-specific schedules. A delay in the first appearance and/or in the full differentiation of one or more pairs of appendages in respect to a more posterior pair can be found in all of the species but, again, with species-specific differences. The most unusual heterochronic pattern involves the full differentiation of trunk appendages in *D. typicus* where, starting with the third post-maxillipedal pair of appendages (the same which also anticipates differentiation in *D. remanei*), differentiation proceeds in a cephalic direction through three developmental stages.

Even so, in the case of retrograde morphogenesis, Izawa (1991) has described a few examples within copepods where more anterior segments are younger, not older than more caudal ones, and Ferrari (1993) has documented the fairly widespread occurrence, in the same group, of a more precocious differentiation of more caudal legs, followed by that of more anterior ones. In Ferrari's words, there are exceptions to the rule that anterior is older among serially homologous segments!

TAXONOMIC AND PHYLOGENETIC IMPLICATIONS OF SEGMENTAL HETEROCHRONY

Heterochrony is also probably a major cause of the frequent mismatch between larval and adult patterns of apparent affinities within a given group. To quote Williamson (1982, p. 93):

An . . . example of the relative independence of adult and larval evolution is seen at the generic level in the Pandalidae (Caridea). Within the family we can postulate adult trends toward the loss of thoracic exopods, epipods, and arthrobranchs and larval trends towards the loss of exopods and abbreviation of development, but the adult and larval changes have not been synchronous. The result is that grouping of species based on zoeal characters would split one adult genus and lump several others.

Heterochronic effects on body segmentation have often been proposed as key steps in the origin of higher groups. A traditional example involves myriapods and insects (reviewed in Dohle, 1988). Heterochrony is also advocated by a new hypothesis of insect origins from euthycarcinoids, a less well known group of fossil arthropods. Indeed, McNamara and Trewin (1993), prompted by the discovery of an unusual euthycarcinoid from the Late Silurian of Western Australia (*Kalbarria brimmellae*) have recently sketched a model of euthycarcinoid origins from myriapods and subsequent origins of hexapods from euthycarcinoids, based on paedomorphic loss of appendages. On the other hand, recent molecular evidence seems to shake the traditional belief in a close affinity between insects and myriapods; the former could even turn out to be closer to crustaceans than to any of the myriapod groups (Ballard *et al.*, 1992; Averoff and Akam, 1993; Whitington, 1993). But in that case, too, heterochrony could prove to have played a major role in the origin of the legless abdomen on insects, as in the origin of Maxillopoda, one of the major clades within Crustacea (Newman, 1983).

ACKNOWLEDGMENTS

This work has been supported by grants from the Italian National Research Council (CNR) and the Italian Ministry of the University and the Scientific and Technical Research (MURST) to A. Minelli. We are most grateful to Ken McNamara for his invitation to contribute this chapter.

REFERENCES

Ambros, V. and Moss, E.G., 1994, Heterochronic genes and the temporal control of *C. elegans* development, *Trends in Genetics*, **10**: 123–127.

André, H.M., 1989, The concept of stase. In H.M. Andr and J.-C. Lions (eds), *L'ontogenèse et le concept de stase chez les Arthropodes*, Agar Publishers, Wavre: 3–16.

Averoff, M. and Akam, M., 1993, *HOM/Hox* genes of *Artemia*: implications for the origin of insect and crustacean body plans, *Curr. Biol.*, **3**: 73–78.

Ballard, J.W.O., Olsen, G.J., Faith, D.P., Odgers, W.A., Rowell, D.M. and Atkinson, P.W., 1992, Evidence from 12S ribosomal RNA sequences that onychophorans are modified arthropods, *Science*, **258**: 1345–1348.

Bates, M. and Martinez Arias, A. (eds), 1993, *The development of* Drosophila melanogaster, Cold Spring Harbor Laboratory Press.

Benesch, R., 1969, Zur Ontogenie und Morphologie von *Artemia salina* L., *Zool. Jahrb. Anat.*, **86**: 307–458.

Blower, J.G., 1985, *Millipedes*, Synopses of the British Fauna (New Series), **35**. London.

Bock, W.J., 1989, The homology concept: its philosophical foundation and practical methodology, *Zoologische Beiträge*, N.F. **32**: 327–353.

David, J.-F., 1989, Le cycle biologique de *Chamaesoma brolemanni* Ribaut & Verhoeff, 1913 (Diplopoda, Craspedosomatida) en forêt d'Orléans (France), *Bull. Mus. natn. Hist. nat., Paris*, (4) **11**, sect. A: 585–590.

Dirsh, V.M., 1937, Postembryonic growth in *Pachyiulus flavipes* C.L. Koch (Diplopoda), *Zool. Zh.*, **16**: 324–335 [in Russian].

Dohle, W., 1988, *Myriapoda and the ancestry of insects*. The Charles H. Brookes memorial lecture. The Manchester Polytechnic in collaboration with the British Myriapod Group, Manchester.

Dohle, W. and Scholz, G., 1988, Clonal analysis of the crustacean segment: the discordance between genealogical and segmental borders, *Development*, **104**, suppl.: 147–160.

Eernisse, D.J., Albert, J.S. and Anderson, F.E., 1992, Annelida and Arthopoda are not sister taxa: a phylogenetic analysis of spiralian metazoan morphology, *Syst. Biol.*, **41**: 305–330.

Enghoff, H., Dohle, W. and Blower, J.G., 1994, Anamorphosis in millipedes (Diplopoda): The present state of knowledge and phylogenetic considerations, *Zool. J. Linn. Soc.*, **109** (1993): 103–234.

Ferrari, F.D., 1993, Exceptions to the rule of development that anterior is older among serially homologous segments of postmaxillipedal legs in copepods, *J. Crust. Biol.*, **13**: 763–768.

Field, K.G., Olsen, G.J., Lane, D.J., Giovannoni, S.J., Ghiselin, M.T., Raff, E.C., Pace, N.R. and Raff, R.A., 1988, Molecular phylogeny of the animal kingdom, *Science*, **239**: 748–753.

Freeling, M., Bertrand-Garcia, R. and Ninha, N., 1992, Maize mutants and variants altering developmental time and their heterochronic interactions, *BioEssays*, **14**: 227–326.

Geoffroy, J.J., 1984, Particularités du développement post-embryonnaire du Diplopode Craspédosomide cavernicole *Opisthocheiron canayerensis*, *Mém. Biospéol.*, **11**: 211–220.

Grandjean, F., 1938, Sur l'ontogénie des acariens, *C.R. Acad. Sci. Paris*, **206D**: 146–150.

Grandjean, F., 1951, Les relations chronologiques entre ontogenèse et phylogenèse d'après les petits caractères discontinus des Acariens, *Bull. biol. France Belg.*, **85**: 269–292.

Held, L.I., 1992, *Models for embryonic periodicity*, Monographs in developmental biology **24**. Karger, Basel.

Hertzel, G., 1984, Die Segmentation des Keimstreifes von *Lithobius forficatus*, L. (Myriapoda, Chilopoda), *Zool. Jahrb., Anat.*, **112**: 369–386.

Huys, R. and Boxshall, G.A., 1991, *Copepod evolution*, The Ray Society, London.

Itow, T., 1986, Inhibitors of DNA synthesis change the differentiation of body segments and increase the segment number in horseshoe crab embryos (Chelicerata, Arthropoda), *Roux's Arch. Dev. Biol.*, **195**: 323–333.

Izawa, I., 1991, Evolutionary reduction of body segments in the poecilostome Cyclopoida (Crustacea, Copepoda), *Plankton Society of Japan*, Special Volume: 71–88.

Korn, H., 1982, *Annelida*. In F. Seidel (ed.), *Morphogenese der Tiere. Erste Reihe: Deskriptive Morphogenese*, Gustav Fischer Verlag, Stuttgart, New York. Lief. 5: H-1, 599 pp.

Lawrence, P.A., 1966, Gradients in the insect segment: the orientation of hairs in the milkweed bug *Oncopeltus fasciatus*, *J. exp. Biol.*, **44**: 607–620.

McNamara, K.J. and Trewin, N.H., 1993, A euthycarcinoid arthropod from the Silurian of Western Australia, *Palaeontology*, **36**: 319–335.

Martinez Arias, A. and Lawrence, P.A., 1985, Parasegments and compartments, *Nature*, **31**: 639–642.

Menshenina, L.L., 1990, Some correlations of limb development in *Euphausia* larvae (Euphausiacea), *Crustaceana*, **58**: 1–16.

Minelli, A., in press, Some thoughts on homology, 150 years after Owen's definition. *Proceedings of a workshop held at the Nat. Hist. Museum in Milan*, June 1993.

Minelli, A. and Bortoletto, S., 1988, Myriapod metamerism and arthropod segmentation, *Biol. J. Linn. Soc.*, **33**: 323–343.

Minelli, A., Negrisolo, E. and Fusco, G., in press, Developmental trends in the post-embryonic development of lithobiomorph centipedes, *Proc. IX Int. Congress Myriapodology*, Paris, 25–30 August 1993.

Minelli, A. and Peruffo, B., 1991, Developmental pathways, homology and homonomy in metameric animals, *J. evol. Biol.*, **3**: 429–445.

Nagy, L.M. and Carroll, S., 1994, Conservation of wingless patterning functions in the short-germ embryos of *Tribolium castaneum*, *Nature*, **367**: 460–463.

Newman, W.A., 1983, Origin of the Maxillopoda; urmalacostracan ontogeny and progenesis. In F.R. Schram (ed.), *Crustacean Issues, 1. Crustacean Phylogeny*, Balkema, Rotterdam: 105–119.

Patel, N.H., Condron, B.G. and Zinn, K., 1994, Pair-rule expression patterns of *even-skipped* are found in both short- and long-germ beetles, *Nature*, **367**: 429–434.

Sander, K., 1976, Specification of the basic body pattern in insect embryogenesis. In J.E.Treherne, M.J.Berridge and V.B. Wigglesworth (eds), *Advances in insect physiology*, **12**, Academic Press, New York: 125–238.

Scholz, G., 1992, Cell lineage studies in the crayfish *Cherax destructor* (Crustacea Decapoda): germ band formation, segmentation and early neurogenesis, *Roux's Arch. Dev. Biol.*, **202**: 36–48.

Schram, F.R., 1986, *Crustacea*, Oxford University Press, Oxford.

Siewing, R., 1963, Zur Problem der Arthropodenkopfsegmentierung, *Zool. Anz.*, **171**: 429–468.

Silvestri F., 1949, Segmentazione del corpo dei Colobognati (Diplopodi), *Boll. Lab. Ent. Agr. Portici*, **9**: 115–121.

Slack, J.M.W., Holland, P.W.H. and Graham, C.F., 1993, The zootype and the phylotypic stage, *Nature*, **361**: 490–492.

Vachon, M., 1953, Commentaires à propos de la distinction des stades et des phases du développement post-embryonnaire chez les Araignées, *Bull. Mus. natn. Hist. nat. Paris*, **2**, 23: 294–297.

Weisz, P.B., 1946, The space-time pattern of segment formation in *Artemia salina*, *Biol. Bull.* **91**: 119–140.

Weisz, P.B., 1947, The histological pattern of metameric development in *Artemia salina*, *J. Morphol.*, **81**: 45–95.

Whitington, P.M., 1993, Conservation vs. change in early axonogenesis in arthropod embryos—a comparison between myriapods, crustaceans and insects. In W. Kutsch and O. Breidbach (eds), *The nervous system of invertebrates: a comparative approach*, Int. Symp. Univ. Konstanz, 4.–7.10.1993: 17–18.

Williamson, D.I., 1982, Larval morphology and diversity. In D.E. Bliss (ed.), *The biology of Crustacea*, Academic Press, New York, London: 43–110.

Wood, W.B. and Edgar, L.G., 1994, Patterning in the *C. elegans* embryo, *Trends in Genetics*, **10**: 49–54.

Zrzavý, J., 1992, Morphogenesis of antennal exoskeleton in Heteroptera (Insecta): from phylogenetic to ontogenetic pattern, *Acta entomol. Bohemoslov.*, **89**: 205–216.

Chapter 4

SEXUAL DIMORPHISM: THE ROLE OF HETEROCHRONY

Kenneth J. McNamara

INTRODUCTION

The evolution of sex has long fascinated biologists. When Darwin (1869) wrote on the "remarkable sexual relations" of the flower of the *Primula*, he could not understand:

> why nature should thus strive after the intercrossing of distinct individuals. We do not even in the least know the final cause of sexuality; why new beings should be produced by the union of the two sexual elements, instead of by a process of parthenogenesis.

Darwin thought that:

> The whole subject is as yet hidden in darkness.

In his book *The Descent of Man and Selection in Relation to Sex* Darwin (1871) set the tone for over a century's research on the evolution of sex. For what Darwin set out to explain was why males and females of many animals and plants often looked so different from each other. After all, why should natural selection favour males and females in particular species that were different from each other in some respect? How did it make the species in some way "fitter"? For instance, in some species adult males are consistently larger than the females (e.g., many terrestrial herbivorous mammals (Jarman, 1983)). Yet in other groups (e.g., in orb-weaving spiders (Vollrath and Parker, 1992)) it is the other way round. In some species, as well as body size differences between the sexes, there are differences in the morphology of each sex as a whole, or of certain structures. Furthermore, specific structures may be present on one sex and not on the other,

Evolutionary Change and Heterochrony. Edited by Kenneth J. McNamara © 1995 The Editor and Contributors. Published in 1995 by John Wiley & Sons Ltd.

such as antlers and horns in the males of some species of mammals and beetles. Such differences between species are known as *sexual dimorphism*. Over the last century, research into the evolution of sexual dimorphism has centred almost exclusively on sexual selection—the benefits to survival of the species contributed by the combination of different traits possessed by the two sexes. However, barely any effort has gone into explaining how such differences arise in the first place.

Within the last decade there has been a limited realisation that differences in the growth rates between sexes within a species, or differences in the age at which maturity is reached by each sex, have been important factors in the evolution of sexual dimorphism. Such dimorphism cannot be achieved by extrinsic selective pressures alone. Intrinsic factors are just as important. Although, as such, differences in growth rates and timing of maturation are heterochronic mechanisms, there has been little attempt to articulate them in terms of our current understanding of heterochrony. An example of this is Jarman's (1983) excellent review of sexual dimorphism in large terrestrial herbivorous mammals. While recognising the significance and importance of differences in maturation times and growth rates, it is not explicitly articulated as heterochrony, and therefore its phylogenetic significance is perhaps somewhat understated as a consequence.

In recent years a small number of publications have looked at the impact of heterochrony on sexual dimorphism, such as Shea (1983), Ravosa (1991), German *et al.* (1994) and Ravosa and Ross (1994) in primates; Griffith (1991) in reptiles; Livezey and Humphrey (1986); Livezey (1989) and Björklund (1991) in birds—see also Livezey, Chapter 9 herein. McKinney and McNamara (1991) have briefly addressed the question of the effect of heterochrony on sexual dimorphism. But most current tracts dealing with the evolution of sex have given intrinsic factors such as heterochrony very short shrift. Thus in Stearns' (1986) edited volume on the evolution of sex, the only references to sexual dimorphism discuss the selective advantages of particular cases, not how the dimorphisms might have evolved in the first place. Similarly, a recent analysis of sexual dimorphism in bird tail length (Winquist and Lemon, 1994) centred almost exclusively on sexual selection, rather than on the inherent mechanism that generated the dimorphism. However, any review of sexual dimorphism, particularly in the animal kingdom, should, I believe, discuss the role that heterochrony has played in its evolution. Indeed, I would go so far as to say that, arguably, most, if not all, cases of sexual dimorphism that involve differences to body size or shape, or of the size of shape of particular organs or structures, have arisen due to heterochrony.

In this chapter I will present a broad overview to show that sexual dimorphism, in vertebrates such as mammals, birds, reptiles and fishes, as well as in a number of invertebrate groups, both aquatic and terrestrial, can be accommodated within modern heterochronic theory. As such, this dimorphism can be explained in all groups simply in terms of relative differences in the timing of onset, offset or rate of development of characters between sexes. Whereas in phylogenetic studies of heterochrony, we look at variability in the timing and rate of development between ancestor and descendant, between sexes we can look at differential development between sexes. Thus in the case of a species in which the male is smaller than the female, having achieved sexual maturity relatively earlier, we can say that the male is relatively progenetic compared with the female. Alternatively, we can say that the female is relatively hypermorphic, compared with the male. Such relative heterochrony has its underlying phylogenetic basis in the fact that these differences between the sexes are the product of the two sexes undergoing variable degrees of heterochronic change between ancestral and descendant species. An understanding of the polarity of heterochronic change between species will allow the polarity of differential change between sexes to be assessed.

Perhaps one of the major reasons why the role of heterochrony in the evolution of sexual dimorphism has been so little appreciated is that it has always been seen (not surprisingly really) as an adult characteristic. Yet sexual dimorphism in adults arises from differences in development, not only in the adult and juvenile phases, but in some instances even during very early embryonic development, where differences may exist.

So, to investigate the part played by heterochrony in sexual dimorphism we must look not only at the "tail end" of development, i.e., differences in the timing of onset of maturity and the transition from juvenile to adult, and its resultant somatic effects, but also at early embryonic development. Although such morphological changes are a reflection of changing levels of sex hormones, recent work on the patterns of early embryonic development of the sexes (Mittwoch, 1993) is revealing that even at very early embryonic stages of development in mammals, even before differentiation of the genitalia (and thus before expression of sex hormones) there are inherent differences between males and females that arise from differential growth rates.

To unravel the underlying mechanisms that determine the extent and the form of sexual dimorphism, it is necessary to understand the hormonal control of both sexual development, by sex hormones (testosterone, oestrogen), as well as somatic growth, by growth hormones (somatotrophin). For it is the interplay between these two endocrinal systems, and their differential timing of expression between sexes, that underlies the evolution of sexual dimorphism. Without such hormonal differences between sexes the peacock would not have evolved its splendid tail, and the deer its staggering array of antlers. And in many species, in particular some invertebrates, where the extent of dimorphism is extreme, the species would not have evolved without the dimorphism. For, in many cases, the dimorphism itself is the adaptive advantage possessed by the species to enable it to occupy its specific niche.

Hence, if we are to comprehend fully the evolution of sexual differences within species, we must expand our vision to encompass not only the extrinsic factors that are selecting particular dimorphs, but also the intrinsic factors that have led to their phenotypic evolution in the first place.

SEXUAL DIMORPHISM IN VERTEBRATES

Although many of the classic studies on sexual dimorphism, and in particular its relationship to mating systems, have been carried out on birds (e.g., Lack, 1968; Selander, 1972; Wiley, 1974), within vertebrates as a whole sexual dimorphism is common in all groups. While this dimorphism often takes the form of differences in coloration (in particular, in birds and fishes), many species in all groups of vertebrates demonstrate dimorphism in body size. The degree of dimorphism varies from slight to extreme. In some fishes (see below) this may be as much as an order of magnitude difference in body size. And as with body size differences *between* species being a function of heterochrony (McKinney, 1990; McKinney and McNamara, 1991), so too are such body size differences between sexes *within* species attributable to heterochrony.

Terrestrial herbivorous mammals

One group which demonstrates the variability in the degree of sexual dimorphism particularly well is terrestrial herbivorous mammals. In his review of sexual dimorphism in this group, Jarman (1983) has provided a wealth of data that demonstrates that in many terrestrial herbivorous mammals the bimaturism that is so characteristic of many of these animals (in other words, distinct differences between males and females in age at sexual maturity) is the product of either differences in timing of maturation, or of relative rates of growth. Such bimaturism in birds has been considered by Wiley (1974) as part and parcel of sexual selection, insofar as it is an important corequisite, with a number of ecological factors. But as with many other studies of sexual dimorphism the phenotypic effects of the differences in maturation times, or of growth rates, has been somewhat understated, when compared with extrinsic factors of selection.

Yet these extrinsic factors would have little effect on the species if male and female maturation times were the same. What Jarman's (1983) data does clearly show is that overwhelmingly where there are dimorphic differences in body size, males are usually the larger. This, of course, must have an appreciable adaptive significance. It also results in a number of incidental benefits, notably enhanced weapon growth (i.e., horns, antlers) and the suppression of the reproductive efforts of younger males. While it is possible that a prolonged juvenile phase in males, relative to that of the female, is advantageous, it could alternatively be argued that it is of adaptive significance for females to mature earlier than males, so enhancing their chances of breeding. Most likely it is a combination of the benefits achieved from both earlier female maturation and delayed maturation in males, resulting in a relatively larger body size and consequent improved combative capabilities, that produces the "fittest" species.

From the growth data of 107 bovids that Jarman (1983) analysed, males of three-quarters of the species were shown to be relatively larger than the females. Yet at birth body weights of males and females were generally about the same. What Jarman discovered from his analysis was that within bovids there are three types of growth patterns to maturity:

1. In small species, such as duikers and dik-diks, males and females attain their adult weight at the same time early in life (Kellas, 1955; Dittrich, 1967; Hutchinson, 1970) Thus here there is no heterochrony between the sexes, as regards body size, and consequently minimal sexual dimorphism (Figure 4.1).
2. In a second group of medium-sized animals, males of species such as impala, waterbuck and lechwe reach maturity one to several years after females (Robinette and Child, 1964; Spinage, 1967; Howells and Hanks, 1975). They then retain this weight for the rest of their lives. Juvenile growth appears to be comparable between males and females, indicating that relative to females the males are hypermorphic (Jarman, 1983: Figure 7A,B; see Figure 4.1 herein).
3. In the third group, the males have essentially indeterminate growth, whereas the females may or may not have indeterminate growth. But significantly the females have a slower growth rate. In heterochronic terms, the females are neotenic, compared with the males (Jarman, 1983: Figure 8). Examples are the African buffalo (Sinclair, 1977) and the American bison (D. Lott *in litt* to Ralls, 1977).

Similar growth patterns to these three categories occur in cervids (Jarman, 1983). However, in macropods Jarman (1983) could distinguish only two groups: one of small species in which female and male growth rates and maturation times are comparable; and a second group of medium- to large-sized species, where females grow at a lower rate than males during the juvenile phase (Jarman, 1983: Figure 16D). Like the large bovid group, this medium to large macropod group demonstrates neotenic growth in the females, relative to the males.

From his analysis of terrestrial herbivorous mammals, Jarman (1983) established a nomenclature to describe the types of dimorphism. These groupings can be readily accommodated within heterochronic theory, not surprisingly, because what they represent are the effects of the operation of different heterochronic processes. The first group recognised by Jarman comprises those species where there are no differences in either growth rates of timing of maturation between the sexes—heterochrony was not a factor. This Jarman has termed *homomorphism*. Where differences in maturation times occur between sexes (i.e., relative hypermorphosis or progenesis have operated), Jarman calls it *determinate heteromorphism*. The term "heteromorphism" lends itself quite admirably to heterochronic theory, for if we consider heterochrony as representing the impact of "different times", then the resultant effect will be heteromorphism—"different shapes".

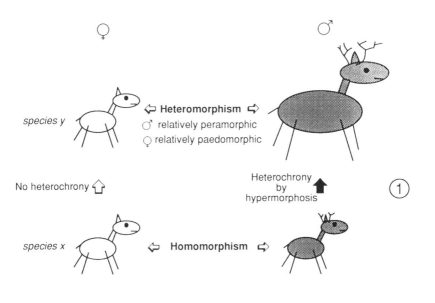

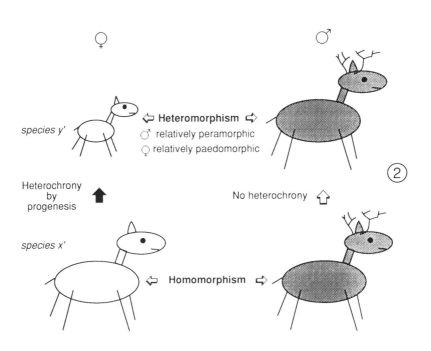

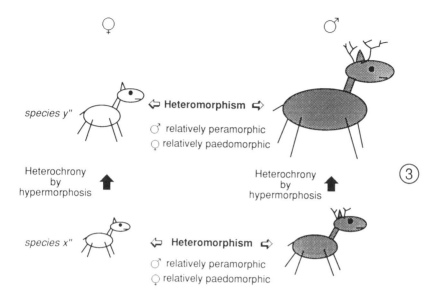

Figure 4.1   Three possible evolutionary scenarios leading to sexual dimorphism in a terrestrial herbivorous mammal by the effect of differential intersexual heterochrony. In Scenario 1 there is stronger selection pressure for larger males—combative reproductive selection. In Scenario 2 there is selection pressure favouring a smaller female—earlier maturation promoting earlier reproduction. Scenario 3 implies equal selection pressure on large body size for both males and females. There could be two types of Scenario 3: one in which the extent of hypermorphosis is the same in the two species; and another where there is greater hypermorphosis in males than in females. In a long-term phylogenetic sense, Scenario 1 is likely to presage Scenario 3. A fourth, but unlikely, scenario, would see evolution from a homomorphic ancestral pair, the female evolving by progenesis, and the male by hypermorphosis. This would produce extreme heteromorphism.

Jarman (1983) further recognised two types of heteromorphism: *telic heteromorphism* and *ecdochic heteromorphism*. In the former, males mature from one to three years after the females, i.e., they are relatively hypermorphic. Furthermore, the presence or absence of weapons or ornaments (e.g., horns) distinguishes subadults from adults. Growth of such structures ceases at the onset of maturity. Species that show the latter (ecdochic heteromorphism) are those that demonstrate greatest sexual dimorphism in body size. It is seen most in medium-sized mammals in which the males reach sexual maturity more than three years after the females. Moreover, weapons and ornaments continue to grow after maturity, with the result that distinct age classes can be categorised within adult males, over and above pure body size differences. For example, red deer are ecdochic heteromorphs, the number of tines in the antler increasing each year, through much of the adult male's life. If horns are present in females of ecdochic heteromorph species, then they cease to grow after maturity is reached.

The third major group of dimorphism in terrestrial herbivorous mammals categorised by Jarman (1983) includes those species in which at least one sex has indeterminate growth. Although not common in mammals, such a growth strategy does occur in some of the very large species, such as the African elephant and in some larger bovids and macropods. This growth strategy Jarman calls *holobiotic heteromorphism*. In such species, the sexual dimorphism does not arise from relative hypermorphosis or progenesis; in other words it is not maturation age determined, but is due to different growth rates. Thus if, as is generally the case in species of this group, males grow with a faster growth rate (i.e., they show relative acceleration) then they attain a larger body size as adults and the disparity increases throughout their life.

What these differences between mammalian groups in modes of attaining dimorphism reveal is that dimorphism in structures other than body size (e.g., antlers) need not necessarily be a simple allometric consequence of body size. Size and complexity will also be a function of the type of heteromorphism, i.e., whether or not the structure ceases growth at the onset of maturation or not. In other words, in telic heteromorphism juvenile allometries will be critical. In ecdochic heteromorphism adult allometries, as well as juvenile ones, will be a factor.

In the mammals that show ecdochic heteromorphism why should body growth, but not horn development, cease at maturity? Jarman (1983) has argued that perhaps there is a physiological limit to body size, particularly if the species are inhabiting a seasonal, fluctuating environment. There is probably selection for cessation of growth at maturity because the number of females in a male's group will eventually be limited by external factors. Therefore it becomes profitless to attain a larger body size. However, since courtship and mating will be very important to reproductive success, the larger the weapons the better. Viewed in terms of telic and ecdochic heteromorphism the relatively larger antlers possessed by the Pleistocene "Irish elk", *Megaloceros giganteus*, may reflect ecdochic heteromorphism, rather than just a simple extension of ancestral allometries, as Gould (1974) has argued.

From a heterochronic perspective the continuation of growth of structures such as antlers can be interpreted as a local peramorphosis by hypermorphosis of this structure, relative to body size itself. In holobiotic heteromorphs, the rank of the male is directly related to body size. Highest frequencies of mating are achieved by highest ranked males. In such a situation there is nothing to be gained by stopping growth at the onset of maturity. Thus by accelerating growth in the males at a faster rate than in the females, and becoming relatively larger, the male is adaptively fitter, in terms of its likely reproductive success. The same selection pressure does not apply to the female; thus a lower growth rate, which is energetically more efficient, is adaptively more successful.

Primates

The general trend of greater degree of sexual dimorphism with greater body size has been explained in a quantitative genetic model by Leutenegger and Cheverud

(1985) and has been similarly recorded in primates (O'Higgins and Dryden, 1993). Thus in large apes the degree of size difference is greater than in smaller apes. Sexual dimorphism in skull shape in *Gorilla* and *Pongo* has been interpreted by Shea (1983) as a function of hypermorphosis, and the resultant scaling effects of extrapolating ancestral allometries. Interestingly, in the chimpanzee, *Pan troglodytes*, there is a significant difference in size between adult males and females, yet there is little shape difference between the two (O'Higgins and Dryden, 1993). Extrapolation of allometries, by males being relatively hypermorphic to the females, would be expected to result in shape changes arising from extending allometries.

O'Higgins and Dryden (1993) have suggested that the reason for the lack of variation in shape between the sexes may be a result of the fact that the intersexual size differences are relatively small. Thus the relatively peramorphic nature of the male is insufficient to have any appreciable shape effect. Another possibility may be that the allometric coefficients are generally not very great in *Pan*. In the larger *Gorilla* and *Pongo* the shape differences between the females and males are significant, particularly in the face and braincase shapes. This is more than likely due to the greater degree of size dimorphism, whether or not the allometric coefficients are any different from those of *Pan*.

Even between the two species of chimpanzee, *Pan paniscus* and *Pan troglodytes*, there are significant differences in the degree of sexual dimorphism, particularly of certain morphological characters. Shea (1983, 1989) has discussed how the facial growth of *P. paniscus* is neotenic, compared with that of *P. troglodytes*, resulting in a relative reduction in sexual dimorphism in the gnathic and facial region. Significantly, there is less dimorphism in the canine teeth in *P. paniscus*. The consequence of this is that there appears to be lower male–male and male–female aggression in this species, compared with *P. troglodytes*, along with increased female bonding and increased food sharing. Subtle differences between species in the intersexual growth rates can therefore lead to appreciable behavioural differences.

In their study of the role of heterochrony in sexual dimorphism in the pigtailed macaque, *Macaca nemestrina*, German *et al.* (1994) found that the sexual dimorphism was not a consequence of simple rate differences between the sexes, but due to a "complex interaction of timing and rate over the entire period of growth". As they point out, growth is essentially non-linear in nature, making elucidation of the underlying processes difficult. What German *et al.* (1994) found was that there appeared to be two distinct growth spurts during development, one before the animals were one year old, the other after about two years (Figure 4.2). Adult males are larger and heavier than the females. This in part is a consequence of the larger growth spurt in juvenile males at the earlier growth spurt. Thus the sexual dimorphism manifested in adults begins to appear quite early in ontogeny. The first phase of male growth extends for a longer period than the female, indicating the operation of sequential hypermorphosis in the males (McKinney and McNamara, 1991). So at the end of this first phase of growth the males are already larger than females of the same age. In the second phase of

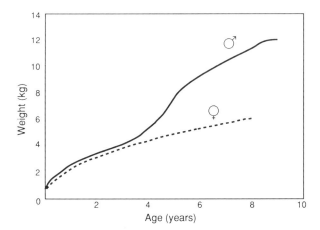

Figure 4.2   Growth data (weight and years) for male (solid line) and female (broken line) pigtailed macaque (*Macaca nemestrina*). Amended from German *et al.* (1994: Figures 5 and 6). Two phase growth curve of the males results in attainment of much larger body weight than females.

growth the males also grew at a faster rate as well as for a longer period. Thus in addition to a second phase of sequential hypermorphosis, acceleration was also operative. The result of these two phases of differential growth between the males and females was thus pronounced sexual dimorphism. A similar phenomenon has been reported in long-tailed macaques (Ravosa, 1991), rhesus macaques (Watts, 1985; 1986) and in howler monkeys (Ravosa and Ross, 1994). In chimpanzees sexual differences appear before adolescence (Watts, 1985; 1986), indicating acceleration in male development, relative to females. Baboons would seem to be dimorphic in body weight from birth, this dimorphism becoming more pronounced during adolescence (Coelho, 1985; Glassman and Coelho, 1987). In proboscis monkeys (*Nasalis larvatus*) the male peramorphosis is the product of only hypermorphosis (Ravosa, 1991).

The toque macaque (*Macaca sinica*) from Sri Lanka also shows a complex pattern of growth (Cheverud *et al.*, 1992). During the juvenile period females have two phases of growth, males three. Sexual dimorphism is minimal in infants and young juveniles, up to two and half years old. However, after that sexual dimorphism, arising from sequential hypermorphosis in the males, starts to become apparent. Skeletal limb growth ceases in juvenile females at about five and a half years of age, but continues in males until about seven and a half years. Muscle mass shows a similar pattern of sequential hypermorphosis in the males, muscle growth ceasing in females at about eight years, but in males not until 12 years. Males also undergo an adolescent growth spurt by acceleration. Thus, as Cheverud *et al.* (1992) point out, while much of the sexual dimorphism is a result of bimaturism, there are also relative differences in growth rates in older juveniles, which contribute significantly to final adult dimorphism. In his review of sexual dimorphism in 45 species of primates, Leigh (1992) noted that dimorphism in many primate species arises from a combination of both bimaturism and rate differences. Combinations of relative hypermorphosis and acceleration in males will therefore lead to accentuated dimorphism, greater than can be achieved by the operation of one process alone. This argues for strong selection pressure on larger male body size in primates.

In *Homo sapiens* sexual dimorphism is largely a result of differential timing in onset of sexual maturity. However, as I discuss below, Mittwoch (1993) has shown that even at very early developmental stages growth rates between male and female foetuses are different, with males growing at a slightly faster rate. Rates of post-embryonic juvenile growth in humans vary slightly between the sexes. Body size differences, between the sexes arise, in general, due to the delay in onset of maturity in males. This is about two years later in males than in females, on average. Males thus continue at the faster juvenile growth rates for longer than females, as in many other primates. In heterochronic parlance human males are hypermorphic, compared with females. The relatively paedomorphic features retained by females include small faces, reduced cranial sinuses, decreased cranial robusticity and cresting and relatively larger brains (Shea, 1989).

Reptiles

Sexual dimorphism, involving both differences in overall body size, as well as size and shape of particular morphological features, has also been documented in many reptilian groups. The effect of heterochrony on sexual dimorphism in a group of skinks, the *fasciatus* group of the genus *Eumeces*, from North America, has been studied by Griffith (1991). While heterochrony has clearly been a controlling factor in generating the morphological differences between the species, the heterochronic mechanisms that generated them are significantly different from those that contributed to the evolution of sexual dimorphism in many mammals (see above).

Griffith (1991) found that the five species within the *fasciatus* group are sexually dimorphic, both in overall body size and in relative head size. The greatest sexual dimorphism occurs in *E. laticeps*, which is also the largest species. This parallels the situation found in mammals, with larger species demonstrating greater sexual dimorphism. *E. laticeps* is peramorphic, not only hatching at a larger size than the other species, but growing at a faster rate. In females and juveniles the head allometries are not significantly different, and have a negative coefficient. However, an increase in rate of growth of the head in males after reaching sexual maturity results in them having a relatively larger head, the allometric coefficient becoming positive. Such an increase in growth rate at maturity is in contrast to the situation in mammals and many other groups of organisms where growth rates are usually strongly attenuated at the onset of maturity. The result in *Eumeces* is thus an allometric divergence between males and females that commences at the onset of sexual maturity. Although life spans and maturation times are identical in males and females, the increase in male allometric coefficients produces pronounced sexual dimorphism in both body size and relative head size. Juvenile head width shows negative allometry in the four small species, but in the larger *E. laticeps* it is near isometric.

As with the situation in terrestrial herbivorous mammals, the larger species show greater dimorphism, particularly in head size, in this case because the allometric coefficients are greater, not because the existing allometries are extended. In comparison with other species *E. laticeps* is peramorphic. This was produced by acceleration; time of onset of maturation is the same as in the other species, 21 months. The relative amounts of male and female growth, and adult and juvenile growth do not vary significantly between the species. Thus the greater dimorphism in *E. laticeps* is a consequence of its accelerated growth, compared with other species.

The different method of attaining sexual dimorphism in these species, compared with terrestrial mammals, is a direct consequence of two factors: male territoriality and a rigid seasonal breeding system. As with many other reptiles, larger male body size plays an important role in territoriality. However, to have "achieved" this by hypermorphosis, i.e., by delaying the onset of sexual maturity, would have been maladaptive. Griffith (1991) has noted how the maturation time in the *fasciatus* group coincides with annual seasonal changes. Reproduction occurs the second spring after hatching. There is a rigid seasonal breeding system with a relatively short breeding period, just a few weeks in late spring. Thus the only way for males to achieve a large adult size, yet still be able to breed on cue, is to accelerate growth following breeding.

In a review of sexual dimorphism in six species of Australian elapid snakes, Shine (1978) found that males generally mature at a slightly younger age than females. The result is that the females attain a larger body size than the males. Shine argues that the positive selection pressure on female hypermorphosis arises from the fact that female fecundity is proportional to body size. In males, however, reproductive success is not correlated with body size. Shine reports that similar bimaturism is apparent in a range of other snakes, such as colubrids, viperids and critalids.

Fishes

Some fishes show extreme sexual dimorphism. Females of the deepsea ceratiid fish (seadevils) are characterised by their large size, up to 145 cm in length. However, males are extremely small, never exceeding 16 cm in length, and relatively paedomorphic. In some species, such as *Ceratias holboelli*, the paedomorphic male attaches to the underside of the female (Pietsch *in* Smith and Heemstra, 1986). In *Photocorynus*, the small, 15 cm long, male is attached to a protruding structure on the head of the 1 m long female. Without adequate growth data it is not possible to be certain whether the male paedomorphosis arose by neoteny or progenesis.

Other, less dramatic, examples of sexual dimorphism in fishes are clearly the result of heterochrony. Hutchins (1992) has shown that 57 species of monacanthid fishes show some form of sexual dimorphism in body and fin shape. Significant variation occurs in 21 species. For example, females of *Eubalichthys mosaicus* much more closely resemble the juveniles in body shape than they do the males (Figure 4.3). This is because the males undergo a greater degree of ontogenetic change than the females, making them relatively peramorphic. Whether this is because they mature later than the females or whether they have accelerated growth is not known. There are also substantial male/female differences in

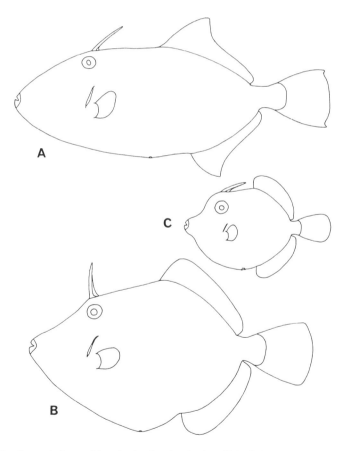

Figure 4.3   Sexual dimorphism in the leatherjacket *Eubalichthys mosaicus*, showing the relative paedomorphic similarity of body shape of the female (B), compared with the juvenile (C), in contrast to the body shape of the male (A). Reproduced from Hutchins (1992, Figure 1), with permission of the Trustees of the Western Australian Museum.

elements of the skeletal structure, such as the pelvis. Again, the female resembles the juvenile. What is particularly interesting is that Hutchins (1992) considers that the difference in fin shape between the male and female explains differences between the two in the way that they swim. So again here we have an example where developmental changes between sexes can induce behavioural differences too.

SEXUAL DIMORPHISM IN INVERTEBRATES

Molluscs

Sexual dimorphism is common in many groups of marine invertebrates, ranging from extreme to subtle differences. Many of these have evolved by heterochrony. A typical example is the galeommatacean bivalve *Pseudopythina rugifera*. Females of this species attain a valve length of up to 15 mm and are ectocommensal on either a species of mudshrimp, or one of two species of polychaete worms. Within their mantle cavity they often house a minute, but sexually mature, male that reaches no more than 1.25 mm in length (O Foighil, 1985). However, these males do not remain as paedomorphic males for their entire life span. Following fertilisation of the host female, they continue their development outside the female, becoming first hermaphrodites, then females. Similar paedomorphic males have been described in other members of the Galeommatacea.

In many marine molluscs varying degrees of more straightforward sexual dimorphism induced by heterochrony occur. In all cases the males are paedomorphic relative to the females. In the prosobranch gastropod family Eulimidae (Waren, 1983) all species with separate sexes show a range of sexual dimorphism, males varying between 0.1 and 0.7 times the size of the females. In other eumilids the smaller male will occur parasitically on the female. In *Thyca crystallina*, the tiny paedomorphic male clings to the right side of the foot of the female. Whereas the female has a strongly ornamented shell, the male shell is smooth and very thin. Whether the male is relatively progenetic or neotenic is not known.

However, this degree of sexual dimorphism pales into insignificance when compared with the extraordinary extent of dimorphism that occurs in a group of prosobranch gastropods that parasitise holothurians: *Enteroxenos, Gasterosiphon, Entocolax, Entoconcha, Thyonicola* and *Diacolax*, (Lützen, 1968). For example, in *Enteroxenos oestergreni* the female is endoparasitic in the intestine of the synaptid holothurian *Stichopus tremulus*. The male is so precociously developed that for a long time the species was considered to be hermaphroditic. However, Lützen showed that the so-called testis of *E. oestergreni* is not a male gonad, but an extremely paedomorphic, presumably progenetic, male. After having lost its larval shell the male enters the female through a ciliated tubule connecting the female to its host's oesophagus. The paedomorphic male then attaches itself to a special male receptacle and expands into what is really little more than a testis. In such species the parasitic occurrence of the male within the female ensures an increased chance of fertilisation of the eggs in a relatively unstable habitat. Thus the strongest selection pressure ensuring optimum survival of the species is the paedomorphic state of the male, enabling localised fertilisation and optimum fecundity.

Spiders

One of the classic examples of sexual dimorphism in body size is portrayed by spiders. There is great variability in the extent of dimorphism, not only between different families, but also even within single species (Vollrath and Parker, 1992), with males typically being smaller. Body size dimorphism occurs more often in sedentary, web-building spiders, and much less in active, roaming hunters. Vollrath and Parker (1992) have demonstrated how in a typical orb spider, such as *Nephila*, males are relatively progenetic, maturing earlier than females. This results in females attaining a larger adult body size than the males. The selective advantage of such differential timing in onset of maturation times is thought to be attributable to larger female size resulting in higher fecundity. Furthermore, the males benefit because being more mobile they suffer higher levels of predation. Consequently, by maturing relatively earlier, male juvenile mortality rates are reduced (Vollrath and Parker, 1992).

Beetles

The males of many species of beetles develop either greatly enlarged front legs, mandibles (Figure 4.4), horns (Figure 4.5) or spear- or fork-like structures on the head or prothroax (Darwin, 1871; Crowson, 1981). There is a close analogy between the development of horns in some herbivorous mammals and in some species of beetles, both in terms of their development and selective advantage. As Crowson (1981) has pointed out, such features are the product of positive allometry, being more prominently developed in larger individuals. Even within males this character may be dimorphic. Cook (1987) has shown how in the dung beetle *Onthophagus ferox* the sexual dimorphism involves, in part, the presence of horns on the males, but not on the female; while there are also two distinctive male morphs: one with small body size and no horns, the other larger and with horns. Such male dimorphism, arising from allometric scaling of horn size, occurs in a number of other beetles, illustrating the formation of such horns occurs as a result of hypermorphosis or acceleration, in conjunction with strong positive allometry in the structure (Eberhard, 1982).

The size of the beetle, and consequently whether or not it develops particularly large dimorphic structures, depends, in part, on the availability of resources. In

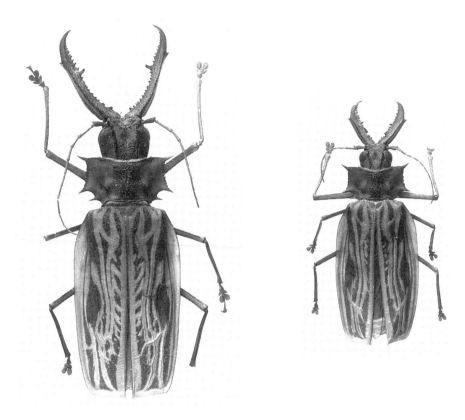

Figure 4.4   Sexual dimorphism in the beetle *Macrodontia cervicornis* from Guyana. The male (left) attains a larger body size than the female (right). Extension of existing positive allometries, particularly of the mandibles, results in their relatively greater size in the male than in the female. Both at two-thirds size.

the forked fungus beetle *Bolitotherus cornutus*, for instance, those individuals that as larvae feed on more nutritious fungi emerge as adults at a larger size, and attain a larger adult size. As a consequence males develop larger horns than those feeding on less nutritious fungi (Brown and Rockwood, 1986). Furthermore, Brown and Rockwood found that the developmental state of the fungus can affect growth rates. Beetle larvae growing in younger fungi typically grow faster and attain a larger body size than those feeding on older, less nutritious, fungi.

Heterochrony resulting in sexual dimorphism in beetles can also be global in its effects. In a number of groups, such as the Drilidae, Phengodidae, Lampyridae (glow-worms), Stylopidae and Rhipidiinae, adult females are relatively paedomorphic, retaining a number of larval characters, including winglessness.

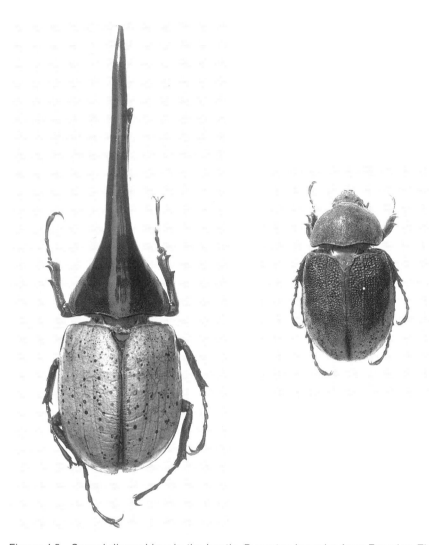

Figure 4.5 Sexual dimorphism in the beetle *Dynastes hercules* from Ecuador. The male (left) has evolved a massive cephalic horn, probably by acceleration, whereas the female (right) has not. Both at three-quarters size.

EFFECT OF HORMONES ON SEXUAL DIMORPHISM

When dealing with heterochrony affecting somatic growth we need to understand the underlying endocrinal control that is responsible for the timing and rate of development. Any changes to these will be a reflection of changes to timing of initiation, amount or cessation of production of hormones that control growth and maturation. Some authors (Privratsky, 1981; Shea, 1989) have argued, for instance, that paedomorphic changes arising from neoteny can be accounted for by changes to the production of growth hormones. However, heterochrony arising from changes to the timing of growth offset, perturbations to which can produce either progenesis or hypermorphosis, is often related to the timing of expression of sex hormones. This coincides in many cases, particularly in many mammals, with a reduction in production of growth hormones. Sex hormones also have the effect of triggering growth in secondary sexual characters. So, in order to understand the role of heterochrony in generating sexual dimorphism we need to understand the role of these various endocrine systems, and the degree to which they interact, in order to understand the processes that lead to the evolution of sexual dimorphism.

Vertebrates

The most important hormone influencing post-natal growth in mammals is known as insulin-like growth factor 1 (IGF-1). Production of this in various parts of the body is under the control of the growth hormone somatotropin. This hormone prepares the targeted structures for response to IGF-1. The amount of growth hormone formed in the pituitary gland is itself under neuroendocrine control (Ludecke and Tolis, 1985; Raiti and Tolman, 1986). Shea *et al.* (1990) have shown how elevating both growth hormone and IGF-1 levels in transgenic mice results in overall increased growth of many skeletal elements. However, these changes were in accordance with the extent of pre-exisiting proportionate growth. These changes occurred by an acceleration in growth. Thus, by changing levels of growth hormone, particularly during post-natal and pre-adult growth, acceleration (producing peramorphosis) or neoteny (producing paedomorphosis) will ensue, if hormone levels are increased or decreased, relative to the ancestral condition. Thus any variations that occur between males and females attributable to changes in rates of growth may arise from intersexual differences in growth hormone levels.

Whereas onset of sexual maturity in arthropods arises from cessation of production of juvenile hormone, in vertebrates it arises from increased levels of production of either testosterone, in males, or oestrogen, in females. Frequently, onset of sexual maturity is also accompanied by reduction in production of growth hormone, resulting in reduction, then effective cessation, of somatic growth. As many groups of vertebrates show, differences between males and females arising from changes to the timing of increased sex hormone levels, and concomitant reduction in growth hormones, will produce sexual dimorphism, and either relatively peramorphic or paedomorphic sexes. Yet there may also be intersexual variation in the timing of onset of sex hormone production in the developing embryo, producing relative pre- or postdisplacement.

Shine and Crews (1988) have shown how sexual dimorphism in the garter snake

Thamnophis sirtalis parietalis is under endocrine control. The faster growth rate in females arises from growth inhibition in the males due to testosterone activity. When males are castrated their growth rates are comparable with those of females. This dimorphism mainly affects head size. However, it is expressed long before the onset of sexual maturity, indicating that the hormonal effect on growth is operating *in utero*. Indeed, Shine and Crews (1988) argue that it is the short-term effect of increased androgen levels early in development that determines the lower growth rate. Thus when testosterone levels are high at a critical embryonic phase, growth rates are lower, indicating reduction in growth hormone production. However, the same androgen activity during the post-embryonic, juvenile phase will result in a much greater reduction in growth hormone levels.

Likewise, hormones are thought to be responsible for sexual dimorphism in plumage in fringillid birds (Björklund, 1991). This dimorphism arises as a consequence of female paedomorphosis, male bright plumage being a function of the relatively late maturation of the males. Björklund (1991) considers that the bright plumage of the male can be correlated with the strong development of the gonads and high testosterone levels.

The degree of sexual dimorphism between related species within a clade can vary over time. Using the fossil record of Neogene peccaries, Wright (1993) has shown patterns of five separate paedomorphic events whereby the zygomatic process and the degree of canine dimorphism are reduced. The changes are not a consequence of allometric scaling but are probably neotenic reductions in these structures. Lorber *et al.* (1979) have demonstrated how canine dimorphism in dogs is under endocrinal control, specifically by gonadotropins. Wright (1993) has suggested that such sex hormone control of canine growth may have influenced the varying degrees of sexual dimorphism within peccaries over time. As Wright argues, "Alteration of a character's response to sex hormones . . . could be a common mechanism for producing sexual dimorphism . . . Characters under hormonal control could conceivably be gained and lost in males or females simply by altering hormone titers [*sic*] during critical periods of development."

In utero *sexual dimorphism*

Male development in mammals is induced by specific signals from the foetal testis (Wilson *et al.*, 1981). In males, cells appear in the testis that synthesises testosterone during the first trimester of development. In females, on the other hand, there is little histological change until the second trimester, when ovarian follicles develop. As Jost (1972) put it:

> The early and very fast differentiation of the testis must be contrasted with the very late and slow differentiation of the ovary.

Again, rates and timing of development. Selection for early differentiation of the male gonads and production of their own sex hormones, probably lies with the fact that eutherian mammals develop in a sea of maternal hormones (Mittwoch, 1992; Mittwoch *et al.*, 1993). While Mittwoch *et al.* (1993) have demonstrated that there is an accelerated rate of development of male gonads, compared with females, the earlier onset of gonad differentiation demonstrates that there is also, significantly, a predisplacement in the male, relative to the female. Indeed, if this predisplacement does not occur, then sex-reversed XY females can ensue (Burgoyne and Palmer, 1991). Formation of the testis actually precedes male phenotypic development, whereas in females ovarian formation develops after female phenotypic development has occurred (Wilson *et al.*, 1981). Development of the male phenotype is therefore dependent on the testis being formed in order for testosterone to be produced. In females, differentiation is not dependent on the presence of the ovary and its hormones.

Until recently, the established status quo was that prior to differentiation of the testis in human males, the pattern of development of male and female embryos was identical. However, recent work by Mittwoch (1993) has shown that in different mammalian species even prior to male gonad differentiation, growth rates between male and female embryos vary. Experiments on mice and cattle have shown that XY blastocysts show accelerated development, compared with XX embryos. In humans and rats quantitative sex differences are also discernible prior to testis differentiation. This supports evidence from measurements of skulls of human foetuses (Fog Pedersen, 1980) that showed that at eight to 12 menstrual weeks, females were about one day behind males, and that at birth this had increased to about six to seven days. Thus both before and after gonadal differentiation, males develop at a faster rate than females. As this is set in train prior to testosterone production in males, the cause must be put down to the direct effect of the sex chromosomes.

As Mittwoch (1993) has pointed out, even in the 1960s it was being suggested that chromosomal differences could affect the duration of the mitotic cycle times and the rate of cell differentiation. Mittwoch (1969, 1970) has suggested that the presence of a second X chromosome has the effect of reducing slightly the rate of cell differentiation in females, compared with males. The slightly faster growth rate engendered by Y-chromosomal genes appears, in mice at least, to function as early as the two-cell stage.

Invertebrates

In insects the control of growth has been relatively well understood for some time. It involves changes to the timing of production of certain hormones, in particular the *juvenile hormone* (which inhibits onset of sexual maturity) and *ecdysone* (a hormone which controls growth by determining the timing of moulting). The moulting sequence in arthropods and its morphological expression is thus under endocrine control. Neurosecretory cells present in the brain generate a hormone that stimulates the synthesis and release of the moulting hormone ecdysone.

Separate secretion of the juvenile hormone by endocrine glands modifies the expression of the moult and acts in close conjunction with the ecdysone. For the period early in development when the juvenile hormone is present at sufficiently high levels, the organism will display juvenile morphological characteristics. But when production of this hormone ceases, the animal metamorphoses into the adult phase. Thus any differences betwen sexes in the timing of reduction in juvenile hormone levels can result in size and shape differences between the sexes. For instance, if secretion of this hormone continues one moult later in the male than in the female, the male is likely to be larger, and those characters that have allometric growth will consequently differ between the sexes.

In social insects there is one factor that can greatly influence the timing of onset of sexual maturity, and so affect the size and shape that the adult animal attains, and that is the production by some members of the communities of other hormones—pheromones. These can trigger a cascade of physiological changes that either delay the onset of maturity, such as occurs in honey bees, or bring it on earlier, as occurs in the desert locust *Schistocerca gregaria*.

It has long been know that the mere presence of adult males within high population density colonies of these locusts will result in the precocious onset of sexual maturity in any immature males or females with which they are living (Butler, 1967). Mature females also have a similar ability, but they are less effective than the males in activating maturation. But paradoxically, females that are less than eight days old secrete pheromones that have the opposite effect, as they have the capacity to retard male maturation. The selective advantage of this strategy is clearly that the females do not want the males to be ready to mate before they are. The high population densities often found with these locusts means that there is usually synchronous maturation, any laggards being accelerated along by the merest whiff of a pheromone. The other advantage of this strategy is that it stops potentially precocious maturers from becoming mature at too small a size. It has been discovered that the pheromone that accelerates maturation is an extremely volatile substance that covers their bodies. A mere 1/5000th of the pheromone that is present on the body of any one individual is sufficient to stimulate a response in an immature male (Loher, 1961).

Pheromones can also have an effect on growth rates in molluscs. Heller and Ittiel (1990) have shown how in the land snail *Helix texta* in Israel, juveniles are inhibited from attaining sexual maturity by pheromones produced by adults. However, when adults suffer high levels of predation, the maturation-inhibiting substance is reduced, and the "Peter Pan" snails accelerate their growth, compared with juveniles in a pre-predation population, and mature into adults.

CONCLUSIONS

Sexual dimorphism implicitly involves differing morphologies between the sexes. As such, if the argument is made that the generation of morphological variation between individuals essentially is controlled by changes to the timing and rate of development of somatic elements, then it becomes hardly surprising that sexual dimorphism can be seen to be the product also of heterochrony. In this broad overview it is clear that dimorphism can be manifest at many stages of ontogeny. In some groups, such as mammals, it is present even at the earliest embryonic stages, arising from differential rates of growth between the sexes. But of course it is most pronounced when allometries are continued for longer in one sex than another, or particularly when this is combined with rate differences as well.

The importance of demonstrating the all-pervasive influence of heterochrony in generating sexual dimorphism lies in the clarification that it provides in demonstrating how major phenotypic effects can be generated by heterochrony, within a single gene pool. The lesson for this in the role of heterochrony in speciation is to reinforce the notion that pronounced phenotypic differences between species may also occur with little genetic change.

ACKNOWLEDGMENTS

For discussion of various points in this chapter, and for assistance with literature, I would like to thank Ian Dadour, Nina Jablonski, Ursula Mittwoch, Paul O'Higgins, Matt Ravosa and Brian Shea. Figure 4.3 is reproduced with the permission of the Trustees of the Western Australian Museum and Barry Hutchins. Kris Brimmell kindly produced the photographs of the beetles.

REFERENCES

Björklund, M., 1991, Coming of age in fringillid birds: heterochrony in the ontogeny of secondary sexual characters, *J. Evol. Biol.*, **4**: 83–92.
Brown, L. and Rockwood, L.L, 1986, On the dilemma of horns, *Nat. Hist.* (**7/86**): 54–61.
Burgoyne, P.S. and Palmer, S.J., 1991, The genetics of XY sex reversal in the mouse and other mammals, *Semin. Devel. Biol.*, **2**: 277–284.
Butler, C.G., 1967, Insect pheromones, *Biol. Rev.*, **42**: 42–87.
Cheverud, J.M., Wilson, P. and Dittus, W.P.J., 1992, Primate population studies at Polonnaruwa. III. Somatometric growth in a natural population of toque macaques (*Macaca sinica*), *J. Hum. Evol.*, **23**: 51–77.
Coelho, A.M., 1985, Baboon dimorphism: growth in weight, length and adiposity from birth to 8 years of age. In E.S. Watts (ed.), *Non-human primate models for human growth and development*, Alan R. Liss, New York: 125–159.
Cook, D., 1987, Sexual selection in Dung Beetles I. A multivariate study of the morphological variation in two species of *Onthophagus* (Scarabaeidae: Onthophagini), *Aust. J. Zool.*, **35**: 123–32.
Crowson, R.A., 1981, *The biology of the coleoptera*, Academic Press, London.
Darwin, C., 1869, On the two forms, or dimorphic condition, in the species of *Primula*, and on their remarkable relations. In P.H. Barrett (ed.), *The Collected Papers of Charles Darwin*, University of Chicago Press, Chicago, 1977.
Darwin, C., 1871, *The descent of man and selection in relation to sex*, John Murray, London.

Dittrich, L., 1967. Breeding Kirk's dikdik *Madoqua kirki thomasi* at Hanover Zoo. *Int. Zoo Yearbook*, **7**: 171–173.

Eberhard, W.G., 1982, Beetle horn dimorphism: making the best of a bad lot. *Am Nat.*, **119**: 420–426.

Fog Pedersen, J.F., 1980, Ultrasound evidence of sexual difference in fetal size in first trimester, *Br. Med. J.*, **281**: 1253.

German, R.Z., Hertweck, D.W., Sirianni, J.E. and Swindler, D.R., 1994, Heterochrony and sexual dimorphism in the Pigtailed Macaque (*Macaca nemestrina*), *Am. J. Phys. Anthrop.*, **93**: 373–380.

Glassman, D.M. and Coelho, A.M., 1987, Principal components analysis of physical growth in savannah baboons, *Am. J. Phys. Anthrop.*, **72**: 59–66.

Gould, S.J., 1974, The evolutionary significance of "bizarre" structures: antler size and skull size in the "Irish Elk', *Megaloceras gigantans*, *Evolution*, **28**: 191–220.

Griffith, H., 1991, Heterochrony and evolution of sexual dimorphism in the *fasciatus* Group of the scincid genus *Eumeces*, *J. Herp.*, **25**: 24–30.

Heller, J. and Ittiel, H., 1990, Natural history and population dynamics of the land snail *Helix texta* in Israel (Pulmonata: Helicidae), *J. Moll. Stud.*, **56**: 189–204.

Howells, W.W. and Hanks, J., 1975, Body growth of the impala (*Aepyceros melampus*) in Wankie National Park, Rhodesia, *J. S. Afr. Wildlife Management Assoc.*, **5**: 95–98.

Hutchins, J.B., 1992, Sexual dimorphism in the osteology and myology of monacanthid fishes, *Rec. West. Aust. Mus.*, **15**: 739–747.

Hutchinson, M., 1970, Observations on the growth rate and development of some East African antelopes, *J. Zool., Lond.*, **161**: 431–436.

Jarman, P., 1983, Mating system and sexual dimorphism in large, terrestrial, mammalian herbivores, *Biol. Rev.*, **58**: 485–520.

Jost, A., 1972, A new look at the mechanisms controlling sex differentiation in mammals, *Johns Hopkins Med. J.*, **130**: 38–53.

Kellas, L., 1955, Observations on the reproductive activities, measurements, and growth rate of the dik-dik (*Rhynchotragus kirkii thomasi* Neumann), *Proc. Zool. Soc. Lond.*, **124**: 751–784.

Lack, D., 1968, *Ecological adaptations for breeding in birds*, Methuen, London.

Leigh, S.R., 1992, Patterns of variation in the ontogeny of primate body size dimorphism, *J. Hum. Evol.*, **23**: 27–50.

Leutenegger, W. and Cheverud, J., 1985, Sexual dimorphism in primates: the effect of size. In W.L. Jungers (ed.), *Size and scaling in primate biology*, Plenum, New York: 33–50.

Livezey, B.C., 1989, Phylogenetic relationships and incipient flightlessness of the extinct Auckland Islands Merganser, *Wilson Bull.*, **101**: 410–435.

Livezey, B.C. and Humphrey, P.S., 1986, Flightlessness in Steamer-ducks (Anatidae: *Tachyeres*): its morphological bases and probable evolution, *Evolution*, **40**: 540–558.

Loher, W., 1961, The chemical acceleration of the maturation process and its hormonal control in the male of the desert locust, *Proc. R. Soc. Lond.*, **B153**: 380–397.

Lorber, M., Alvo, G. and Zontine, W.J., 1979, Sexual dimorphism of canine teeth of small dogs, *Arch. Oral Biol.*, **24**: 585–589.

Ludecke, D.K. and Tolis, G. (eds), 1985, *Growth hormone, growth factors and acromegaly*, Vol. 3, *Progress in endocrine research and therapy*, Raven, New York.

Lützen, J., 1968, Unisexuality in the parasitic family Entoconchidae (Gastropoda: Prosobranchia), *Malacologia*, **7**: 7–15.

McKinney, M.L., 1990, Trends in body-size evolution. In K.J. McNamara (ed.), *Evolutionary trends*, Belhaven, London: 75–118.

McKinney, M.L. and McNamara, K.J., 1991, *Heterochrony: the evolution of ontogeny*, Plenum, New York.

Mittwoch, U., 1969, Do genes determine sex? *Nature*, **221**: 446–448.

Mittwoch, U., 1970, How does the Y chromosome affect gonadal differentiation? *Phil. Trans. R. Soc. Lond.*, **B259**: 113–117.

Mittwoch, U., 1992, Sex determination and sex reversal: genotype, phenotype, dogma and semantics, *Hum. Genet.*, **89**: 467–479.

Mittwoch, U., 1993, Blastocysts prepare for the race to be male, *Hum Reprod.*, **8**: 1550–1555.

Mittwoch, U., Burgess, A.M.C. and Baker, P.J., 1993, Male sexual development in "a sea of oestrogen", *Lancet*, **342**: 123–124.

O'Higgins, P. and Dryden, I.L, 1993, Sexual dimorphism in hominoids: further studies of craniofacial shape differences in *Pan*, *Gorilla* and *Pongo*, *J. Hum. Evol.*, **24**: 183–205.

O Foighil, D., 1985, Form, function, and origin of temporary dwarf males in *Pseudopythina rugifera* (Carpenter, 1864) (Bivalvia: Galeommatacea), *The Veliger*, **27**: 245–252.

Privratsky, V., 1981, Neoteny and its role in the process of hominization, *Anthropologie*, **19**: 219–230.

Raiti, S. and Tolman, R.A. (eds), 1986, *Human growth hormone*, Plenum, New York.

Ralls, K., 1977, Sexual dimorphism in mammals: avian models and unanswered questions, *Am. Nat.*, **111**: 917–938.

Ravosa, M.J., 1991, The ontogeny of cranial sexual dimorphism in two Old World monkeys: *Macaca fascicularis* (Cercopithecine) and *Nasalis larvatus* (Colobinae), *Int. J. Primatol.*, **12**: 403–426.

Ravosa, M.J. and Ross, C.F., 1994, Craniodental allometry and heterochrony in two howler monkeys: *Alouatta seniculus* and *A. palliata*, *Am. J. Primatol.*, **33**: 277–299.

Robinette, W.L. and Child, G.F.T., 1964, Notes on biology of the lechwe (*Kobus leche*), *The Puku*, **2**: 84–117.

Selander, R.K., 1972, Sexual selection and dimorphism in birds. In B.G. Campbell (ed.), *Sexual selection and the descent of man, 1871–1971*, Aldine, Chicago: 180–230.

Shea, B.T., 1983, Allometry and heterochrony in the African apes, *Am. J. Phys. Anthrop.*, **62**: 275–289.

Shea, B.T., 1989, Heterochrony in human evolution: the case for neoteny reconsidered, *Yearbook Phys. Anthrop.*, **32**: 69–101.

Shea, B.T., Hammer, R.E., Brinster, R.L. and Ravosa, M.J., 1990, Relative growth of the skull and postcranium in giant transgenic mice, *Genet. Res., Camb.*, **56**: 21–34.

Shine, R., 1978, Growth rates and sexual maturation in six species of Australian elapid snakes, *Herpetologica*, **34**: 73–79.

Shine, R. and Crews, D., 1988, Why male garter snakes have small heads: the evolution and endocrine control of sexual dimorphism, *Evolution*, **42**: 1105–1110.

Sinclair, A.R.E., 1977, *The African buffalo: a study of resource limitation of populations*, University of Chicago Press, Chicago.

Smith, M.M. and Heemstra, P.C., 1986, *Smiths' Sea Fishes*, Macmillan, Johannesburg.

Spinage, C.A., 1967, Ageing the Uganda defassa waterbuck *Kobus defassa ungandae*, *E. Afr. Wildlife Jl*, **5**: 1–17.

Stearns, S.C. (ed.), 1986, *The evolution of sex and its consequences*, Birkhäuser Verlag, Basel.

Vollrath, F. and Parker, G.A., 1992, Sexual dimorphism and distorted sex ratios in spiders, *Nature*, **360**: 156–159.

Waren, A., 1983, A generic revision of the family Eulimidae (Gastropoda, Prosobranchia), *J. Moll. Stud. Suppl.*, **13**: 1–96.

Watts, E.S., 1985, Adolescent growth and the development of monkeys, apes and humans. In E.S. Watts (ed.), *Non-human primate models for human growth and development*, Alan R.Liss, New York: 41–65.

Watts, E.S., 1986, Evolution of the human growth curve. In F. Falkner and J.M. Tanner (eds), *Human growth*, Vol. 3, Plenum, New York: 153–166.

Wiley, R.H., 1974, Evolution of social organization and life history patterns among grouse, *Qt. Rev. Biol.*, **49**: 201–227.

Wilson, J.D., George, F.W. and Griffin, J.E., 1981, The hormonal control of sexual development, *Science*, **211**: 1278–1284.

Winquist, T. and Lemon, R.E., 1994, Sexual selection and exaggerated male tail length in birds, *Am. Nat.*, **143**: 95–116.

Wright, D.B., 1993, Evolution of sexually dimorphic characters in peccaries (Mammalia, Tayassuidae), *Paleobiology*, **19**: 52–70.

Part Two

HETEROCHRONY AND MORPHOLOGY

Chapter 5

HETEROCHRONY AND THE EVOLUTION OF LAND PLANTS

Volker Mosbrugger

INTRODUCTION

The phenomenon of heterochrony, as it is understood today, has been basically known at least since the fundamental embryological studies of Karl Ernst von Baer (e.g., von Baer, 1828) and Wilhelm Hofmeister (e.g., Hofmeister, 1851). Ernst Haeckel's term "heterochrony", however, has been used in numerous different ways (Osche, 1966; Zimmermann, 1953; see Gould, 1988 for a review) but has gained its present-day meaning and popularity from Gould's (1977) book *Ontogeny and Phylogeny* and subsequently from the publication of Alberch *et al.* (1979). The concept of heterochrony is very simple and refers to "changes in the relative time of appearance and rate of development for characters already present in ancestors" (Gould, 1977, p. 2). Transferred to a genetic level, heterochrony probably means that genetic variations or mutations may affect not only structural genes but also regulatory genes—an evolutionary concept that has been unanimously accepted since the first modern gene regulation concepts were developed (e.g., Jacob and Monod, 1961). For instance, Riedl (1977) has always advocated a cybernetic or system theory of evolution, and thus emphasized that one of the easiest ways of changing a system is by changing its cybernetics.

In this chapter I will review the relevance of heterochrony to the evolution of land plants from the viewpoint of a palaeobotanist. I will discuss three examples: life cycles, telome theory and stelar evolution, to illustrate that heterochrony does indeed provide an excellent theoretical concept to describe many crucial steps in the evolution of land plants. Finally, I will briefly mention some of the reasons why heterochrony is particularly common in land plants. It has to be emphasized, however, that the relevance of heterochrony for understanding the evolution of land plants has long been recognized (e.g., Takhtajan, 1954; 1969; Zimmermann, 1959) although an increasing interest in heterochrony can recently be observed in

Evolutionary Change and Heterochrony. Edited by Kenneth J. McNamara © 1995 The Editor and Contributors. Published in 1995 by John Wiley & Sons Ltd.

both botany and palaeobotany (Gallardo *et al.*, 1993; Guerrant, 1988; Jones, 1993; McLellan, 1993; Rothwell, 1987; Sattler, 1992).

Semantic problems are not the focus of this chapter, and thus the term heterochrony is used in a very broad sense as defined above. Alberch *et al.* (1979) have developed a complex nomenclature discriminating between the heterochronic processes and their morphological results (see Preface). In this paper, however, I will only use the descriptive morphological terms such as peramorphosis and paedomorphosis which I consider very useful, but I will not refer to the putative underlying processes such as neoteny, acceleration, etc. Personally, I doubt whether these latter terms can really be used properly for fossil material because we have no adequate control over the ontogenetic processes and timing of gene activities in fossils. Obviously, size is a very unreliable proxy for measuring ontogenetic time, one reason being the mosaic evolution or heterobathmy of almost all organisms. In addition, heterochrony has to be considered from a hierarchical perspective (Raff and Wray, 1989) and most heterochronic evolutionary changes will involve a combination of several heterochronic processes. Hence, without detailed control over the ontogeny (including the timing of gene activities) of the organisms it may not be possible to discriminate reliably between the various postulated heterochronic processes.

LIFE CYCLES OF LAND PLANTS

Traditionally, botanists and palaeobotanists recognize four major groups of "true" land plants (Stewart and Rothwell, 1993; Taylor and Taylor, 1993):

1. Psilophytes (including Rhyniopsida, Zosterophyllopsida, Trimerophytopsida).
2. Mosses (including Bryopsida, Marchantiopsida, Anthocerotopsida).
3. Pteridophytes (including Lycopsida, Sphenopsida, Filicopsida, Progymnospermopsida).
4. Seed plants (including gymnosperms and angiosperms).

Beside these there are a few land-living algae, cyanobacteria and fungi, which may also live on land, but are no longer considered to be "true" land plants. Admittedly, cladists do not like these four groups of land plants because the psilophytes and pteridophytes, and possibly even the mosses and seed plants, are paraphyla and not monophyla. Together, however, these four groups of land plants form a monophylum (Graham, 1993), and all have basically the same type of haplodiplontic life cycle (Figure 5.1): there is a diplontic sporophyte producing spores which give rise to a gametophyte which then produces gametes; after fertilization the zygote forms a new sporophyte. Variations of this life cycle characterize each of the four groups of land plants.

The psilophytes are the first of the plant groups that colonized the land and are well-documented from the Upper Silurian to the Devonian. Their life cycle is

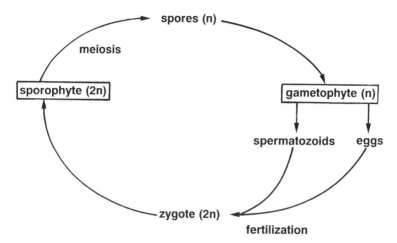

Figure 5.1 Structure of the life cycle of land plants (n = haploid; 2n = diploid).

possibly the most primitive and has only recently become known from compression–impression and structurally preserved material (Remy, 1982; Remy and Remy, 1980; Remy *et al.*, 1993; Schweitzer, 1981). For instance, *Sciadophyton* is a gametophyte commonly found as compression–impressions in the Lower Devonian, and possibly represents the haplontic stage of various Lower Devonian psilophytes such as *Zosterophyllum*, *Stockmansella* and *Huvenia*. It consists of star-like, repeatedly branched, creeping axes which obviously contain a vascular bundle (Figure 5.2). In later ontogenetic stages the creeping axes turn upright and form cup-like structures at their ends which contain the gametangia. Fertilization took place on these cup-like structures, and the zygote developed into a sporophyte which in its early stages looks very similar to the gametophyte. In later ontogenetic stages the creeping axes of the young sporophyte also turned upwards and some of them formed sporangia at their ends.

Structurally preserved psilophyte gametophytes are known only from the classical Lower Devonian Rhynie locality in Scotland (Remy and Remy, 1980; Remy, 1982; Remy *et al.*, 1993). Various types of gametophytes (*Lyonophyton*, *Kidstonophyton*, *Langiophyton*) have been described. On the basis of their anatomical structures they can be referred to sporophytes such as *Aglaophyton*, *Nothia* and *Horneophyton*. All gametophytes, however, show a similar growth habit with upright axes terminating in a cup-like gametangiophore thus resembling the basic growth habit of *Sciadophyton* gametophytes. Another persistent character of these gametophytes is the existence of a vascular bundle and of stomata. Obviously, in psilophytes the sporophytes and gametophytes possess a similar morphological and anatomical complexity: the gametophyte is only slightly smaller than the sporophyte and forms cup-like gametangiophores instead of sporangia at the terminations of the axes.

In mosses we observe the same alternation of a gametophyte and sporophyte

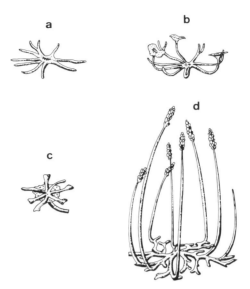

Figure 5. 2   Gametophyte and sporophyte of early land plants. a–c, Various ontogenetic stages in the development of the gametophyte *Sciadophyton* from the Lower Devonian (a, early star-like stage; b, "adult" stage with cup-like gametangiophores; c, gametangiophore with a developing sporophyte); d, *Zosterophyllum* from the Lower Devonian, which may represent one of the sporophytes of *Sciadophyton*. (After Schweitzer, 1981.)

but here the gametophyte is obviously the dominant generation: the sporophyte is little more than a sporangium with its stalk fixed to the gametophytic moss plant. Hence the life cycle of mosses differs in two respects from the life cycle of psilophytes and all other land plant groups: firstly, gametophyte and sporophyte look totally different and the gametophyte is by and large the dominant generation; secondly, gametophyte and sporophyte are not independent generations, instead the sporophyte is parasitizing the gametophyte.

Another situation again is found in pteridophytes such as ferns, club mosses and horsetails. The sporophyte represents the dominant generation whereas the gametophyte remains very small, and very different morphologically and anatomically from the sporophyte. In fact, the gametophyte of pteridophytes is always only a few millimetres in size and may represent a one cell layer thick algae-like thallus as in many ferns, or a club-like or irregular shaped, mostly saprophytic structure as in club mosses. Vascular elements and stomata are absent in almost all pteridophyte gametophytes but in contrast to mosses sporophyte and gametophyte generations are still independent.

In seed plants the gametophytic generation is further reduced and remains within the initial meiospores: the female gametophyte corresponds to what plant anatomists have called the "embryosac" of the developing seed and is surrounded not only by the spore membrane but also by the sporangium wall (called nucellus) and the integuments; the male gametophyte is represented by the multi-nuclear pollen grain and the pollen tube. Thus, in seed plants the gametophyte has become a very reduced generation which in its energy resources depends entirely on the sporophyte.

Obviously, the four basic land plant groups differ considerably in their life cycles and these differences can easily be explained by heterochrony, i.e. evolutionary changes in relative rate and timing of development: as compared with the psilophytes the sporophyte of mosses is paedomorphic, whereas in pteridophytes and seed plants the gametophyte generation is paedomorphic. Thus the life cycles of land plants represent the most classic example of heterochrony in land plants, the basic principles of which have been known to all botanists since at least the discovery of the life cycle in mosses, pteridophytes and seed plants by Hofmeister (1851).

TELOME THEORY

The groups of land plants considered so far differ not only in their life cycle but also in many other characters. The telome theory of Zimmermann (1959) tries to explain the evolutionary origin of many of these other differences. The basic idea of the telome theory is that most new structures of more advanced land plants, such as leaves or the integument of seeds, were derived from telomes (i.e., naked plant axes) of the primitive psilophytes by a combination of only a few so-called "elementary processes" which are planation, webbing, incurvation, reduction and overtopping (Figure 5.3). For instance, large leaves or megaphylls are considered to have originated from planation and webbing of telomes; small leaves or microphylls are interpreted as the result of a reduction of telomes or of larger leaves. Obviously the telome theory is a concept based on heterochrony and Zimmermann has always pointed out that the elementary processes are those

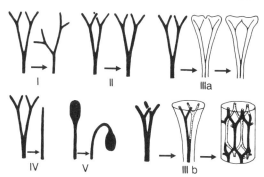

Figure 5.3   The five elementary processes of W. Zimmermann's telome theory. I, Overtopping. II, Planation. IIIa, Webbing in two dimensions. IIIb, Webbing in three dimensions. IV, Reduction. V, Incurvation. (After Zimmermann, 1959.)

occurring during ontogeny. The term "hologeny" has been introduced by Zimmermann to emphasize that evolutionary changes always include ontogenetic changes in the development of characters.

For many years the telome theory was considered to be a powerful concept to explain the evolution of land plants. Important structures of land plants explained by the telome theory include, for instance, the microphylls of lycopods and horsetails, the characteristic sporangiophore of horsetails, the cupule and integument of seed plants, and the evolution of the stele types. More recent data, however, suggests that many of the telome theory's interpretations of evolutionary pathways are wrong, or have to be modified, but—interestingly enough—the new interpretations still refer to heterochronic processes. This will be illustrated using the evolution of microphylls and of the seed integument as examples.

Evolution of microphylls

According to telome theory, microphylls as they exist, for instance, in lycopods and horsetails, originated from a reduction of a branched telome truss (Zimmermann, 1959). Analysis of various genera of Lower Devonian psilophytes of the Zosterophyllopsida group indicates, however, that at least in some cases microphylls may also originate from enations (these are leaf- or hair-like outgrowths of the epidermis and subepidermal cell layers and have no vascular supply). According to various authors (e.g., Schweitzer, 1980), this is clearly illustrated by the Lower Devonian genera *Sawdonia*, *Asteroxylon* and *Drepanophycus*. In *Sawdonia* the axes are covered with enations; the axes of *Asteroxylon* are also covered with enations but vascular bundles or initial leaf traces go directly from the stem bundle to the base of the enations; finally, in *Drepanophycus* the "enations" are fully vascularized thus representing true microphylls.

Functionally this evolution of microphylls is obviously related to a size increase of the genera, and it has been demonstrated that the enations and microphylls evolved in order to maintain a high surface to volume ratio despite an increase in size (Mosbrugger, 1992; Mosbrugger and Roth, 1993). If a structure increases in size isometrically (i.e., without changing its shape) then its surface to volume ratio will continuously drop for simple geometrical reasons. Photoautotrophic plants, however, have to maintain a high surface to volume ratio and hence have to increase their "flatness" (LaBarbera and Vogel, 1982) when increasing in size: the formation of enations, such as those in *Sawdonia*, is one possible solution to the necessity of increasing the plant's flatness. When the plant further increases in size, the enations will also increase in size. Computer simulations of water transport in parenchyma (Roth *et al.*, 1994a) have shown that once these enations attain a size of only a few millimetres a vascular supply becomes essential. This vascular bundle may develop in the enations according to the same regulatory process as it originates in shoots so that, apart from the size increase of the enation, no specific genetic information should be required.

Hence, the first microphylls may have evolved (at least in some lycopods) by peramorphosis of enations as a consequence of a size increase in land plants. That many of the earliest and most primitive land plants were capable of producing enations has been demonstrated by Remy and Hass (1988).

Evolution of seed integument

Thus, with respect to the microphyll evolution, the telome theory is probably wrong in at least some cases but the alternative interpretation involves heterochronic changes nonetheless. A similar situation applies to the evolution of the seed integument. According to classical telome theory the integument of seeds originated from the fusion of telomes surrounding the original (mega-)sporangium (Figure 5.4). This view was supported by Lower Carboniferous ovules with lobed integuments which seemed to represent various stages in the fusion of surrounding telomes (Figure 5.5); however, up until now no fossils have been found to show the gradual evolution of the initial stage with a central sporangium surrounded by sterile telomes. A more likely interpretation of the evolution of the seed integument was recently proposed by Remy and Hass (1991) (Figure 5.6). They pointed out that cup-like structures with or without lobes were frequently formed at the end of telomes of early land plant gametophytes and sporophytes. In fact, the gametangia of all gametophytes are situated on cup-like telome endings (see

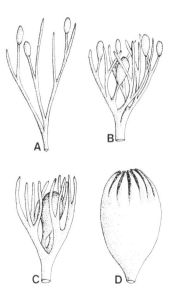

Figure 5.4   Evolution of the seed integument from surrounding sterile telomes according to the telome theory. A, Telome truss; B, shift of a megasporangium towards the centre of the telome truss; C, megasporangium surrounded by sterile telomes; D, fusion of the sterile telomes (after H.N. Andrews from Stewart and Rothwell, 1993).

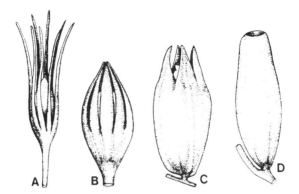

Figure 5.5   Lower Carboniferous ovules with increasingly fused integumentary lobes (after A.G. Long from Stewart and Rothwell, 1993).

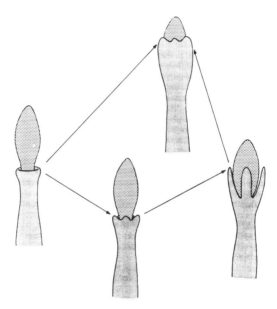

Figure 5.6   Possible pathways of the evolution of the seed integument according to Remy and Hass (1991).

above). Similarly, in the sporophytes of some genera (e.g., *Huvenia*) the sporangia are also situated on cup-like enlargements of the telome endings (Hass and Remy, 1991). It therefore seems reasonable to assume that the integument surrounding the sporangium or nucellus represents a peramorphosed homologue of a cup-like telome ending supporting a sporangium.

EVOLUTION OF WATER TRANSPORT SYSTEMS IN THE AXES OF EARLY LAND PLANTS

As a third example illustrating the importance of heterochrony in the evolution of land plants the evolution of water transport systems is considered. Land plants continuously lose water due to transpiration. To compensate for this water loss they take up water from the soil and transport it to the sites of transpiration. This water transport along the plant axis occurs in a stele or vascular bundle which consists of numerous elongated and cylinder-like, dead tracheid cells. The earliest and most simple type of such a vascular bundle is the so-called protostele in which the water-transporting tracheid-cells are arranged in a central cylinder. Later on, more complex stele types such as siphonosteles, actinosteles, etc. are developed (Figure 5.7).

To understand the functional aspect of this stelar evolution a Finite Element computer modelling approach has been used (Roth *et al.*, 1994a,b). The water transport in cylindrical plant axes was simulated as a fluid flow through a porous medium following Darcy's law. It was assumed that there was a continuous transpiration loss at the surface of the axis and that the water reservoir in the soil was unlimited. From the computer models it emerged that the protostele is probably most effective in small-sized axes (up to a few millimetres in diameter), and that the evolution of the protostele may be understood in terms of a self-regulating process in which the water permeability in an initially homogeneous axis is increased at those sites with the lowest fluid flow. The advantage of the central protostele is that the photosynthesizing parenchyma can be placed at the

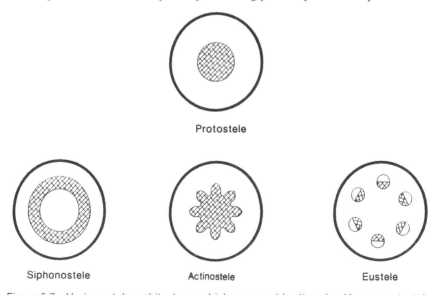

Figure 5.7   Various stele architectures which presumably all evolved from a protostele.

axis periphery, and that the water transporting system is isolated from the transpiring surface thus minimizing water loss. It is important, however, that the distance between the protostele and the transpiring surface remains small (less than 1 mm) because the transpiration stream has to bridge this distance by a parenchymatous water transport which is much slower than the transport through tracheids. In fact the distance from the vascular bundle to the transpiring telome surface proved to be a crucial factor in determining the water transport efficiency of a stele.

Hence, when the plant axis increases in diameter the protostele also has to increase its diameter in order to maintain a small distance of the bundle from the transpiring surface. If, however, the protostele increases in such a way that the distance from the bundle to the surface remains constant then the cross-sectional area of the vascular system increases more rapidly than the transpiring surface (the cross-sectional area of the vascular bundle varies in proportion to the radius squared of the bundle; the transpiring surface, however, varies with the radius of the axis). This explains why the protostele structure is abandoned in thicker axes and why stele types such as siphonosteles, actinosteles, polysteles or eusteles are developed: they all represent more dissected stele types and allow for a peripheral arrangement of the vascular system without the formation of too much dead xylem tissue.

As a whole the simulation study indicates that all types of vascular bundles may have originated by basically the same self-regulation process (vascular tissue is formed where it is actually needed): in smaller axes this process will lead to a protostele; in thicker axes siphonosteles, actinosteles, polysteles, etc., may develop. Thus, the evolution of a protostele and then of siphonosteles, actinosteles, etc., is related to the size increase of plants and may be understood as the result of a peramorphosis, although heterochronic processes alone cannot account for the observed evolutionary sequence. The example illustrates two important facts. Firstly, evolutionary changes presumably can never be explained by heterochronic processes alone; rather in all evolutionary lineages heterochronic processes are combined (although to a different degree) with non-heterochronic genetic changes. Secondly, heterochrony, i.e. "changes in the relative time of appearance and rate of development for characters already present in ancestors", may not only affect the expression of morphological characters but also the expression of regulation processes (e.g. the regulation process governing the development of the vascular bundle) and hence lead to morphologically very different structures. Both phenomena, the common mixture of heterochronic and non-heterochronic processes as well as the heterochronic alteration of regulation processes, may obscure the overall effect and importance of heterochrony. Thus the evolution of structures, which are morphologically very different from each other such as proto- and siphonostele, may nevertheless be the result of heterochrony.

WHY IS HETEROCHRONY SO IMPORTANT IN LAND PLANT EVOLUTION?

The examples have been chosen to demonstrate that heterochrony is indeed a crucial phenomenon in the evolution of land plants. In fact, many of the key characters of the four land plant groups can be explained in terms of heterochrony. The land plants obviously have several features in common which make heterochronic changes particularly easy. First of all, many land plants—in contrast to many animals—have an indeterminate growth and no really definite adult morphology thus favouring the formation of peramorphosis. In addition, this indeterminate growth is based on a modular construction. For instance, early land plants consist of numerous telomes which can be varied independently and combined in various ways. More advanced land plants also consist of only a few modules, such as shoots, roots, and leaves, so the flowers of angiosperms are nothing other than a branch with leaves modified by heterochrony. The modularity of land plants is actually the reason why they can so easily be modelled in a computer as so-called Lindenmayer structures (Prusinkiewicz and Lindenmayer, 1990). In this respect it is interesting to remember that land plants also have "homeotic genes" (MADS-box genes of Schwarz-Sommer *et al.*, 1990) and that now numerous homeotic mutants of angiosperms are known (e.g., Meinke, 1992; Rasmussen and Green, 1993; Schwarz-Sommer *et al.*, 1990). It seems clear that indeterminate growth, modularity and the existence of homeotic genes make heterochronic changes particularly easy.

Meristemes and meristemoids are another characteristic feature of land plants favouring heterochronic changes. In multicellular animals typically all cells with the exception of reproductive cells become differentiated during ontogeny. In land plants, however, embryonic cells persist as meristemes or (smaller) meristemoids throughout the lifetime of the plants. Such meristemes may exist at all the telome or shoot apices but also the epidermis cells may remain largely embryonic or meristematic, as is the case for instance, in the Lower Devonian psilophyte *Horneophyton* (Hass, 1991). Correspondingly, plants have a non-Weismannian development and do not segregate a germ line as do many animals (Buss, 1987). Thus, in plants all cell lineages can differentiate into reproductive cells so that somatic mutations can be accumulated during the lifetime of an organism and transmitted to the next generation (Mosbrugger and Roth, 1994).

SUMMARY

Heterochrony has played a crucial role in the evolution of land plants. This is demonstrated by the life cycles of various land plant groups and by the evolution of microphylls, seed integument and stele types. The effect of heterochronic processes can be difficult to recognize because heterochronic and non-heterochronic evolutionary changes are commonly mixed, and because heterochrony may affect the ontogenetic development not only of morphological characters but also of regulation processes. The specific construction of land plants with indeterminate growth, ontogenetic persistence of meristemes, modular organization, existence of homeotic genes and non-Weismannian development may explain why heterochronic changes are particularly common in the evolution of land plants.

REFERENCES

Alberch, P., Gould, S.J., Oster, G.F. and Wake, D.B., 1979, Size and shape in ontogeny and phylogeny, *Paleobiology*, **5**: 296–317.

Baer, K.E. von, 1828, *Entwicklungsgeschichte der Thiere: Beobachtung und Reflexion*, Borntraeger, Königsberg.

Buss, L., 1987, *The evolution of individuality*, Princeton University Press, Princeton.

Gallardo, R., Dominquez, E. and Munoz, J.M., 1993, The heterochronic origin of the cleistogamous flower in *Astragalus cymbicarpos* (Fabaceae), *Am. J. Bot.*, **80**: 814–823.

Gould, S.J., 1977, *Ontogeny and phylogeny*, Harvard University Press, Cambridge, Ma.

Gould, S.J., 1988, The uses of heterochrony. In M.L. McKinney (ed.), *Heterochrony in evolution: a multidisciplinary approach*, Plenum Press, New York: 1–13.

Graham, L., 1993, *Origin of land plants*, John Wiley and Sons, New York.

Guerrant, E.O., 1988, Heterochrony in plants—the intersection of evolution, ecology and ontogeny. In M.L. McKinney (ed.), *Heterochrony in evolution: a multidisciplinary approach*, Plenum Press, New York: 111–133.

Hass, H., 1991, Die Epidermis von *Horneophyton lignieri* (Kidston & Lang) Barghorn & Darrah, *N. Jb. Geol. Paläont. Abh.*, **183**: 61–85.

Hass, H. and Remy, W., 1991, *Huvenia kleui* nov. gen., nov. spec.—ein Vertreter der Rhyniaceae aus dem höheren Siegen des Rheinischen Schiefergebirges, *Argumenta Palaeobot.*, **8**: 141–168.

Hofmeister, W., 1851, *Vergleichende Untersuchungen der Keimung, Entfaltung und Fruchtbildung höherer Kryptogamen*, Leipzig.

Jacob, F. and Monod, J., 1961, Genetic regulatory mechanisms in the synthesis of proteins, *J. Mol. Biol.*, **3**: 318.

Jones, C.S., 1993, Heterochrony and heteroblastic leaf development in two subspecies of *Cucurbita argyrosperma* (Cucurbitaceae), *Am. J. Bot.*, **80**: 778–795.

LaBarbera, M. and Vogel, S., 1982, The design of fluid transport systems in organisms, *Am. Sci.*, **70**: 54–60.

McLellan, T., 1993, The roles of heterochrony and heteroblasty in the diversification of leaf shapes in *Begonia dregei* (Begoniaceae), *Am. J. Bot.*, **80**: 796–804.

Meinke, D.W., 1992, A homeotic mutant of *Arabidopsis thaliana* with leaf cotyledons, *Science*, **258**: 1647–1650.

Mosbrugger, V., 1992, Constructional morphology as a useful approach in fossil plant biology, *Cour. Forsch.-Inst. Senckenberg*, **147**: 19–29.

Mosbrugger, V. and Roth, A., 1993, Self-organization, size, and the evolution of early land plants, *N.Jb. Geol. Paläont. Abh.*, **190**: 267–278.

Osche, G., 1966, Grundzüge der allgemeinen Phylogenetik. In L. v. Bertalanffy and F. Gessner (eds), *Handbuch der Biologie*, Akademische Verlagsgesellschaft, Frankfurt a.M.: 817–906.

Prusinkiewicz, P. and Lindenmayer, A., 1990, *The algorithmic beauty of plants*, Springer Verlag, Berlin.

Raff, R.A. and Wray, G.A., 1989, Heterochrony: developmental mechanisms and evolutionary results, *J. evol. Biol.*, **2**: 409–434.

Rasmussen, N. and Green, P.B., 1993, Organogenesis in flowers of the homeotic green pistillate mutant of tomato (*Lycopersicon esculentum*), *Am. J. Bot.*, **80**: 805–813.

Remy, W., 1982, Lower Devonian gametophytes: relation to the phylogeny of land plants, *Science*, **215**: 1625–1627.

Remy, W., Gensel, P. and Hass, H., 1993, The gametophyte generation of some early Devonian land plants, *Int. J. Plant Sci.*, **154**: 35–58.

Remy, W. and Hass, H., 1988, Wandlung und Plastizität von Organisationsplänen am Beispiel altdevonischer Pflanzen, *Eclogae geol. Helv.*, **81**: 911–914.

Remy, W. and Hass, H., 1991, Gametophyten und Sporophyten im Unterdevon, *Fakten und Spekulationen*, **8**: 193–223.

Remy, W. and Remy, R., 1980, *Lyonophyton rhyniensis* nov. gen. et nov. spec., ein Gametophyt aus dem Chert von Rhynie (Unterdevon, Schottland), *Argumenta Palaeobot.*, **6**: 37–72.

Riedl, R., 1977, A systems-analytical approach to macro-evolutionary phenomena, *Quart. Rev. Biol.*, **52**: 351–370.

Roth, A., Mosbrugger, V. and Neugebauer, J., 1994a, Efficiency and evolution of water transport systems in higher plants—a modelling approach. I. The earliest land plants, *Phil. Trans. Roy. Soc. Lond*, **B345**: 137–152.

Roth, A., Mosbrugger, V. and Neugebauer, J., 1994b, Efficiency and evolution of water transport systems in higher plants—a modelling approach. II. Stelar evolution, *Phil. Trans. Roy. Soc. Lond*, **B345**: 153–162.

Rothwell, G., 1987, The role of development in plant phylogeny: a paleobotanical perspective, *Rev. Palaeobot. Palynol.*, **50**: 97–114.

Sattler, R., 1992, Process morphology: structural dynamics in development and evolution, *Can. J. Bot.*, **70**: 708–714.

Schwarz-Sommer, S., Huijser, P., Nacken, W., Saedler, H. and Sommer, H., 1990, Genetic control of flower development by homeotic genes in *Antirrhinum majus*, *Science*, **250**: 931–936.

Schweitzer, H.J., 1980, Über *Drepanophycus spinaeformis*, *Bonner Paläobot. Mitt.*, **7**: 1–29.

Schweitzer, H.J., 1981, Der Generationswechsel der Psilophyten, *Bonner Paläobot. Mitt.*, **8**: 1–19.

Stewart, W.N. and Rothwell, G., 1993, *Paleobotany and the evolution of plants*, 2nd Ed., Cambridge University Press, Cambridge.

Takhtajan, A., 1954, *The origin of angiospermous plants*, Moscow [in Russian].

Takhtajan, A., 1969, *Flowering plants—origin and dispersal*, Oliver and Boyd, Edinburgh.

Taylor, T.N. and Taylor, E.L., 1993, *The biology and evolution of fossil plants*, Prentice Hall, Englewood Cliffs, New Jersey.

Zimmermann, W., 1953, *Evolution—Geschichte ihrer Probleme und Erkenntnisse*, Verlag Karl Alber, Freiburg i.Br.

Zimmermann, W., 1959, *Die Phylogenie der Pflanzen*, 2nd Ed., G. Fischer, Stuttgart.

Chapter 6

HETEROCHRONY IN THE DEVELOPMENT OF THE AMPHIBIAN HEAD

Rainer Schoch

INTRODUCTION

The ancestry of the three living amphibian orders, the anurans (frogs), urodeles (salamanders), and limbless gymnophionans, has been much debated (e.g., Parsons and Williams, 1963; Reig, 1964; Carroll and Currie, 1975; Carroll and Holmes, 1980; Duellman and Trueb, 1986; Trueb and Cloutier, 1991). Firstly, living amphibians meet all the criteria for three essentially separate clades, having fundamentally different morphologies and life strategies. Secondly, all of them differ distinctly from the majority of Palaeozoic and Mesozoic ruling amphibians: the stegocephalians. The latter discrepancy is seen most clearly in their different size, but it involves a great many other, more subtle details. Although accepted as a monophyletic group, the exact placement of modern amphibians in the phylogenetic tree has been questioned (Parsons and Williams, 1963; Reig, 1964; Rage and Janvier, 1982; Duellman and Trueb, 1986; Milner, 1988; Bolt, 1991; Trueb and Cloutier, 1991).

A convincing series of characters appears to indicate their origin in a clade of animals that shows increasing terrestriality, combined with decreasing body size: the Carboniferous to Triassic superfamily Dissorophoidea (Bolt, 1969; 1977; Milner, 1988; Trueb and Cloutier, 1991). Interestingly, some advanced dissorophoids share the most significant skeletal synapomorphy that unites modern amphibians: the pedicellate teeth. In most frogs, salamanders and gymnophionans the teeth are subdivided, allowing the pointed end to be retracted into the buccal cavity while the tongue is drawn back into the mouth (Duellman and Trueb, 1986; Bolt, 1969, 1977). The phylogenetic position of gymnophionans remains problematic due to their specialised status (although reasonably complete skeletons of a Jurassic protocaecilian have been described

Evolutionary Change and Heterochrony. Edited by Kenneth J. McNamara © 1995 The Editor and Contributors. Published in 1995 by John Wiley & Sons Ltd.

by Jenkins and Walsh, 1993). Therefore they are not considered further in this study.

Frequently, a paedomorphic origin for salamanders and frogs has been postulated (e.g. Bolt, 1977, 1979; Milner, 1988, 1990). The aim of this was to search for an explanation of the striking differences between stegocephalians and extant forms, rather than pursuing interests in heterochrony itself (exceptions are Boy, 1993b; de Ricqles, 1975). In order to analyse the role that heterochrony has played in amphibian evolution, I have investigated the transition from basic stegocephalians towards anurans and urodeles, focusing, in particular, on the development of the head region.

PHYLOGENETIC ANALYSIS

The group from which anurans and urodeles most probably originated is the order Temnospondyli (for definition see Smithson, 1985). This order forms the great majority of Palaeozoic and Triassic amphibians, and in conventional definition (i.e., excluding modern amphibians) survived up to the Cretaceous (Milner, 1990). Most of the famous, large labyrinthodonts belong here, such as the heavy *Eryops* from the Texas Red-beds, the giant *Mastodonsaurus*, and the large, presumably neotenic, flattened plagiosaurs.

The cladogram in Figure 6.1 follows in many details the hypotheses of Milner (1988, 1990), Boy (1972, 1981), and Bolt (1991). Altogether 27 characters have been determined in the temnospondyl clade including the proposed common ancestry of anurans and urodeles. These refer exclusively to skull morphology and focus on characters that can be observed as ontogenetic traits. The definitions of the characters are listed in Appendix 6.1.

HETEROCHRONIES TOWARD URODELES AND ANURANS

Ontogenetic data: the fossil record of temnospondyl amphibians

Most species included in the analysis above are known to have different ontogenetic stages, and some of them are recorded in detail, so that a comparison with the ontogenies of extant species (anurans and urodeles) is possible. Ontogenetic series exist in *Dendrerpeton* (Milner, 1980), *Sclerocephalus* (Boy, 1972, 1988; Schoch, in prep.), *Micromelerpeton* and related taxa (Boy, 1972; Milner, 1987; Werneburg, 1989), *Apateon* and related taxa (Boy, 1974; Werneburg, 1989; Schoch, 1992), *Amphibamus* (Milner, 1980, and personal observation), and finally *Tersomius* (Carroll, 1964; Bolt, 1977; Boy, 1980). The treatment of extant species here concentrates on forms regarded as primitive, such as the hynobiid salamanders and the ascaphid and discoglossid anurans, respectively (Lebedkina, 1964, 1968, 1979; Schmalhausen, 1968).

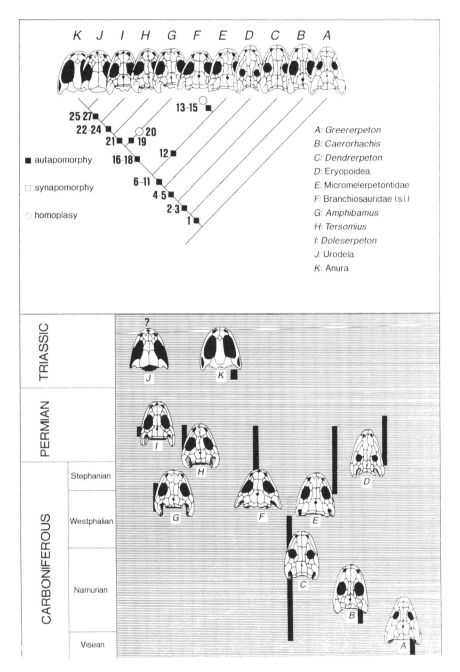

Figure 6.1 Hypothetical phylogeny of temnospondyl amphibians. Only ontogenetically traceable characters of the head region have been considered. For character definition see Appendix 6.1.

Trends in the early evolution of temnospondyl amphibians

The earliest tetrapods closely resemble one another, even in the stem groups of different clades, so that any phylogenetic interpretation must be treated cautiously. One of the best known species is the Visean *Greererpeton*, a heavy, aquatic animal with a superficial resemblance to a crocodile (Figure 6.2; Smithson,1982). Its elongate head was strongly ossified and hardly fenestrated (just small openings for nares, eyes, and the pineal organ). The palate was likewise strongly ossified, bearing three rows of different teeth: marginally about 60 medium-sized teeth on each side, accompanied by large tusks somewhat inside the palate, and finally a battery of small teeth in the central palate region (on the pterygoids, see Figure 6.2). In addition, the posterior skull lacked what is called a tympanic notch (Milner, 1990). This is a curved bony frame, bordering the tympanum *in vivo*. *Greererpeton* was hence an obvious predator of large, aquatic prey, lacking a developed impedance-matching ear (the common ear ossicle, the stapes, was here a supporting element, unable to transmit vibrations). Instead, it possessed an effective lateral sense system (documented by open canals in the skull roof). Predation focused on the capture and entrapment of large fishes. In the back of the palate, there was a joint allowing little movement of the marginal parts of the skull with respect to the braincase and their supporting bones. Depressing the lateral portions slightly, prey could probably be swallowed more easily, being fixed by the battery of small teeth (however, the skull roof did not allow much movement, appearing to be the remnant of a more elaborate feeding mechanism in the crossopterygian ancestors that had several intracranial joints).

More advanced than *Greererpeton* is *Caerorhachis* from the Namurian coal measures of Nova Scotia (Holmes and Carroll, 1977). It has several terrestrial adaptations, lacks the lateral line canals, and is considerably smaller than *Greererpeton*. The loss of the lateral sense may indicate an increasing degree of terrestriality (as do some features in the postcranial skeleton), but in general the absence of lateral line sulci can have different causes (Boy, 1972). The smaller size may be due to juvenility of the only known specimen or true paedomorphosis. The palate displays small round openings, the number of large teeth is reduced, and instead the bony surface is covered by small denticles that are directed backward. The origin of the fenestrae is of considerable interest, because they appeared several times in tetrapods, but never reached the dimensions and characteristic shape of those in temnospondyls, anurans, and urodeles. These so-called interpterygoid vacuities (see Figure 6.2) have been correlated with an increasing importance of the eyes and their musculature (Boy, 1972). Because the skull was rather flat, there was not much space, if the eyes and their support expanded by local acceleration.

A more terrestrial, and far better known temnospondyl, the Visean to Westphalian *Dendrerpeton*, demonstrates these trends. This animal has a deep

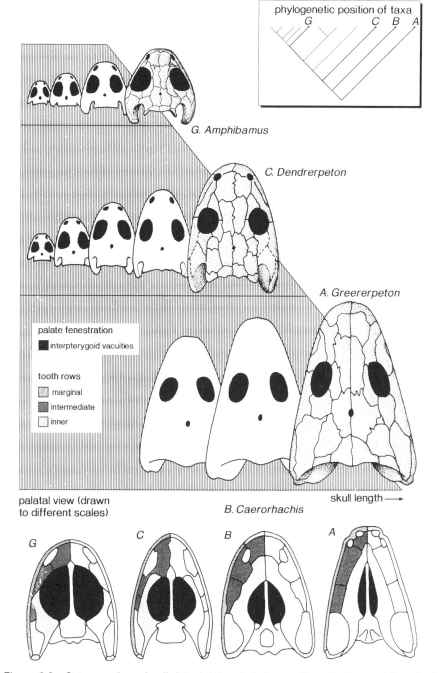

Figure 6.2   Ontogenetic series (left to right) and phylogenetic clade (upward direction) of selected basic temnospondyls. Note decreasing body size (paedomorphoclade) and eye enlargement (peramorphoclade). Below note the paedomorphic reduction of inner and intermediate tooth rows and the peramorphic enlargement of the interpterygoid vacuities, being influenced by accelerated eye growth (see text for functional interpretation).

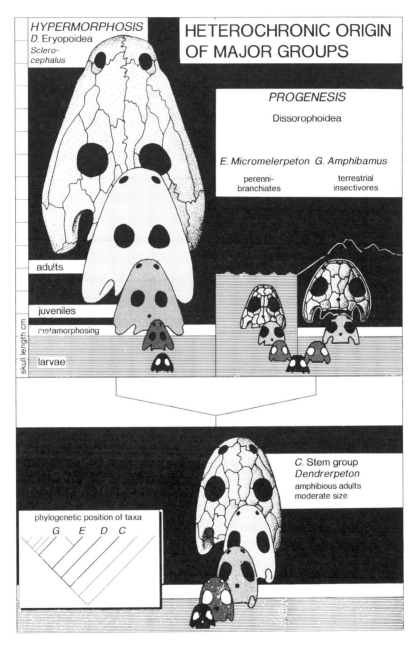

Figure 6.3   Ontogenetic series of selected species of major temnospondyl clades, assigned to the putative stem-group representative *Dendrerpeton*. Note that both *Micromelerpeton* and *Amphibamus* are believed to be progenetic, the former neotenic, the latter accelerated. Gill-breathing life period drawn with striped, lung-breathing period with black background (fossil evidence exists). Data from Boy (1972, 1988), Carroll (1967), Godfrey *et al*. (1987), Milner (1980, 1982a, b).

notch in which the tympanum was situated, connected with an impedance-matching ear (Milner, 1980). The interpterygoid vacuities in the palate are much broader than in *Caerorhachis*, due to a narrowing of the three palatal tooth rows. Also, the tusks have been reduced in size, and the intermediate row (palatine and ectopterygoid) is distinctly shorter, concentrating on the support of the large teeth. All of the aforementioned changes appear to be the result of local neoteny. The spaces in the palate have reached an extension that may not be sufficiently explained by the size increase of the eyes and their muscles. As Milner (1988) mentioned, salamanders and frogs manage part of their respiration by raising the eyes and the soft dermal floor that covers the interpterygoid vacuities. A second feature, shared exclusively by anurans and urodeles (among living species) may be equally significant. Large palatal openings permit a depression of the eyes in order to assist in swallowing prey (Duellman and Trueb, 1986). The crucial indication that such a mechanism may have existed in large temnospondyl ancestors is the observation that some frogs contain denticles on the dermal floor that facilitate prey handling behaviour in the swallowing process. Similar denticles can be found in both eryopoids and dissorophoids (advanced temnospondyls) on a mosaic of polygonal plates that covered the palatal vacuities (*Sclerocephalus*: Boy, 1988; *Amphibamus*: Carroll, 1964).

The most successful groups that have obviously evolved from a *Dendrerpeton*-like ancestor are the large piscivorous Eryopoidea (Stephanian–Permian) and the tiny, probably insectivorous, Dissorophoidea (Westphalian–Lower Triassic; Milner, 1988; Trueb and Cloutier, 1991) (see Figure 6.3). The former is characterised by an aquatic or amphibious life style, a strong increase of body size throughout life (reaching up to three metres), and a long-snouted, flattened head. Metamorphosis occurred early in ontogeny, but a variety of bones that formed were either retarded with respect to *Dendrerpeton*, or even failed to ossify at all (in particular, the braincase and parts of the girdles and extremities). These animals inhabited stable environments, such as permanent lakes and river shores, and were never in need of changing habitat. This was by definition a *K*-selective regime, allowing undisturbed growth periods, enabling eryopoids to focus on large prey. Hypermorphosis almost certainly prolonged the growth period, for it is unlikely that they grew to giants in the same time that it took the tiny *Dendrerpeton* to become an adult. Milner (1990) described this as acceleration. Some bone ossification shows postdisplacement, and the development of the extremities was slowed down or suppressed (probably by local neoteny and progenesis) because they were not required in aquatic locomotion (except for the hand that enlarged, possibly by acceleration, to a hydrofoil). Nevertheless, eryopoids were a divergent group, and some, such as *Eryops* and *Zatrachys*, that display somewhat different life strategies, returned to more terrestrial habitats.

The Dissorophoidea underwent a decrease in body size, with the head becoming proportionally larger (a localised peramorphosis), and lived in terrestrial habitats. Feeding focused on small prey, presumably insects. The middle ear, already well-established in *Dendrerpeton*, must have been highly improved, because the tympanical frame, now often completely surrounded by bones, was large and the stapes predestined for an effective transmission of airborne vibrations. Likewise, eyes and brain enlarged, and the jaw musculature apparently required less space than in the fish-eating ancestors. The Westphalian *Amphibamus* already possessed many advanced features, most significant being pedicellate, bicuspid teeth in the juvenile stage of development (Bolt, 1977), two synapomorphies of salamanders and frogs. This derived mechanism obviously prevents the teeth from breaking while the tongue is being retracted into the mouth, and the shift of these teeth from the juvenile stage in *Amphibamus* to the adult stage in the advanced *Doleserpeton* has been regarded as evidence for progenesis (Bolt, 1979). Most of the trends leading to the origin of the Dissorophoidea are a consequence of an adaptation to foraging within an unstable and dry, probably seasonal, environment. The correlation of such regimes with *r*-selection suggests an accelerated growth rate, especially of the brain and the highly developed sense organs, but also involving the extremities and girdles to accomplish refined terrestrial locomotion. Overall growth was then limited by progenesis (see also Milner, 1990).

Small scale trends and reversions: neoteny and progenesis in branchiosaurid larvae?

Within the well-known clade of branchiosaurids and their relatives, heterochrony is especially obvious. This involves very small scale trends, and in several cases it is unclear whether intra- or infraspecific changes are being observed. Morphological data on species of *Apateon* from the Rotliegend (Autunian, Early Permian) in southwestern Germany suggest differences in developmental timing within a single population leading to heterogenies in feeding periods. According to these data, one larval type specialised in small invertebrate prey capture, utilising a particular kind of suction feeding connected with a branchial filter apparatus. Another type, presumably with accelerated growth, developed stronger jaws at an early growth stage, feeding on larger animals and perhaps even having cannibalistic tendencies (documented by fossils, Boy, pers. comm.). The small putative suction feeding and plankton-filtering animals (*Apateon pedestris*) retained a weakly ossified skull table with certain sutures that appear to have been movable to some degree (personal observations). This could have facilitated a volume increase of the buccal cavity while swallowing an amount of water containing prey. Also, the dentition of the median palate was retarded here. In the other larval type (*Apateon caducus*) the quadratojugal contacted the maxilla early, and soon a strengthening of the cheek was gained (Figure 6.4). In addition, the teeth of the margin became conical and slightly bent, and in the palate a central teeth field appeared. Biting of larger prey was apparently possible. Figure 6.4 (modified from Boy, 1987a) shows several lineages that illustrate the above mentioned trends with various subtle differences between

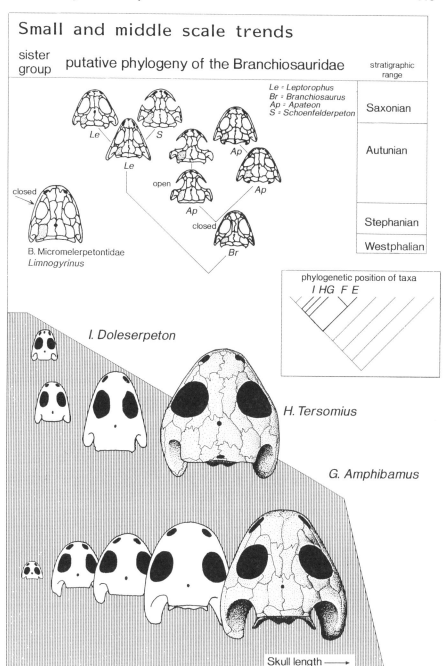

Small and middle scale trends

sister group · putative phylogeny of the Branchiosauridae · stratigraphic range

Le = *Leptorophus*
Br = *Branchiosaurus*
Ap = *Apateon*
S = *Schoenfelderpeton*

Saxonian

Autunian

Stephanian

Westphalian

Le
S
Le
closed
open
Ap
Ap
Ap
closed
Br

B. Micromelerpetontidae
Limnogyrinus

phylogenetic position of taxa
I HG F E

I. Doleserpeton

H. Tersomius

G. Amphibamus

Skull length ⟶

Figure 6.4 Advanced dissorophoid evolution. Above, branchiosaurid radiation (with polarity reversions of two characters indicated and tested by outgroup comparison); below, mosaic heterochronoclade in terrestrial dissorophoids (progenesis and acceleration). Data from Bolt (1969), Boy (1987b), Carroll (1964), Daly (1973), Milner (1982a).

the lineages. Though the tree is based on cladistic analysis and partially with stratigraphical congruence, many convergences appear which often display polarity reversions.

Clearly, most species within that clade retained their gills (fossil evidence exists) which was obviously an adaptation to a prolonged aquatic feeding style. Interpretation of the life cycle of branchiosaurids, however, appears to be much more difficult, dealing with more detailed data. It is evident that they migrated after having reached a certain size to habitats that were rarely preserved (Boy, 1978). The rarity of large fossils suggests that many species did not metamorphose. The small larvae lived in fairly stable environments, ranging from isolated water bodies to shallow lakes with fluvial connection or even deeper-sited aquatic milieus (Boy, 1978). The sedimentary facies, often containing black shales, indicates plankton-rich water, and in some cases branchiosaurids were the unique vertebrates in their environment, probably living in considerable population densities.

An explanation for this complex phenomenon may be as follows. Most branchiosaurids reached metamorphosis at larger body size than other stegocephalians (Boy, 1988, 1993; Werneburg, 1989), if they metamorphosed at all. Also, the environments were fairly stable in that the water bodies were not in danger of drying out, combined with rich food resources and relatively few or no competing species. Progenesis obviously eliminated the terrestrial life period here. This appears to have occurred in order to profit from a favourable ontogenetic niche, namely the plankton-rich water, rather than in response to an unpredictable environment (for which clear evidence is lacking). Neoteny of certain developmental processes may be concluded, accompanying the truncated development; this would also fit with most characteristics of the environment (Milner (1990) also mentioned neoteny as a possible factor in branchiosaurid evolution, while Boy (1993b) assumed progenesis and *r*-selection). Focusing on a smaller scale, the described intrapopulational differences (as heterochronic morphotypes) might indicate how branchiosaurids could have managed the problem of high population density: by pre- or postdisplacing the change of feeding style, accompanied by a variety of additional heterochronies in many growth fields (jaw bones, teeth, proportions of the head). This interpretation is strengthened by the observation of similar morphotype strategies in the Recent axolotl, *Ambystoma tigrinum* (Duellman and Trueb, 1986). Branchiosaurids appear thus to be an excellent example of dissociated and mosaic heterochrony, being subject to indirect size selection.

Advanced dissorophoid evolution: miniaturisation

A crucial evolutionary step appears to have occurred towards *Doleserpeton*, which was considered close to the ancestry of modern amphibians (Bolt, 1969, 1991; Trueb and Cloutier, 1991; and others cited therein), and at least as close to the ancestry of frogs (Carroll and Holmes, 1980). Following Bolt (1977), the most similar relative of *Doleserpeton* is *Tersomius*, which may have arisen from an *Amphibamus*-like animal (Figure 6.4 and Appendix 6.1). If we consider a transition from an ancestor similar to *Amphibamus lyelli* through *Tersomius* to *Doleserpeton*, the decrease in body size during the second step is remarkable. In comparison with an equally-sized specimen of *Tersomius*, *Doleserpeton* displays two peculiarities: (1) the eyes are distinctly larger (diminishing both the space between the eyes and behind them), and (2) the snout region is relatively elongated. These allometries may be interpreted as follows. The miniaturisation of body size in *Doleserpeton* reached a value, beyond which essential organs could not shrink further. A similar case of size decrease in modern plethodontid urodeles was discussed by Hanken (1984).

Doleserpeton was a terrestrial amphibian that did pass through a metamorphosis (Bolt, 1977), presumably smaller in size and younger in age than its relatives *Amphibamus* or *Tersomius*. That means there was precocious onset of development, accompanied by the predisplacement of braincase ossification (which appears logical due to the strong positive allometry of the brain and sense organs and the correlated weakening and opening of the skull bones). Having hardly reached the size of a late *Amphibamus* (or branchiosaurid) larva, *Doleserpeton* terminated growth, hence being a progenetic dwarf with a variety of interesting differences in timing and rate of development and growth. The crucial change was obviously a minimisation of body size with the other heterochronies being by-products of it (Milner, 1988). The ecological background of this direct size selection is difficult to analyse and provisionally discussed below.

The origins of salamanders and frogs

From the foregoing analysis, *Doleserpeton annectens* can be considered to represent the most convincing candidate for analysing the ancestry of both anurans and urodeles (Bolt, 1969, 1991). In both groups, some or all bony elements surrounding the eyes have become lost, together with the very posterior part of the median skull table (tabular and postparietal, see Figure 6.5). *Triadobatrachus massinoti* from the early Triassic of Madagascar has retained a tympanic notch very similar to dissorophoids that has become lost in all urodeles. Also, the prefrontals have been conserved, whereas all other above mentioned elements have disappeared. Additionally, frontals and parietals are fused to a single ossification (data from Rage and Roček, 1989), although this is not universal for Recent anurans. In the snout, the lacrimal has become lost in anurans, as has the prefrontal, which is only known in *Triadobatrachus*. In urodeles, the snout region has undergone no significant change (lacrimal and prefrontal are retained), but again the region behind the eyes lacks mineralisations. This is evident in the earliest complete salamander fossil, *Karaurus* from the lower Jurassic of Kazakhstan, and is universal among living species. The

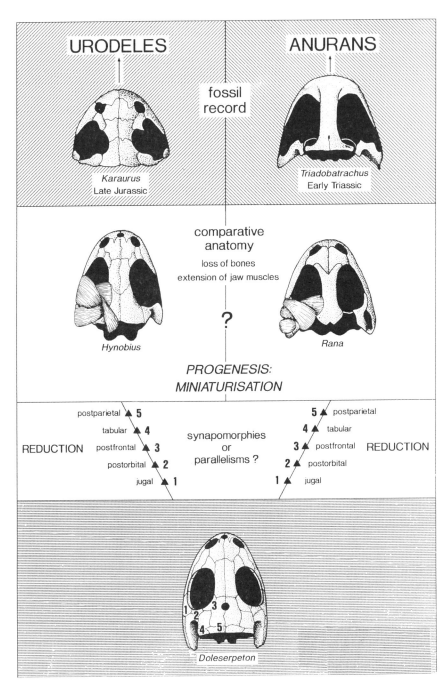

URODELES

ANURANS

fossil record

Karaurus
Late Jurassic

Triadobatrachus
Early Triassic

comparative anatomy

loss of bones
extension of jaw muscles

?

Hynobius

Rana

PROGENESIS:
MINIATURISATION

REDUCTION

postparietal ▲ 5

tabular ▲ 4

postfrontal ▲ 3

postorbital ▲ 2

jugal ▲ 1

synapomorphies
or
parallelisms ?

5 ▲ postparietal

4 ▲ tabular

3 ▲ postfrontal

2 ▲ postorbital

1 ▲ jugal

REDUCTION

Doleserpeton

Albanerpetontidae are not considered here owing to the results of Milner's (1988) study. The opened skull region behind the eyes not only appears similar in anurans and urodeles, but is, in its peculiarities, a unique character among tetrapods (Bolt, 1991, p. 199). Nonetheless, Benton (1990) is right in arguing that the reduction of elements is not sufficient evidence on its own for a close relationship, since a superficially similar reduction must have occurred many times in early tetrapods. However, dissection of salamander and frog heads shows in both groups a similar attachment of large jaw muscles to the borders of frontals, parietals, squamosals and the underlying auditory capsules. Studying these muscles in their ontogeny and function, Iordansky (1990a,b, 1991, 1992) showed that anurans and urodeles share many homologous jaw muscles (as well as skull kinetics connected with these muscles), and hence inferred a close common ancestry for them (he revised the study of Carroll and Holmes (1980) who had concluded a separate origin for the urodeles).

Whether salamanders and frogs derived independently or together from a *Doleserpeton*-like ancestor is not the case in point here. Rather I shall focus on the implications of skull remodelling that occurred to each of them. The loss of temporal and eye-surrounding bones must have been caused by a relative enlargement of the jaw musculature. What is known about the feeding style of dissorophoids, however, does not appear essentially different from that of modern salamanders and frogs (Bolt, 1977). Bearing in mind that pedicellate teeth were already present and the ancient stegocephalian feeding style rearranged into an insectivorous one, the expansion of the jaw muscles as observed might not have been expected in animals of about dissorophoid size.

If one considers the trend of miniaturisation in *Doleserpeton*, a new perspective arises. Should this trend have continued further, would not a complete reduction of the affected bones be a consequence? *Doleserpeton* skulls measure between 12 and 17 mm. Bolt (1977) assumed that completely grown individuals would have been slightly larger than the largest specimens found. The mean skull length of urodeles is 10 mm, that of anurans 15 mm (Benton, 1990). It may thus be concluded that the ancestors of both groups did indeed go through a further miniaturisation that finally reached the minimal length of the jaw muscles. The disappearance of most dermal bones in Recent animals occurred in regions where the muscles take over to attach onto the skull surface. Most of these bones fail to mineralise by an extreme postdisplacement. This could have been easily arranged

---

Figure 6.5 Miniaturisation hypothesis of anuran and urodele origin. Progenesis and acceleration are suggested as having reduced the indicated bones (1–5) of advanced dissorophoid amphibians, permitting the jaw musculature to insert on the margins of the remaining skull roof (data from Iordansky, 1991, 1992). The decisive proof of miniaturisation has yet to be found: a dissorophoid somewhat advanced in morphology in respect to *Doleserpeton* and of smaller size. Data from Bolt (1969), Ivakhnenko (1978), Milner (1988), Rage and Roček (1989).

ontogenetically, because these bones were evidently among the last to mineralise (data on branchiosaurids in Boy, 1974 and in Schoch, 1992). In addition, modern amphibians reach metamorphosis far earlier with respect to overall bone connection than is known in stegocephalians (Milner, 1988; Lebedkina, 1979; Boy, 1974; Schoch, 1992).

If the hypothesis of salamander and frog origin by miniaturisation is true, then the processes involved should have been a further progenetic reduction of body size, accompanied by a precocious onset of development, and a local postdisplacement and neoteny of a variety of bony elements. To test this hypothesis, more convincing near-ancestor candidates, which should be as small or smaller than *Doleserpeton*, must be searched for, especially in the Permian. Furthermore, more ontogenetic sequences, particularly those displaying bone formation, of related and outgroup taxa need to be described (see the cladogram in Figure 6.1 and the data in Appendix 6.1).

The ecological background influencing the hypothetical direct size selection has been outlined by Bolt (1979). According to him, the increasing aridity of many regions in the late Permian and Triassic must have favoured the selection of small amphibians by its restricted water capacities and fewer favourable localities. Possibly, salamander and frog ancestors focused on a life as naked dwellers in the vicinity of small water sites, whereas relatives of *Tersomius* prevented desiccation by strong dermal armour, allowing these animals to keep a moderate size of up to half a metre (DeMar, 1968).

DISCUSSION

A variety of problems and restrictions arose in the course of this study. These result, in part, from the incompleteness of the fossil record (Figure 6.1). For instance, the ontogenetic record of the extinct taxa generally concentrates on few growth stages, which mainly depend on the sedimentary facies (ontogenetic habitat changes were presumably frequent, see Boy, 1978, 1987b). In cases of excellent preservation and continuity of habitat, many growth stages can be studied (Boy, 1972, 1974; Milner, 1982a; Schoch, 1992 and in prep.), but in some cases the paedomorphic retention of gills and termination of growth (branchiosaurids, micromelerpetontids) preclude comparison with adults of related taxa. The latter point is of considerable importance, since many characters have been erected previously that cannot be observed in their developmental change. Hence wrong phylogenetic conclusions are easily made. Consequently, one should restrict phylogenetic analyses to ontogenetically traceable characters in cases such as those described above, although the analysis certainly becomes more difficult or even impossible in some clades.

A second problem is demonstrated in the multiple polarity reversals in branchiosaurid larvae. Paradoxically, the fossil record is so detailed here that problems arise in distinguishing between species and mere ecologically specialised variants. The crucial insurmountable obstacle here is the greatly restricted data set of fossil vertebrates which contains at best information about all of the skeleton and directly correlated soft-anatomical details (an unrepresentative set of muscles, cartilages and vessels). These reversals of polarity cannot be excluded in other reconstructed "lineages" here, as well; more secure appear the trends at the intermediate level, such as from *Amphibamus* to *Doleserpeton* (here polarity reversals also exist, but the anatomical details of the convergent organs are decisive).

Finally, the hypothesis of anuran and urodele origin has to be qualified in two important respects. Firstly, the focus on the head region does, unavoidably, provide a narrow perspective, particularly in the search for ecological implications. Secondly, the hypothesis about the origin of the anuran head omits the premetamorphic tadpole stages. We are still far from understanding the evolutionary transition from a dissorophoid larva to an anuran tadpole.

Hanken (1984) described the consequences of miniaturisation in plethodontid salamanders, and summarised them in three observations formulated to be of general validity: paedomorphic morphology (by progenesis), increased variability and evolutionary novelty. Branchiosaurids allow similar observations: their variability was remarkable (Figure 6.4) and the prolonged aquatic period was obviously a successful innovation. A more important novelty should have been the skull remodelling of salamander and frog ancestors: the variation in skull morphology of fossil and modern anurans and urodeles is extensive, if compared with the stegocephalous ancestral skull (Duellman and Trueb, 1986).

The analysis of heterochronies in temnospondyl amphibian evolution reveals a complicated story of evolutionary change in timing of development at all phylogenetic scales. Dissociated and mosaic heterochronoclines dominate, and moreover the case of branchiosaurid larvae demonstrates a variety of problems that do not appear at larger scales due to their more restricted data set.

ACKNOWLEDGMENTS

I thank Dr Kenneth J. McNamara (Perth) and Dieter Korn (Tübingen) for reading and commenting on the manuscript and for interesting discussions. Drs Juergen A. Boy (Mainz), Wolfgang Maier, Wolf-Ernst Reif, Frank Westphal, and Matthias Kröner (all Tübingen) helped with comments and literature, and Drs Andrew R. Milner (London), John Long (Perth), and Per Ahlberg (London) stimulated thinking. Dr Fischer (Humboldt Museum, Berlin) kindly permitted me to study material in his care. Dieter Korn assisted in the conceptual and technical realisation of the figures, Dr Raimund Feist provided hospitality during the Montpellier meeting, and Professor Volker Mosbrugger organised financial support.

APPENDIX 6.1

1. Pterygoids curved medially, framing the fairly large interpterygoid vacuities (compare Smithson, 1982 and Holmes and Carroll, 1977).

2. Vomers participate in the margin of the interpterygoid vacuities, separating pterygoids anteriorly (compare Milner, 1980 and Godfrey *et al.*, 1987).
3. Shortened ectopterygoid (compare Holmes and Carroll, 1977 and Milner, 1980).
4. Stapes with elongate, rod-like, anterioposteriad compressed shaft; exoccipital enlarged to give bilobed occipital condyle; reduced basioccipital still participating in condyles (Milner, 1990).
5. "Loss" of intertemporal by fusion with supratemporal (Godfrey *et al.*, 1987).
6. Shortened snout region.
7. Decreased body size.
8. Laterally exposed palatine; reduction of jugal anteriorly (Bolt, 1969).
9. Palatine, ectopterygoid, and pterygoid reduced to narrow rods of bone, expanding the interpterygoid vacuities marginally (Milner, 1990).
10. Medial quadratojugal process (Boy, 1972).
11. Dorsal quadrate process (Bolt, 1969).
12. Partial separation of ectopterygoid and palatine from the maxilla (compare Boy, 1972: Figure 5).
13. Specialised gill-rakers (Boy, 1972).
14. Shortening of posterior skull table (Boy, 1972).
15. Ectopterygoid reduced to small bar (Boy, 1972).
16. Broadening of tympanic notch; shortening of squamosal-supratemporal contact.
17. Palatine and ectopterygoid further narrowed, partly or completely separated by pterygoid.
18. Marginal teeth with bicuspid crowns (Milner, 1988).
19. Extreme broadening of tympanic notch; narrowing of supratemporal and tabular.
20. Shortening of posterior skull table.
21. Loss of prefrontal–postfrontal contact; narrowing of frontals (Bolt, 1969).
22. Loss of ectopterygoid (Bolt, 1969).
23. Pedicellate teeth (Bolt, 1969).
24. Vomerine and palatal fangs replaced by clumps or lines of small teeth (Bolt, 1977).
25. Neurocranium and dermatocranium forming a single structure wrapped around the brain (Milner, 1988).
26. Loss of sclerotic ring (Milner, 1988).
27. Reduction to one coronoid (Milner, 1988).

REFERENCES

Benton, M.J., 1990, Reptiles. In K.J. McNamara (ed.), *Evolutionary trends*, Belhaven, London: 279–300.
Bolt, J.R., 1969, Lissamphibian origins: possible protolissamphibian from the Lower Permian of Oklahoma, *Science*, **166**: 888–891.
Bolt, J.R., 1977, Dissorophoid relationships and ontogeny, and the origin of the Lissamphibia. *J.Paleont.*, **51**: 235–249.
Bolt, J.R., 1979, *Amphibamus grandiceps* as a juvenile dissorophid: evidence and implications. In Nitecki, M.H. (ed.), *Mazon Creek fossils*, Academic Press, New York: 529–563.
Bolt, J.R., 1991, Lissamphibian origins. In H.P. Schultze and L. Trueb (eds), *Origin of the higher groups of tetrapods*, Comstock, Ithaca and London: 194–222.
Boy, J.A., 1972, Die Branchiosaurier (Amphibia) des saarpfälzischen Rotliegenden (Perm, SW-Deutschland), *Abh.hess.L.-A.Bodenforsch.*, **65**: 1–137.
Boy, J.A., 1974, Die Larven der rhachitomen Amphibien (Amphibia: Temnospondyli; Karbon-Trias), *Paläont. Z.*, **48**: 236–268.

Boy, J.A., 1978, Die Tetrapodenfauna (Amphibia, Reptilia) des saarpfälzischen Rotliegenden (Unter-Perm; SW-Deutschland). 1. *Branchiosaurus, Mainzer geowiss. Mitt.*, **7**: 27–76.

Boy, J.A., 1980, Die Tetrapodenfauna (Amphibia, Reptilia) des saarpfälzischen Rotliegenden (Unter-Perm; SW-Deutschland). 2. *Tersomius graumanni* n.sp., *Mainzer geowiss. Mitt.*, **8**: 17–30.

Boy, J.A., 1981, Zur Anwendung der Hennig'schen Methode in der Wirbeltierpaläontologie, *Paläont. Z.*, **55**: 1–2, 87–107.

Boy, J.A., 1987a, Studien über die Branchiosauridae (Amphibia: Temnospondyli; Ober-Karbon-Unter-Perm). Systematische Übersicht, *Neues Jahrb. Geol. Paläont. Abh.*, **174**: 75–104.

Boy, J.A., 1987b, Die Tetrapoden-Lokalitäten des saarpfälzischen Rotliegenden (?Ober-Karbon-Unter-Perm; SW-Deutschland) und die Biostratigraphie der Rotliegend-Tetrapoden, *Mainzer geowiss. Mitt.*, **16**: 31–65.

Boy, J.A., 1988, Über einige Vertreter der Eryopoidea (Amphibia: Temnospondyli) aus dem europäischen Rotliegenden (?höchstes Karbon-Perm). 1. *Sclerocephalus, Paläont. Z.*, **62**: 107–132.

Boy, J.A., 1993a, Über einige Vertreter der Eryopoidea (Amphibia: Temnospondyli) aus dem europäischen Rotliegenden (?höchstes Karbon-Perm) 4. *Cheliderpeton latirostre, Paläont. Z.*, **67**: 1–2.

Boy, J.A., 1993b, Evolution bei den Amphibien, *Veröff. Übersee-Mus., Nat. Wiss.*, **11**: 27–40.

Carroll, R.L., 1964, Early evolution of the dissorophid amphibians, *Bull. Mus. Comp. Zool. Harvard*, **131**: 161–250.

Carroll, R.L., 1967, Labyrinthodonts from the Joggins Formation, *J. Paleont.*, **41**: 111–142.

Carroll, R.L. and Currie, P.J., 1975, Microsaurs as possible apodan ancestors, *Zool. J. Linn. Soc.*, **57**: 229–247.

Carroll, R.L. and Holmes, R., 1980, The skull and jaw musculature as guides to the ancestry of salamanders, *Zool. J. Linn. Soc.*, **68**: 1–40.

Daly, E., 1973, A lower Permian vertebrate fauna from southern Oklahoma, *J. Paleont.*, **47**: 562–589.

DeMar, R., 1968, The Permian labyrinthodont *Dissorophus multicinctus*, and adaptations and phylogeny of the family Dissorophidae, *J. Paleont.*, **42**: 1210–1242.

Duellman, W.E. and Trueb, L., 1986, *Biology of amphibians*, McGraw Hill, New York.

Godfrey, S.J., Fiorillo, A.R. and Carroll, R.L., 1987, A newly discovered skull of the temnospondyl amphibian *Dendrerpeton acadianum* Owen, *Can. J. Earth Sci.*, **24**: 796–805.

Hanken, J., 1984, Miniaturization and its effects on cranial morphology in plethodontid salamanders, genus *Thorius* (Amphibia, Plethodontidae). I. Osteological variation, *Biol. J. Linn. Soc.*, **23**: 55–75.

Holmes, R. and Carroll, R.L., 1977, A temnospondyl amphibian from the Mississippian of Scotland, *Bull. Mus. Comp. Zool.*, **147**: 489–511.

Iordansky, N.N., 1990a, *Evolution of cranial kinesis in lower tetrapods*, Nauka, Moscow (in Russian).

Iordansky, N.N., 1990b, *Evolution of complex adaptations (jaw apparatus of Amphibians and Reptiles)*, Nauka, Moscow [in Russian].

Iordansky, N.N., 1991, Jaw musculature and the problem of phylogenetic relations of the Lissamphibia, *Zool. Zh.* **70**: 50–62 [in Russian].

Iordansky, N.N., 1992, Jaw muscles of the Urodela and Anura: some features of development, functions, and homology, *Zool. Jb. Anat.*, **122**: 225–232.

Ivakhnenko, M.F., 1978, Urodelans from the Triassic and Jurassic of Soviet Central Asia, *Paleont. JL*, **3**: 362–368.

Jenkins, F.A and Walsh, D.M., 1993, An early Jurassic caecilian with limbs, *Nature*, **365**: 246–250.

Lebedkina, N.S., 1964, *Razvite pokrovnikh kostey osnovaniya cerepa khvostatykh amfibiy sem'a Hynobiidae*, Nauka, Moscow.

Lebedkina, N.S., 1968, *Razvite kostey kryshy cerepa amfibiy*, Nauka, Leningrad.

Lebedkina, N.S., 1979, *Evol'uziya cerepa amfibiy*, Nauka, Moscow.

Milner, A.R., 1980, The temnospondyl amphibian *Dendrerpeton* from the Upper Carboniferous of Ireland, *Palaeontology*, **23**: 125–141.

Milner, A.R., 1982a, Small temnospondyl amphibians from the Middle Pennsylvanian of Illinois, *Palaeontology*, **25**: 635–664.

Milner, A.R., 1982b, A small temnospondyl amphibian from the Lower Pennsylvanian of Nova Scotia, *J. Paleont.*, **56**: 1302–1305.

Milner, A.R., 1987, The Westphalian tetrapod fauna: some aspects of its geography and ecology, *J. Geol. Soc. Lond*, **144**: 495–506.

Milner, A.R., 1988, The relationships and origin of living amphibians. In M.J. Benton (ed.), *The phylogeny and classification of the tetrapods. 1: Amphibians, Reptiles, Birds*, Syst. Assoc. Special Vol. **35A**: 59–102.

Milner, A.R., 1990, The radiations of temnospondyl amphibians. In P.D. Taylor and G.P. Larwood (eds), *Major evolutionary radiations*, Clarendon, New York.

Parsons, T.S. and Williams, E.E., 1963, The relationships of the modern Amphibia: a reexamination, *Q. Rev. Biol.*, **38**: 26–53.

Rage, J.-C. and Janvier, P., 1982, Le problème de la monophylie des amphibiens actuels, a la lumière des nouvelles données sur les affinités des tétrapodes, *Géobios, Mem. spéc.*, **6**: 65–83.

Rage, J.-C. and Roček, Z., 1989, Redescription of *Triadobatrachus massinoti* (Piveteau, 1936) an anuran amphibian from the early Triassic, *Palaeontographica A*, **206**: 1–2, 1–16.

Reig, O.A., 1964, El problema del origen monofilético o polifilético de los amfibios, con consideraciones sobre las relaciones entre Anuros, Urodelos y Ápodos, *Ameghiniana*, 3: 191–211.

Ricqles, A. de, 1975, Quelques remarques paléo-histologiques sur le problème de la néoténie chez les stégocéphales, *C.N.R.S*, **218**: 351–363.

Schmalhausen, I.I., 1968, *The origin of terrestrial vertebrates*, Academic Press, New York and London.

Schoch, R.R., 1992, Comparative ontogeny of early Permian branchiosaurid amphibians from southwestern Germany. Developmental stages, *Palaeontographica A*, **222**: 43–83.

Smithson, T.S., 1982, The cranial morphology of *Greererpeton burkemorani* Romer (Amphibia: Temnospondyli), *Zool. J. Linn. Soc.*, **76**: 29–90.

Smithson, T.S., 1985, The morphology and relationships of the Carboniferous amphibian *Eoherpeton watsoni* Panchen, *Zool. J. Linn. Soc.*, **85**: 317–410.

Trueb, L. and Cloutier, R., 1991, A phylogenetic investigation of the inter- and intrarelationships of the Lissamphibia (Amphibia: Temnospondyli). In: H.P. Schultze and L. Trueb (eds), *Origins of the higher groups of tetrapods*, Comstock, Ithaca and London: 223–316.

Werneburg, R., 1989, Labyrinthodontier (Amphibia) aus dem Oberkarbon und Unterperm Mitteleuropas. Systematik, Phylogenie und Biostratigraphie, *Freiberger Forsch.-H.*, C **436**: 7–57.

Chapter 7

HETEROCHRONY IN THE EVOLUTION OF ENAMEL IN VERTEBRATES

Moya M. Smith

INTRODUCTION

Teeth and denticles are classic examples of structures with a precise shape, position and specific tissue composition that result from a cascade of epithelial–mesenchymal interactions occurring at every stage of development, from initiation to completion of the tissue product. In the example discussed here, enamel is a tissue type found in the dermal armour and oral teeth of all vertebrates, in which the biological mineralized structure is preserved in both fossil and extant forms, covering the 500 million years of geological history of vertebrates. Because enamel and enameloid are the hardest biological mineralized tissues they are always the best preserved in the fossil record and therefore available for detailed analysis of tissue microstructure. Also, because these biological hard tissues retain a precise record within them of variation in cellular secretory activity, both extant and fossil examples can be compared and developmental information extrapolated from the mature tissue. This information, on rates of cell secretion, relative timing and direction of growth, and influence of cell or matrix on crystal orientation, can be obtained from the spacing and direction of the incremental lines and the specific arrangement of the crystallites into ordered groups.

Perhaps because this is an example of a developmental system producing one tissue, and is simple and comprehensive, in that it operates in all vertebrates, it is amenable to an analysis of how heterochrony may have influenced the tissue diversity of enamel. Also, because dermal denticles and oral teeth are made of enamel and dentine, we can examine the subtle changes in timing of secretion by the two cell layers, epithelium and mesenchyme, that contribute to the differences between the two main tissue types, enamel and enameloid. Now that molecular

Evolutionary Change and Heterochrony. Edited by Kenneth J. McNamara © 1995 The Editor and Contributors.
Published in 1995 by John Wiley & Sons Ltd.

data is accumulating to provide evidence of the dialogue between these cells during tooth development (see below), changes in timing of events can be explained as the cells' ability to make, or respond to, growth factors and how these directly influence the timing of expression of homeobox gene products. These recent studies of the molecular controls operating at all stages of tooth development allow a heterochronic mechanism to be proposed through which evolution could operate.

All examples of dermal armour and teeth with a covering of enamel or enameloid over dentine will be considered, as products of a single modifiable morphogenetic system (Schaeffer, 1977). The developmental unit of this system is an epithelial–mesenchymal grouping of cells formed into a dental cap (ectoderm), dental papilla and dental follicle (ectomesenchyme) which, through a cascade of dependent interactions, produces the odontode (Figure 7.1). Odontodes are covered by enamel or enameloid. Both are acellular tissue products of this series of epithelial–mesenchymal interactions, and are shiny, translucent, and extra hard. These developmental processes are assumed to be homologous for the dermal skeleton and the oral teeth, both dependent upon neural crest derived mesenchyme for induction and differentiation of the dental ectoderm into ameloblasts, and dental mesenchyme into odontoblasts, of the tooth germ (Smith and Hall, 1990, 1993).

The developmental programme of the tooth germ constrains phenotypic expression within a causal and dependent sequence, but it is proposed that flexibility through a change in relative timing of the onset of differentiation allows a qualitative change in the tissues produced. This is achieved by a shift in timing of the relative differentiation of ameloblasts and odontoblasts in producing enamel and dentine. The distinctions between enamel and enameloid, although of necessity determined by microstructural characteristics in fossil forms, are based on developmental differences known for the extant forms. The former is a secretion product entirely of epithelial origin and the latter a product of mixed epithelial and mesenchymal activity. These differences, due to a change in timing of cell activity, can be detected by the presence and direction of incremental lines and by the arrangement of crystals into protoprisms/prisms or woven/parallel crystal bundles (Smith, 1989). A dissociation of timing of synthetic activity between the two cell types (ameloblasts and odontoblasts), such that enamel proteins are secreted after the dentine has begun to mineralize, results in appositional growth of enamel away from the junction with dentine and an epithelial product without any mesenchymal contribution. An earlier time of secretion of enamel proteins to occur simultaneously with those from the odontoblasts, results in enameloid as a product formed from both cell types because it forms before dentine matrix has calcified. This tissue is without appositional lines of growth and forms on the papilla side of the basal lamina (Figure 7.4b).

This ontogenetic change, which produces different phenotypes, is proposed as an example of heterochrony in evolution, in which the timing of differentiation of

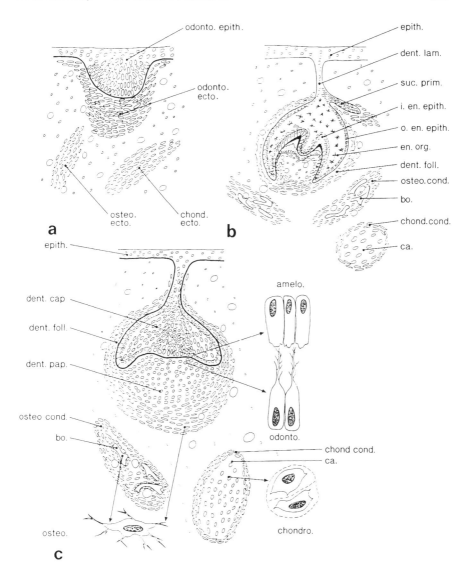

Figure 7.1 Three developmental periods of a mammalian tooth germ recognized morphogenetically as bud (a), cap (c) and bell stages (b), showing the odontogenetic components of the epithelium and ectomesenchyme. The bud stage (a) consists of a thickening of basal epithelial cells and condensed ectomesenchyme. At the cap stage (c) ectomesenchyme is recognized as dental papilla and dental follicle. In the centre of the dental cap is the site of the first cusp where the early signalling molecules are found. At the bell stage (b) cusps develop in positions where cell division stops first, cell differentiation occurs, and dentine and enamel are produced. (From Smith and Hall, 1993, Figure 1, the caption to which provides details of abbreviations; with permission of Plenum Press.)

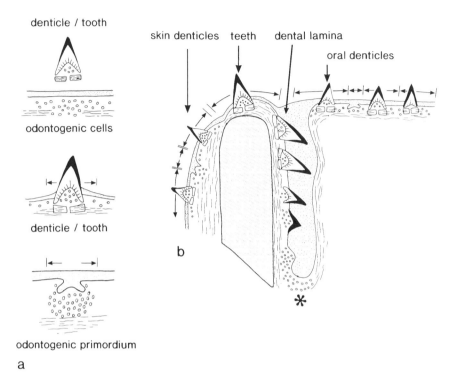

Figure 7.2   Skin denticles and oral teeth are compared as homologous developmental units in this shark model (b), where germs of odontodes (odontogenic primordium denticles a) initiate denticles in a region in the epithelium postulated to be free from inhibition zones around existing denticles (arrows). Free end of dental lamina away from any inhibition zones for denticles or teeth, at which ectomesenchyme can be induced to form a new tooth (asterisk). (From Smith and Hall, 1993, Figure 2; with permission of Plenum Press.)

ameloblasts is displaced in the descendant to an earlier stage relative to the differentiation of odontoblasts, producing enameloid, and that this is a change from the primitive condition, in which enamel is produced. This theory differs from generally accepted ones, in which the developmental mechanism that produces enameloid is considered to be the primitive condition for vertebrates. This proposition is reached from an analysis of the fossil record which reveals that enamel occurs in cladistically primitive groups, found to be also the earliest occurring vertebrates in the fossil record. The switch between enamel and enameloid has occurred many times in the vertebrate record and can occur independently in the dermal skeleton and the teeth in individual taxa.

TEETH AS DERIVATIVES OF EPITHELIAL-MESENCHYMAL INTERACTIONS

Homology between teeth and dermal denticles

Teeth and skin denticles exist in all groups of vertebrates, except birds, in both extant and fossil forms (see Smith and Hall, 1990 for a review). These have long been assumed to be similar structures, although it was Schaeffer's (1977) review of the dermal skeleton in fishes that first clearly linked common structure with shared morphogenetic mechanisms. He proposed the concept that "the dermal skeleton develops from a single, modifiable morphogenetic system" and envisaged that regulation of morphogenetic change was through interaction in this morphogenetic system and that it was significant for evolution of the dermal skeleton. Smith and Hall (1993) proposed that the odontogenic part of the "morphogenetic system" is the equivalent of the tooth germ and formed of the dental primordium of interactive dental cap ectoderm and dental papillary and follicular ectomesenchyme (Figure 7.1). This system forms the developmental basis of the *odontode*, the structure common to all ornamented dermal skeletons and oral teeth (Figure 7.2).

Based on Schaeffer's proposals, Reif (1982) in his "odontode regulation theory", proposed that regulatory processes are the means of varying differentiation programmes of odontodes. Smith and Hall (1993) commented that these regulatory processes would involve cell or matrix-based regulatory molecules, and proposed that part of the differentiation programme could be suppressed or altered in its relative timing through heterochrony in evolution to change tissue composition, and alter phenotypes.

Development of tooth germs, dependence on neural crest

What level of control do we expect to find intrinsic to the cells and to what extent is the epithelium or the ectomesenchyme prespecified, or alternatively dependent upon cell determined epigenetic factors to initiate differentiation? Timing of differentiation of cells could be predetermined as part of the cells' intrinsic ability to specify the times of differentiation (see Chapter 1). Although the process of cell division and differentiation is intrinsic to the cells "extrinsic influences are as important a component in control as are the genes which produce the products of the cell" (Hall, 1994). Some of the molecular controls for either intrinsic or extrinsic mechanisms operating in the tooth germ will be discussed in the next section.

The embryonic origin of cells that contribute to the tooth germ is well known: ectoderm or endoderm of the primitive mouth for the dental epithelium and bilaterally arranged dorsal ectoderm of the neural plate for the cranial neural crest which gives rise to the dental papilla and dental follicle. Potential neural crest cells initially lie within the dorsal neural epithelium of the neurula. As it invaginates and they transform into mesenchymal cells they migrate away from the developing neural tube (Hall and Hörstadius, 1988). All of the craniofacial skeleton, including teeth, is dependent upon neural crest for its differentiation. The role of cranial neural crest in inducing teeth and contributing cells to make dentine is firmly established and has been expertly reviewed by Gaunt and Miles (1967), Lumsden (1987) and Hall and Hörstadius (1988). Although evidence that the role of neural crest in tooth development is not just evocative, but includes the contribution of cells differentiating within the tooth germs, was provided by the experiments of Chibon (1966, 1967) on amphibians, it was not until more recently that direct evidence for the neural crest contribution to teeth in mammals was provided by Lumsden's (1987) explant experiments.

Recent work has demonstrated that precursor cells of the premigratory cranial neural crest are both position and type specified. They generate progeny of only one cell type and are also topographically restricted to one of the branchial arches (Schilling and Kimmel, 1994). This implies that the cells have positional information (rostro-caudal, medio-lateral) carried with them to their postmigration destination. Schilling and Kimmel found that the progenitors of these cell types were spatially separated in the premigratory, segregated crest, and suggested that cells may be fate specified before the morphogenetic movements that take them to their destination. If this is also true for the progenitor cells of dental tissues in the visceral skeleton then odontogenic cells would be prespecified at their sites in the segregated, premigratory neural crest.

Since it is known that odontogenic potential is very precisely localized in the cranial neural folds in amphibians (Chibon, 1966, 1967), it is highly likely that these premigratory cells are also type specified before they reach their destination. These locations are cranio-rostral regions expressed as a series of 30° segments, starting from the medial axis and progressing through 180°, and segments covering 30–100° correlate with tooth formation. Cells from the 30–70° regions produce teeth of the upper jaws, cells of the 70–100° regions teeth of the lower jaws. Lumsden (1987) suggested that "a degree of commitment is acquired by the crest before migration" but that "this commitment is not complete", there is a requirement for interaction with epithelium before odontogenesis is effected. Pharyngeal endoderm rather than stomodeal ectoderm has been implicated as the embryonic source of epithelium providing the initial signal for commitment of cranial neural crest to odontogenesis in urodele amphibians (Cassin and Capuron, 1979). They concluded that endoderm promotes the differentiation of cranial neural crest into odontoblasts. Graveson (1993) has also found, in isochronic and heterochronic recombinations of neural crest with ecto/endoderm as explants from amphibian neurulae, that ectoderm alone will not induce teeth from neural crest, but that endoderm will. Whichever of the two embryonic cell layers is found to be essential for tooth specification, a fundamental question remains. What is the mechanism of the initial embryonic patterning in the determination of tooth morphology and tissue specification? Examination of some of the data now emerging from molecular studies of tooth development in the following section provides clues to this problem.

Molecular data on dental epithelial–mesenchymal interactions

A basis for explanation of tissue diversity in the composition of teeth will include molecular control mechanisms as reviewed by Smith and Hall (1993, p. 406). This data is beginning to accumulate for mammalian teeth, an excellent model system for both *in vivo* and *in vitro* studies, and assuming that there are common bases for differentiation and morphogenesis of vertebrate teeth we can use this information to predict the molecular basis for heterochrony in enamel development and evolution. Hall (1994) has discussed the variety of ways in which epigenetic control occurs during tissue interactions in developing systems, including those between epithelium and mesenchyme, and given examples of how the site of gene action between either of these cell layers is determined using mutants, and the three possible ways in which tissue component signals may arise between epithelium and mesenchyme. These tissue components include retinoic acid acting as a morphogen, hormones mediating between epithelium and mesenchyme, cell-to-cell adhesion molecules (CAMS), substrate-to-cell adhesion molecules (SAMS), and growth factors.

Smith and Hall (1993, p. 407) have commented that differentiation can be manipulated by "different mixes of intrinsic and epigenetic controls and these mixes can change through ontogeny". Further that "differentiation of odontoblasts and ameloblasts can be coupled or uncoupled, or coregulated or independently regulated". It is proposed that there are five classes of interdependent molecules involved in molecular control of odontode differentiation (Smith and Hall, 1993), providing enormous potential for developmental and evolutionary modulation. These molecules found to be involved in tooth differentiation are cell surface molecules such as syndecan, extracellular matrix components (tenascin, fibronectin), growth factors (epidermal growth factor (EGF)), transforming growth factor (BMP-2,4), homeobox genes (*msx-1, -2*) and retinoids.

Progressive determination of the dental epithelium and mesenchyme at classic, recognized histological stages of tooth germ morphodifferentiation has now been described in molecular terms using immunohistochemical (Hurmerinta *et al.*, 1986; Vainio *et al.*, 1989) and *in situ* hybridization techniques (Mackenzie *et al.*, 1991, 1992; Vainio *et al.*, 1993) for analysis of patterns of distribution of molecules relative to recognized developmental events (Figure 7.3). The early stages of tooth induction, morphogenesis and morphodifferentiation have been shown to be controlled by these molecules during epithelial–mesenchymal interactions, but the later stages involving odontoblast and ameloblast differentiation have been less extensively studied. However, good data on some of the molecules has been reported to relate their differential spatial and temporal

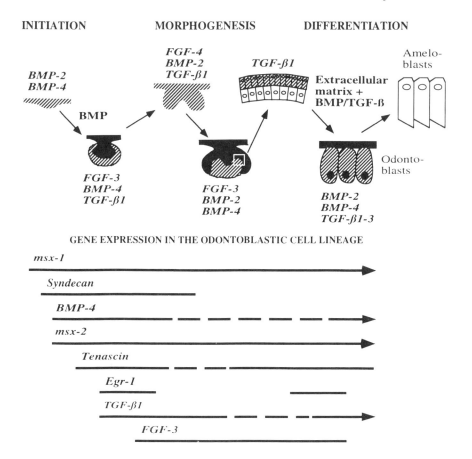

Figure 7.3   Schematic summary of the expression of genes encoding for transcription factors and growth factors in the dental epithelium during initiation of the tooth germ, and morphogenesis and differentiation into ameloblasts as enamel production commences. Below is a representation of the periods of expression of the transcription factors (*msx-1, msx-2, Egr-1*), growth factors (BMP-4, TGF-β1, FGF-3) and structural molecules (syndecan, tenascin) during determination and differentiation of the odontoblastic cell lineage in the mouse. The relative timing of these events will play a crucial role in heterochrony. (Figure provided by Professor Irma Thesleff, modified from Thesleff *et al.* (1995).)

occurrence with odontoblast and ameloblast differentiation. Some of these concern growth factors, such as epidermal growth factor (EGF), small peptides acting as key molecules in regulating embryonic development, and the differential expression patterns of their receptors (Partanen, 1990; Partanen and Thesleff, 1987, 1989; Thesleff *et al.*, 1987, 1989). Partanen and Thesleff (1987, 1989) have shown that EGF will stimulate cell proliferation and inhibit dental cell differentiation, first in the dental epithelium of the bud stage, but at the cap stage and onwards it is the follicular mesenchyme that is stimulated. The distribution of EGF receptors changes during tooth morphogenesis, but because the distribution of these receptors was not related to the stimulation of proliferation by EGF in these tissues Partanen and Thesleff suggested that epithelial–mesenchymal interactions control the response of the dental tissues to EGF. It could further be suggested that EGF, EGF receptors and intrinsic control, could inhibit the differentiation of odontoblasts till a later stage and produce the heterochronic shift from enamel to enameloid. This is only one example of molecular control of the timing of cell differentiation, but there are many other examples in different categories of cellular control.

Many of these studies show that there is up-regulation of expression of several growth factors, transcription factors and synthesis of structural extracellular molecules of the matrix and of the cell surface during key differentiation events in tooth development. The response of cells to growth factors is also influenced by extracellular matrix molecules. It has been suggested that these bind and store growth factors for local release under cell control (Partanen and Thesleff, 1989). This may also be a significant mechanism in the control of enamel or enameloid production in which matrix molecules are regularly degraded and taken up into the cells. One of the early events preceding odontoblast/ameloblast differentiation is removal of the basal lamina at the junction between the undifferentiated cells. This itself may release bound growth factors, or allow epithelial cell contact with the mesenchymal matrix and their associated bound factors (Figure 7.3). Thesleff *et al.* (1995) have shown how this may work in a series of *in vitro* experiments where dental epithelium and mesenchyme, taken from specific embryonic times, are recombined either alone or with beads soaked in growth factors, used instead of epithelium. The response of mesenchyme was then compared in both situations by revealing induction sites of syndecan, tenascin and cell proliferation. They concluded that the beads only partly mimicked the dental epithelium and that other epithelial signals are needed for effective induction. They further proposed that these additional signals would be fibroblast growth factors (FGFs) and transforming growth factors (TGF-βs).

In particular the newer techniques of *in situ* hybridization are providing significant, interesting information on growth factor-mediated signalling and homeobox gene expression in specification of tooth morphogenesis (Figure 7.3). Vainio *et al.* (1993) have identified bone morphogenetic protein (BMP-4), and now BMP-2 (Thesleff *et al.*, 1994), members of the transforming growth factor β superfamily, as early signals in the dental epithelium that lead to mesenchymal induction of the genes controlling two homeobox containing transcription factors, *msx-1* and *msx-2*. In relation to ameloblast differentiation, Vainio *et al.* found that BMP-4 expression shifted from mesenchyme to epithelial cells prior to this event. Earlier BMP-4 was expressed in the mesenchyme coincident with differentiation into odontoblasts, and the expression of the related BMP-2 three days later, increased in the secretory odontoblasts and was transiently expressed in the ameloblasts during their differentiation, suggesting a role in dental cell differentiation. Vaino *et al.* (1993) further suggested that epithelial–mesenchymal interactions, required for differentiation of these cells, are regulated by BMP gene expression in these cells. Odontoblasts will differentiate if TGF-β and/or BMP-2 are combined with matrix components (heparin sulphate/fibronectin). Also, in the differentiated odontoblasts BMP-2, -4 and TGF-β1, -3 are all expressed and proposed as potential candidates to induce ameloblast differentiation (Bèque-Kirn *et al.*, 1992). These are all molecular examples of a mechanism for control of sequential, reciprocal signalling events between epithelium and mesenchyme that provide the mechanistic basis of heterochrony.

The *hox* gene *EGR-1*, also induced by BMP-4, is reported by Karanova *et al.* (1992) to be rapidly and recurrently activated and down-regulated during determination of cell fate in tooth development. *Msx -1* is expressed maximally in the cap stage of tooth development (Thesleff *et al.*, 1990; Mackenzie *et al.*, 1991; Jowett *et al.*, 1993) and Thesleff *et al.* (1990) considered it as active early in the cascade of molecular changes regulated by epithelial–mesenchymal signals, one of which could be BMP-4, found to be a functional signal molecule autoregulated between epithelium and mesenchyme (Vainio *et al.*, 1993). *Msx-1* and *msx-2* are putative transcription factors that may play a role in regulating the expression of other genes during all stages of differentiation in the tooth germ (Jowett *et al.*, 1993) and as such have been proposed as "master genes" by Thesleff *et al.* (1995). Undoubtedly ongoing research in this field will provide very specific information on how controls work on cell differentiation, and this applied comparatively across the vertebrate taxa will allow differences to be evaluated in a phylogenetic context that postulates heterochrony.

HISTOGENESIS OF ENAMEL AND ENAMELOID

Timing of differentiation

Two early studies of enameloid development focused on the controversial issue of the potential contribution of epithelium and mesenchyme to the enameloid in larval urodele teeth (Smith and Miles, 1969) and in two teleost fishes (Shellis, 1975; Shellis and Miles, 1974; 1976). These studies, and Smith and Hall (1990), have demonstrated that the epithelial cells (ameloblasts) differentiate early and deposit a protein into the enameloid matrix at the same time as the ectomesenchymal cells (odontoblasts) differentiate, secrete a protein and then withdraw to leave behind cell processes responsible for tubule formation in the mature tissue. In these examples enameloid was formed in a matrix jointly secreted by epithelium and ectomesenchyme, and was further matured through resorptive action and secretion of minerals by ameloblasts (Figure 7.4). Since then numerous other studies have documented the dual origin of the enameloid tissue (Shellis, 1978; Meinke, 1982b; Graham, 1984). However, studies on elasmobranch enameloid have produced conflicting evidence on the exact contribution of the epithelium

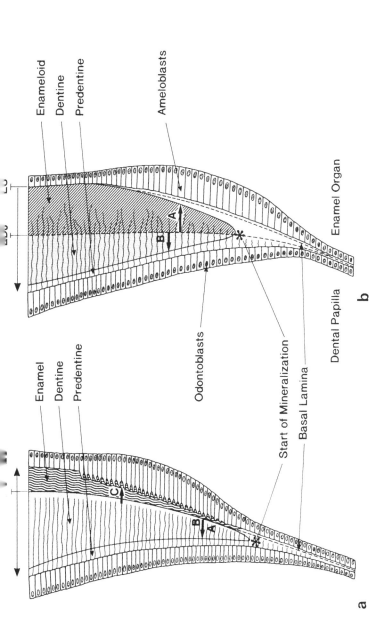

Figure 7.4 Comparison of the relative timing of differentiation of ameloblasts and odontoblasts and the events of matrix production and mineralization, relative to the position of the basal lamina and the enamel/enameloid–dentine junction (EDJ), in a section through a part of the tooth germ of the late bell stage (b, in Figure 7.1). In (a) ameloblast differentiation is after that of the odontoblasts and enamel forms in the direction shown (arrow C) by appositional growth after dentine has begun to calcify in the opposite direction (arrow B). The EDJ forms at the site of the basal lamina and calcification is initiated in the dentine (*) before ameloblast differentiation. In (b) ameloblasts differentiate before odontoblasts produce dentine, both secrete matrix components into the area below the basal lamina, before the mineralization event (*), and the EDJ forms at a site below the basal lamina. Mineralization of the enameloid begins first in the direction shown (arrow A) and the dentine after (arrow B). ES = enamel surface/enameloid surface. (Modified from Smith, 1992.)

(Shellis, 1978; Prostak *et al.*, 1991; Risnes, 1990) and the nature of the proteins (Graham, 1984; Sasagawa, 1991).

Variation in the timing of differentiation of either cell type could disrupt enameloid formation and lead to a failure of its formation (Poole, 1971). Shellis and Miles (1974) and Schaeffer (1977) noted that the consequences of such a delay to epithelial differentiation would result in dentine forming before any epithelial contribution, and that the late-appearing epithelial protein would be deposited as enamel. This dissociation in the timing of secretion of specific matrix proteins would result in two different tissues, enamel and enameloid, in different taxa. Differentiation of the dental epithelium in the shaft of the tooth dissociates from that in the crown, producing enamel in the collar and enameloid in the cap (Smith and Miles, 1969; Figures 7.5, 7.6). In fossil and extant actinopterygian teeth (Smith, 1992) the tooth cap is an extensive area of enameloid and the collar region is enamel distributed to a different extent along the shaft of teeth in each

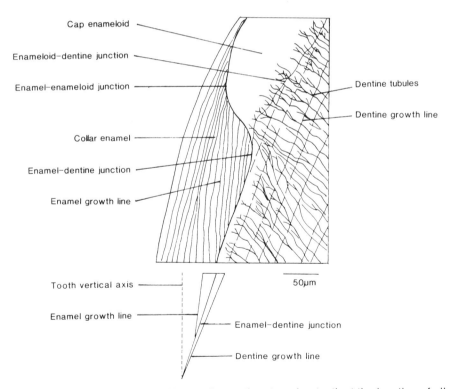

Figure 7.5   Diagram of a small part of an actinopterygian tooth at the junction of all three tissues, cap enameloid, dentine and collar enamel (position of these junctions is shown on the whole tooth in Figure 7.6 (e)). Numerous growth lines are seen in the collar enamel and dentine, but not in enameloid; they are interrupted at the enameloid–dentine junction. (Modified from Smith, 1992.)

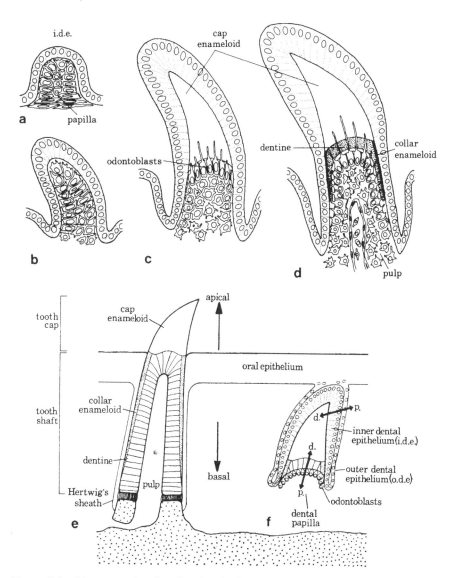

Figure 7.6 Diagrams showing the developing tooth germ (a–d) and the functional tooth (e) and tooth germ *in situ* (f) of a generalized teleost. They show the relative position of cap enameloid, dentine and collar enamel (enameloid) and their sequence of development. Ameloblasts differentiate first (b), enameloid is produced before dentine (c) and collar enamel after both (d). Secretion is from the distal end of both cell types (f). i.d.e., inner dental epithelium; d, distal; p, proximal. (Reproduced by permission of Shellis and Miles (1974, 1976) and the Royal Society.)

taxon (Figure 7.6). Smith concluded, in contrast to current opinion, that enameloid was the later, more derived tissue (Smith and Hall, 1990, p. 345;

Smith, 1992) and enamel the more primitive state for vertebrates. Smith did not agree with Fearnhead (1979) and Kawasaki and Fearnhead (1983) that "there is a logical progression of change from the simple primitive . . . enameloid to more complex mammalian enamels", because this was not supported by tissue distribution in the fossil record and the ontogenetic pathways were not rooted in a phylogenetic tree (Fink, 1982). Also, the concept of enameloid being a simple tissue cannot be upheld from its great diversity and complexity of formation across the taxa.

   Meinke (1982a,b) in studies of the teeth and dermal skeleton of the primitive extant actinopterygian fish *Polypterus*, concluded that "the enamel/enameloid/ dentine system forms a continuum of tissues that have diverged from each other by changes in relative developmental events and matrix production". It was suggested that enamel and ganoine evolved from enameloid via regulatory changes, and did not evolve independently (Meinke and Thomson, 1983). While regulation does affect the timing of differentiation and hence the tissue phenotype, it is envisaged that all three tissues arise in different combinations at different times in evolution in the gnathostome vertebrates, and that enamel and ganoine are almost identical (Smith, 1992; Richter and Smith, 1995). They did not evolve from enameloid but preceded it.

Tissue characters

The essential difference between enamel and enameloid is that enamel is secreted at the interface between mesenchyme and epithelium beginning at the site of the basal lamina and growing centrifugally (Figures 7.4, 7.6). At this time the epithelial cells are no longer in contact with the mesenchyme of the dental papilla because dentine has already formed, as a result of the odontoblasts differentiating earlier than ameloblasts in the programmed development of the tooth germ. The proteins characteristic of enamel, enamelin and amelogenin have different roles, the former in nucleating crystals and the latter in spacing out the initial hydroxylapatite crystals and then controlling their growth by protein disaggre- gation. The secretion product of the ameloblasts is deposited onto an already mineralized surface (dentine) and grows away from the enamel–dentine junction (EDJ). The cell membrane of the ameloblasts controls the orientation of the initial crystallites and a fibrous matrix is never present. Each layer forms by appositional growth of enamel onto the dentine, and each pause in growth is marked by an incremental line in the tissue. Successive enamel layers increase the thickness and also the height of the crown, tapering down onto the dentine surface as another group of ameloblasts differentiates at the lower margin of the tooth germ. This series of lines forms at an acute angle to the enamel–dentine junction (EDJ), and diverges from a similar set of lines in the dentine (Figure 7.5).

   In marked contrast, enameloid forms in contact with the dental mesenchyme cells (odontoblasts) and epithelial cells (ameloblasts), and may be secreted by both into the space created at the junction between them. The ameloblast products are secreted distal to and through the basal lamina and mix with products from the odontoblast processes, before any dentine has formed (Figures 7.4–7.6). This tissue has no previously formed dentine matrix on which to grow but instead forms as a jointly produced unmineralized matrix, beneath the basal lamina, developing in three stages into a progressively mineralized extensive cap to the tooth without the appositional growth lines detected in the enamel (Shellis and Miles, 1974). Enameloid growth is multidirectional with the matrix components secreted from all directions, and only when it has formed at its definitive thickness and shape does it begin to mineralize. Only as this mineralization begins does the predentine begin to form on its inner surface with a stable, insoluble form of collagen as the main matrix component before mineralization begins at a later stage (see direction of arrows B and C in Figure 7.4). Enameloid often contains tubules from the odontoblast cell processes, but their direction is quite random and discordant with those in the dentine. The cell processes from the odontoblasts may help to control the orientation of the crystallites in addition to special matrix components such as "tubular vesicles" derived from odontoblast processes in elamobranch enameloid, and also giant fibres (Prostak *et al.*, 1991; Sasagawa and Akai, 1992), but neither remains in the mature tissue (Figures 7.5, 7.6). Crystallites in enameloids are organized into parallel groups, either all radial and some surface parallel, or into crystal bundles each interlacing forming a woven pattern, the variation depending on function of the tooth, either cutting or crushing. By contrast, in enamel, crystallite orientation is ordered from the distal surface of the ameloblasts into groups of crystals (prisms or protoprisms) arranged with their orientation gradually changing from normal to the enamel surface to up to 60°. Borders between prisms arise where crystallite axes diverge maximally from each other. Protoprisms, with a lesser degree of order than prisms, have been described in many groups of sarcopterygian fishes, amphibians, reptiles and early mammals (Smith, 1989, 1992). There are major changes in crystallite arrangement amongst the taxa, with ordered prism patterns emerging in mammals, each one used as a taxonomic character.

It is clear that there are major differences between the development of enameloid in elasmobranchs and in teleosts (Sasagawa, 1991; Prostak *et al.*, 1991). Graham (1984) has demonstrated that ameloblasts in dogfish are active in synthesizing two different proteins with mammalian antigenic determinants, but these probably may be only one class of protein, the enamelins. He further proposed that the evolution of enamel from enameloid (as he assumed), required the addition of amelogenin to allow the space for the crystals to grow to a larger size. Herold *et al.* (1989) also suggested, from immunohistochemical studies, that enamelin was the only protein present in enameloids, and that evolution involved the development of a gene for amelogenin as well as the extended, later period of expression. In contrast, the theories proposed here suggest the *reduction* to one dominant epithelial protein as an evolved character, linked with the earlier time of secretion of the ameloblasts through heterochrony. Both Graham (1984) and Sasagawa and Akai (1992) proposed that odontoblasts, rather than ameloblasts, produce the predominant amount of protein in enameloid, in sharks much more than in teleosts.

Recent reviews (Sasagawa, 1991; Prostak *et al.*, 1991) place the topic of epithelial participation in enameloid matrix production in dispute. What is becoming clearer is the nature of the major differences that exist between osteichthyan and chondrichthyan enameloid development and structure. Osteichthyan enameloid contains a labile form of collagen secreted mainly from the inner dental epithelium (ameloblasts) (Prostak *et al.*, 1991). In teleosts longitudinal crystal growth is allowed through collagen fibre dissociation (Prostak and Skobe, 1985), and mineralization is initiated at the dentine–enameloid junction. The action of the ameloblasts promotes the high degree of mineralization by absorption of matrix components. Sasagawa and Ferguson (1990) described organic compartments around each crystal in teleost enameloid and assumed it to be enamelin by comparison with its function and position in mammalian enamel. By contrast, chondrichthyan enameloid is mainly mesenchymal in origin (Risnes, 1990; Sasagawa and Akai, 1992) and mineralizes from the outer surface towards the dental papilla. Mineralization is initiated within "tubular vesicles" of odontoblastic origin (Sasagawa, 1991; Sasagawa and Akai, 1992) and may involve giant fibres. The whole process occurs with a basal lamina intact and appears to be unique. The matrix is claimed to lack any affinity for enamel antibodies (Prostak *et al.*, 1991). Sasagawa (1991) summarizes these differences between teleost and shark enameloid and concludes that although the final form is similar, "each has its own peculiar evolutionary development". The proposal in this review, that enameloid is a derived tissue, is enhanced by this concept and shows that in addition to a peramorphic process of predisplacement as a heterochronic event, the gene products may be novel. Current research in molecular biology when applied to these concepts will provide more definitive information.

ENAMEL IN OSTEICHTHYAN FISHES AND EARLY TETRAPODS

Osteichthyans

Amongst the osteichthyan fishes several names have been given to the tissue forming the outer covering to the oral teeth. Although it has been proposed to include all types, not the same as true enamel (tetrapod-type), under the term enameloid (Poole, 1967), many other names are used to describe this tooth-tissue. The tooth crown (cap enameloid) in actinopterygians is referred to as acrodin (Ørvig, 1978c) and the tissue around the neck of the tooth (collar enameloid) as collar ganoine. The microstructure of cap enameloid in non-teleostean actinopterygians has been described by Ørvig (1978 a,b,c) and in teleosts by Reif (1979). Reif compared that of teleosts with enameloid of sharks (Reif, 1977, 1979) and concluded that structural differences between the two allowed them to be distinguished from each other, and that a highly ordered microstructure had evolved secondarily in two teleost families, Characidae and Sphyrenidae, and in modern sharks, Euselachi. He considered that enameloid with a low degree of order is the primitive feature in vertebrates. The information on collar enameloid is far more confused but new data on this tissue in selected fossil actinopterygians, non-teleostean, has been discussed by Smith (1992).

The possibility of "true enamel" occurring in the teeth of actinopterygians has been considered by Smith (1989, 1992). Of considerable importance is the evidence from extant actinopterygians on the development of ganoine in the dermal skeleton during regeneration (Sire *et al.*, 1987; Meunier *et al.*, 1988; Sire, 1994). This reversed prevalent opinions on the developmental origin of this type of covering tissue, showing that it is secreted by epithelium and not by mesodermal cells. They further suggested that ganoine is homologous with enamel and that enamel is a synapomorphy of actinopterygians and sarcopterygians. Smith (1989, 1992) agreed with this conclusion and reported on the microstructure of "collar ganoine" in fossil taxa comparing it with enamel of sarcopterygians. In a survey of ganoine tissue in selected acanthodian and osteichthyan fish, Richter and Smith (1995) concluded that in the majority ganoine showed a microstructure aligned with enamel and that there are no clear criteria to distinguish between them. They suggested that the term ganoine as traditionally adopted is more descriptive than systematically relevant. Ganoine had been described previously as "pseudoprismatic" (Ørvig, 1978c), both in the dermal skeleton and in the collar tissue. This type of arrangement, termed protoprismatic enamel (Smith, 1989), is found in many osteichthyan and reptilian teeth, and is considered to be synonymous with preprismatic enamel in mammal-like reptiles, an arrangement described as one character of true enamel, or monotypic enamel.

Actinopterygians

Whereas most people have recognized that in actinopterygians there are two distinct regions to the tooth, a cap and a collar, the names given to the tissues have revealed a distinct lack of agreement on their homologies. The cap region is a type of enameloid (acrodin of Ørvig, 1978c), whereas the tissue surrounding the base of the tooth below the apical cap, topographically the collar, has been called collar enameloid (Shellis, 1975, 1978), collar ganoine, circumverrucal ganoine (Ørvig, 1978c), and collar enamel (Reif, 1982, p. 343). Since the studies of Ørvig (1978 a,b,c) on the microstructure of the collar region and ganoine in the dermal skeleton, in which he noted a similarity to "pseudoprismatic enamel", the debate about the homology of these two regions has continued.

Apart from Peyer (1968), Reif (1982) is the only person to have suggested that enamel is present in these teeth; a thick layer in chondrostean and holostean genera, but only a very thin layer in teleosts. However, the studies of recent teleosts (*Halichoris, Pagrus*) have all shown that the collar region is a form of modified dentine, and as such an enameloid (Sasagawa and Ishiyama, 1988), although apparently not the same as the cap enameloid. Meinke (1982a) while suggesting that in *Polypterus* the collar region was not ganoine and not enameloid, but a form of dentine, concluded that ganoine was a type of enamel. Smith (1992) concluded that within many fossil taxa the outer layer of the collar region of the tooth is enamel. This is deduced from the arrangement of the crystallites and the position of the incremental lines (Figure 7.5). There is a sharp distinction between this tissue and the cap enameloid, suggesting that the dental epithelium has two different activities in adjacent regions of the tooth germ, possibly separated by an interval of time in development. It is apparent that the collar region differs amongst genera from the major groups of actinopterygians; in extant teleosts it is a modified dentine, enameloid, or bitypic enamel, whereas in some fossil forms of lower actinopterygians it is more similar to monotypic enamel. In the garfish, a primitive extant actinopterygian, Prostak *et al.* (1991) reported that the inner dental epithelium secretes a matrix onto the collar region and that incremental lines are also found. Apparently the ability to form enamel has been lost in more evolved forms of actinopterygian, a suggestion also made by Kemp (1985).

Sarcopterygians

The conclusion that true enamel forms the covering to the tooth in this group of fishes has been reached independently by many workers (Peyer, 1968: Schultze, 1969: Shellis and Poole, 1978: Smith, 1978; 1989: Meinke and Thomson, 1983). The microstructure of the enamel in fossil species from each of the main taxa has been described by Smith (1989) and found to vary amongst them, with at least four different patterns, possibly specific to each genus. Several have a type of arrangement, previously referred to as pseudoprismatic (Peyer, 1968: Schultze, 1969), now termed protoprismatic (Smith, 1989). In particular, it has been shown that porolepids and panderichthyids have a similar arrangement to that described in early mammals as hexagonal columns, or preprisms (Osborn and Hillman, 1979).

Although it is well established that enamel completely covers the whole tooth in sarcopterygians (Smith, 1989), there is no evidence in these teeth of enameloid. Enamel, as an increasingly thicker layer, represents the sole epithelial product in sarcopterygian oral teeth, and it is this distribution rather than its presence per se that should be considered as a synapomorphy of sarcopterygians. The information from the earliest osteichthyan teeth of the Silurian (Janvier, 1978) prompted Smith (1989) to conclude that the reported absence of enamel in *Andreolepis* and *Lophosteus* most likely implies that a very thin layer of enamel was probably present *in vivo* and that enameloid was not, being normally much thicker and more intimately bound to the dentine, and hence, more likely to be retained in fossils. In this case enamel would be a character at the base of of osteichthyans, and also a plesiomorphic character of vertebrates as suggested by Smith *et al.* (1995).

ENAMEL AS A SYNAPOMORPHY OF VERTEBRATES

Amongst previous postulates on types of early vertebrate tissues is one that proposes that a distinction between enamel and dentine is primitive for vertebrates (Dzik, 1986). This leads to the assumption that a sharp junction marks the site of initiation of extracellular deposition of the two tissues and that their appositional growth away from this is recorded as a series of lines set at an acute angle to the junction and divergent to each other. Dzik (1986) also proposed that in cladistically more primitive groups, sclerites contain an enamel homologue as the dominant tissue. This, would include the lamellar crown of conodonts.

Euconodonts as early vertebrates

Recent publications on the conodont animal (Aldridge *et al.*, 1986, 1993) have established their chordate affinities. Those on the element structure have proposed homology between these hard tissues and those of Lower Palaeozoic vertebrates (Dzik, 1986; Sansom *et al.*, 1992, 1994). Many conodonts have a simple two component element consisting solely of a lamellar crown and a basal body, in which opposing sets of divergent growth lines have been recorded recently in the Ordovician conodonts, *Chirognathus*, *Erismodus* and *Pseudooneotodus*, and regarded as evidence of vertebrate tissues (Sansom *et al.*, 1992; Smith *et al.*, 1995). The simple conical elements of *Pseudooneotodus* not only provide evidence of appositional growth but definitive evidence that the tissue type in the crown of these elements is enamel (Sansom, 1992), of a type that is indistinguishable from that of sarcopterygian fishes. These observations on fossil taxa extend the earliest recorded occurrence of enamel considerably, into the Early Ordovician.

Other early vertebrates

The early vertebrates with good preservation of histology from the Ordovician Harding Sandstone in central Australia have examples of both tissue types, enamel in *Eriptychius* and enameloid in *Astraspis* (see Smith and Hall, 1990, for a review of these tissues), although enamel is the only tissue type found in euconodonts that also occur in this region. From this type of fossil evidence it is proposed that the enamel covering to teeth and dermal denticles is a synapomorphy of vertebrates and is the primitive condition, and that diversity of tissue type occurred early on in skeletal evolution.

THEORIES OF HETEROCHRONY

Four possible states can occur in odontodes that reflect heterochrony, i.e., changes in the developmental programme postulated to have effected evolutionary change. These are: (i) an enamel covering separated from the underlying papillary tissue (dentine) by a distinct junction; (ii) an enameloid cap over dentine, without a distinct junction; (iii) no enamel cover to the dentine; (iv) an enameloid cap in the coronal region and basal to this, an enamel collar in the same tooth, both covering dentine.

Traditional ideas have been based on the assumptions that chondrichthyan teeth are primitive for all jawed vertebrates, that the "tetrapod" type of enamel does not occur in more basal clades such as sarcopterygian fishes, and therefore, enameloid is the earlier tissue type and is the only tissue present in the earliest vertebrates. All previous accounts have proposed that evolutionary change occurred by a heterochronic shift to a later time for secretion of the enamel proteins and because of these assumptions, that enamel is the more evolved tissue. The suggestion that enameloid is the primitive tissue, and that it is still represented by a modified mantle dentine zone (the Von Korff layer) immediately beneath the enamel in tetrapod teeth (Fearnhead, 1979; Kawasaki and Fearnhead, 1983), is not supported by the fossil evidence and is based on ontogenetic information which is not rooted in a robust phylogeny. Indeed, the new evidence from the early fossil vertebrate record challenges this assumption and supports a suggestion (Smith and Hall, 1990; Smith, 1992) that secretion of enamel at the later time in histodifferentiation of the tooth germ is the condition in basal vertebrate groups, and that the heterochronic shift is to an earlier time relative to dentine formation, producing enameloid as the more derived tissue.

These traditional ideas of the polarity of tissue evolution in teeth influenced the interpretation of changes to developmental sequences necessary to produce enamel rather than enameloid. Both Poole (1971) and Shellis and Miles (1974) proposed that variation in the timing of secretory activity of either cell type could disrupt enameloid formation to produce enamel as a later type. Smith and Hall (1990, p. 343) proposed that epithelial–mesenchymal interactions would be prime candidates as developmental regulators for changes in evolution through heterochrony.

From the observations on osteichthyan teeth, Smith (1992) produced a new model of heterochronic shift, namely that ameloblast differentiation shifted from a late stage in histodifferentiation of the tooth germ to an earlier one to produce the evolutionary change from primitive enamel to more derived enameloid (Figure 7.4). The only other evidence for a primitive enamel comes from immuno-labelling studies on the keratinous toothlets of the otherwise toothless (anodontogenic) agnathan, the hagfish (Slavkin *et al.*, 1983). They state that there is evidence, from cross-reaction with polyclonal antibodies to mammalian enamel proteins (enamelin), of material between the epithelial cells, that primitively these acraniates had the genes to produce enamel. However, as it is claimed that similar proteins are secreted into enameloids (Herold *et al.*, 1989) and keratinous teeth of extant agnathans are not universally accepted as homologous with phosphatic mineralized teeth (Smith *et al.*, 1995), it does not resolve the problem, even with extant forms, of the earliest tissue type in the vertebrate odontogenic skeleton.

This example of one differentiation event in the causal sequence of development in the tooth germ, producing different tissues in homologous teeth (odontodes), can be explained as a peramorphic process of predisplacement where the initiation of differentiation of ameloblasts is displaced to an earlier stage than that of the ancestral condition. We can only speculate on the selective advantage that this acceleration of development within the odontode of a descendant's ontogeny may have.

PREVIOUS ASSUMPTIONS AND NEW POSTULATES

It is concluded from the fossil data that the differentiation of ameloblasts changed during evolution, from a relatively late time in relation to odontoblast differentiation, to an earlier time in the tooth germ. This conclusion, focused on a peramorphic process of heterochrony, has been made possible by the detailed analysis of phosphatic microfossils of potential vertebrates, and the unequivocal identification of enamel in stratigraphically very early odontode-like forms, probably present in the oral cavity. These fossil forms occur early in the stratigraphic record of vertebrates and belong to one group of conodonts, euconodonts, proposed as a basal chordate in the vertebrate hierarchy (Aldridge *et al.*, 1993). The fossil evidence from the stratigraphically earliest and cladistically most primitive examples shows that the state when enamel is present is the primitive one and the proposition is that this is a synapomorphy for vertebrates.

Control and regulation of this tissue product of an epithelial–mesenchymal morphogenetic system is through molecular coded exchanges between epithelium and neural crest-derived mesenchyme in an odontogenic pentapartite system. Neural crest, and some of the *hox* gene translation products identified as significant in these interactions, are unique developments to vertebrates, as discussed by Smith and Hall (1993). One of the innovative changes, or synapomorphy, of vertebrates is a phosphatic mineralized skeleton of dermal odontodes (either oral or exoskeletal) and proposed to be dependent on neural crest-derived mesenchyme for their development. Genes of the *msx* family are expressed in all vertebrate groups so far examined in skeletogenesis associated with neural crest, and gene duplication of this family is thought to have occurred concurrently with the origin of the vertebrates (Holland, 1991). It is proposed that this is one example of a molecular mechanism through which heterochrony can operate to produce diversification of the tissue types such as those derived from enamel.

The molecular basis of early ontogenetic inductive mechanisms in the tooth germ needs to be examined in selected different species to be able to demonstrate different temporal and spatial expression patterns of signalling molecules and their regulators. This will provide data for the mechanistic basis of heterochrony, linked to the classic histological stages of morphogenesis and histogenesis in the tooth germ the stable reference points referred to by Hall and Miyake (Chapter 1). Now that we have the molecular markers for timing of induction and differentiation in many epithelial–mesenchymal interactions during organogenesis, it should be possible to acquire data for known ancestor and descendent groups to be able to specify the heterochronic events.

ACKNOWLEDGMENTS

I thank the Wellcome Trust for a grant to attend the meeting in Montpellier, France at which the paper was first presented. I am particularly grateful to Irma Thesleff for giving me the diagram for Figure 7.3, for making available material in press, and for her comments on the section on molecular data. Amongst many I acknowledge valuable discussion with Brian Hall and Ivan Sansom. Part of the work on early vertebrates was done through a NERC grant (GR3/8543).

REFERENCES

Aldridge, R.J., Briggs, D.E.G., Clarkson, E.N.K. and Smith, M.P., 1986, The affinities of conodonts—new evidence from the Carboniferous of Edinburgh, Scotland, *Lethia*, **19**: 279–291.

Aldridge, R.J., Briggs, D.E.G., Smith, M.P., Clarkson, E.N.K. and Clark, N.D.L., 1993, The anatomy of conodonts, *Phil. Trans. Roy. Soc.* **B338**: 405–421.

Bèque-Kirn, C., Smith, A., Ruch, J.J., Wozney, J.M., Purchio, A., Hartmann, D. and Lesot, H., 1992, Effects of dentine proteins, transforming growth factor $\beta1$ (TGF $\beta1$) and bone morphogenetic protein (BMP2) on the differentiation of odontoblasts *in vitro*, *Int. J. Dev. Biol.*, **36**: 491–503.

Cassin, C. and Capuron, A., 1979, Buccal organogenesis in *Pleurodeles waltll* Michah (urodele amphibian), study by intrablastocelic transplantation and *in vitro* culture, *J. Biol. Buccale*, **7**: 61–76.

Chibon, P., 1966, Analyse expérimentale de la régionalisation et des capacités morphogénètiques de la crête neurale chez l'amphibien urodèle *Pleurodeles waltll* Michah, *Mém. Soc. Zool. France*, **36**: 1–107.

Chibon, P., 1967, Marquage nucléaire par la thymidine tritée des dérivés de la crête neurale chez l'amphibien urodèle *Pleurodeles waltll* Michah, *J. Embryol. Exp. Morphol.*, **18**: 343–358.

Dzik, J., 1986, Chordate affinities of the conodonts. In A. Hoffman and M.H. Nitecki (eds), *Problematic fossil taxa*, Oxford Monographs on Geology and Geophysics No. 5, Oxford University Press, New York: 240–254.

Fearnhead, R.W.F., 1979, Matrix mineral relationships in enamel tissues, *J. Dent. Res.* **58 (B)**: 909–916.

Fink, W., 1982, The conceptual relationship between ontogeny and phylogeny, *Paleobiology*, **8**: 254–264.

Gaunt, W. A. and Miles, A. E. W., 1967, Fundamental aspects of tooth morphogenesis. In A. E. W. Miles (ed.), *Structural and chemical organization of teeth*, Academic Press, London, 1: 151–197.

Graham, E.E., 1984, Protein biosynthesis during spiny dogfish (*Squalus acanthias*) enameloid formation, *Arch. Oral Biol.*, **29**: 821–825.

Graveson, A., 1993, Neural crest contributions to the development of the vertebrate head, *Am. Zool.*, **33**: 424–433.

Hall, B.K., 1994, Biology and mechanisms of tissue interaction in developing systems. In G.M. Hodges and C. Rowlatt (eds), *Developmental biology and cancer*, CRC Press, Boco Raton: 161–185.

Hall, B. K. and Hörstadius, S., 1988, *The neural crest*, Oxford University Press, Oxford.

Herold, R., Rosenbloom, J. and Granovsky, M., 1989, Phylogenetic distribution of enamel proteins: immunohistochemical localization with monoclonal antibodies indicates the evolutionary appearance of enamelins prior to amelogenins, *Calcif. Tissue Int.*, **45**: 88–94.

Holland, P.W.H., 1991, Cloning and evolutionary analysis of *msh*-like genes from mouse, zebrafish and ascidian, *Gene*, **98**: 253–257.

Hurmerinta, K., Kuusela, P. and Thesleff, I., 1986, The cellular origin of fibronectin in the basement membrane zone of developing teeth, *J. Embryol. exp. Morph.*, **95**: 73–80.

Janvier, P., 1978, On the oldest known teleostome fish *Andreolepis hedei* Gross (Ludlow of Gotland), and the systematic position of the lophosteids, *Koide Geologica*, **3**: 86–95.

Jowett, A.K., Vainio, S., Ferguson, M.W.J., Sharpe, P.T. and Thesleff, I., 1993, Epithelial–mesenchymal interactions are required for *msx 1* and *msx 2* gene expression in the developing murine molar tooth, *Development*, **117**: 461–470.

Karanova, I., Vainio, S. and Thesleff, I., 1992, Transient and recurrent expression of the *Egr-1* gene in epithelial and mesenchymal cells during tooth morphogenesis suggests involvement in tissue interactions and determination of cell fate, *Mech. Dev.*, **39**: 41–50.

Kawasaki, K. and Fearnhead, R.W., 1983, Comparative histology of tooth enamel and enameloid. In S. Suga (ed.), *Mechanisms of tooth enamel formation*, Quint. Pub. Co., Tokyo, Japan: 229–238.

Kemp, N.E., 1985, Ameloblastic secretion and calcification of the enamel layer in shark teeth, *J. Morph.*, **184**: 215–230.

Lumsden, A.G.S., 1987, The neural crest contribution to tooth development in the mammalian embryo. In P.F.A. Maderson (ed.), *Developmental and evolutionary aspects of the neural crest*, John Wiley and Sons, New York: 261–300.

Mackenzie, A., Ferguson, M.W.J. and Sharpe, P.T., 1992, Expression patterns of the homeobox gene, *Hox-8* in the mouse embryo suggest a role specifying tooth initiation and shape, *Development*, **115**: 403–420.

Mackenzie, A., Lemming, G.L., Jowett, A.K., Ferguson, M.W.J. and Sharpe, P.T., 1991, The homeobox gene *Hox 7.1* has specific regional and temporal expression patterns during early murine craniofacial embryogenesis, especially tooth development *in vivo* and *in vitro*, *Development*, **111**: 269–285.

Meinke, D.K., 1982a, A histological and histochemical study of developing teeth in *Polypterus* (Pisces, Actinopterygii), *Arch. Oral Biol.*, **27**: 197–206.

Meinke, D.K., 1982b, A light and scanning electron microscope study of microstructure, growth and development of the dermal skeleton of *Polypterus* (Pisces: Actinopterygii), *J. Zool. Lond.*, **197**: 355–382.

Meinke, D.K. and Thomson, K.S., 1983, The distribution and significance of enamel and enameloid in the dermal skeleton of osteolepiform rhipidistian fishes, *Paleobiology*, **9**: 138–149.

Meunier, F.J., Gayet, M., Geraudie, J., Sire, J.-Y. and Zylberberg, L., 1988, Données ultrastructurales sur la ganoine du dermosquelette des actinopterygiens primitifs, *Mem. Mus. natn. Hist. nat. Paris* (serie C), **53**: 77–83.

Ørvig, T., 1978a, Microstructure and growth of the dermal skeleton in fossil actinopterygian fishes: *Birgeria* and *Scanilepis, Zool. Scr.,* **7**: 33–56.

Ørvig, T., 1978b, Microstructure and growth of the dermal skeleton in fossil actinopterygian fishes: *Boreosomus, Plegmolepis* and *Gyrolepis, Zool. Scr.,* **7**: 125–144.

Ørvig, T., 1978c, Microstructure and growth of the dermal skeleton in fossil actinopterygian fishes: *Nephrotus* and *Colobodus,* with remarks on the dentition in other forms, *Zool. Scr.,* **7**: 297–326.

Osborn, J.W. and Hillman, J., 1979, Enamel structure in some therapsids and Mesozoic mammals, *Calcif. Tissue Int.,* **29**: 47–61.

Partanen, A.-M., 1990, Epidermal growth factor and transforming growth factor-alpha in the development of epithelial–mesenchymal organs of the mouse, *Curr. Top. Devel. Biol.,* **24**: 31–55.

Partanen, A.-M. and Thesleff, I., 1987, Localization and quantitation of 125I-epidermal growth factor binding in mouse embryonic tooth and other embryonic tissues at different developmental stages, *Dev. Biol.,* **120**: 186–197.

Partanen, A.-M. and Thesleff, I., 1989, Growth factors and tooth development, *Int. J. Dev. Biol.,* **33**: 165–172.

Peyer, B., 1968, *Comparative odontology,* University of Chicago Press, Chicago.

Poole, D.F.G., 1967, The phylogeny of tooth tissues; enamel and enameloid in recent vertebrates. In A.E.W. Miles (ed.), *Structural and chemical organization of teeth I,* Academic Press, New York: 119–149.

Poole, D.F.G., 1971, An introduction to the phylogeny of calcified tissues. In A.A. Dahlberg (ed.), *Dental morphology and evolution,* University of Chicago Press, Chicago: 65–80.

Prostak, K., Seifert, P. and Skobe, Z., 1991, Tooth matrix formation and mineralization in extant fishes. In S. Suga and H. Nakahara (eds), *Mechanisms and phylogeny of mineralization in biological systems,* Springer-Verlag, Tokyo: 465–469.

Prostak, K. and Skobe, Z., 1985, The effects of colchicine on the ultrastructure of the dental epithelium and odontoblasts of teleost tooth buds, *J. Craniofac. Genet. and Dev. Biol.* **5**: 75–88.

Reif, W.-E., 1977, Tooth enameloid as a taxonomic criterion: 1. A new euselachian shark from the Rhaetic–Liassic boundary, *N. Jb. Geol. Paläont.,* **1977**: 565–576.

Reif, W.-E., 1979, Structural convergences between enameloid of actinopterygian teeth and of shark teeth, *Scanning Electron Microscopy* **II**, 547–554.

Reif, W.-E., 1982, Evolution of dermal skeleton and dentition in vertebrates: the odontode regulation theory, *Evol. Biol.,* **15**: 287–368.

Richter, M. and Smith, M.M., 1995, A microstructural study of the ganoine tissue of selected lower vertebrates, *Zool. J. Linn. Soc.,* **144**: 173–212.

Risnes, S., 1990, Shark tooth morphogenesis: an SEM and EDX analysis of enameloid and dentine development in various shark species, *J. Biol. Buccale,* **18**: 237–248.

Sansom, I.J., 1992, *The palaeobiology of the Panderodontacea and selected other euconodonts,* Unpublished Doctorate Thesis, University of Durham, U.K.

Sansom, I.J., Smith, M.P., Armstrong, H.A. and Smith, M.M., 1992, Presence of the earliest vertebrate hard tissues in conodonts, *Science,* **256**: 1308–1311.

Sansom, I. J., Smith, M. P. and Smith, M.M., 1994, Dentine in conodonts, *Nature,* **368**: 591.

Sasagawa, I., 1991, The initial mineralization during tooth development in sharks. In S. Suga and H. Nakahara (eds), *Mechanisms and phylogeny of mineralization in biological Systems,* Springer-Verlag, Tokyo: 199–203.

Sasagawa, I. and Akai, J., 1992, The fine structure of the enameloid matrix and initial mineralization during tooth development in the sting rays, *Dasyatis akajei* and *Urolophus aurantiacus, J. Electron. Micros.,* **41**: 242–252.

Sasagawa, I. and Ferguson, M.W.J., 1990, Fine structure of the organic matrix remaining in the mature cap enameloid in *Halichores poecilopterus*, teleost, *Arch. Oral Biol.*, **35**: 765–770.

Sasagawa, I. and Ishiyama, M., 1988, The structure and development of the collar enameloid in two teleost fishes, *Halichores poecilopterus* and *Pagrus major*, *Anat. Embryol.*, **178**: 499–511.

Schaeffer, B., 1977, The dermal skeleton in fishes. In S.M. Andrews, R.S. Miles and A.D. Walker (eds), *Problems in vertebrate evolution*, Linnean Society Symposium No 4, Academic Press, London: 25–52.

Schilling, T.F. and Kimmel, C.B., 1994, Segment and cell type lineage restrictions during pharyngeal arch development in the zebrafish embryo, *Development* **120**, 483–494.

Schultze, H.-P., 1969, Die Faltenzahne der rhipidistiiden Crossopterygier, der Tetrapoden und der Actinopterygier-Gattung *Lepisosteus*, nebst einer Beschreibung der Zahnstruktur van *Onychodus* (structiiformes Crossopterygier), *Palaeontogr. Ital.*, **65**: 63–137.

Shellis, R.P., 1975, A histological and histochemical study of the matrices of enameloid and dentine in teleost fishes, *Arch. Oral Biol.*, **20**: 181–187.

Shellis, R.P., 1978, The role of the inner dental epithelium in the formation of the teeth in fish. In P.M. Butler and K.A. Joysey (eds), *Development, function and evolution of teeth*, Academic Press, London: 31–42.

Shellis, R.P. and Miles, A.E.W., 1974, Autoradiographic study of the formation of enameloid and dentine matrices in teleost fishes using tritiated amino acids, *Proc. R. Soc. Lond. B*, **185**: 51–72.

Shellis, R.P. and Miles, A.E.W., 1976, Observations with the electron microscope on enameloid formation in the common eel (*Anguilla anguilla*: Teleostei), *Proc. R. Soc. Lond. B*, **194**: 253–269.

Shellis, R.P. and Poole, D.F.G., 1978, The structure of the dental hard tissues of the coelacanthid fish *Latimeria chalunnae* Smith, *Arch. Oral Biol.*, **23**: 1105–1113.

Sire, J.-Y., 1994, A light and TEM study of non-regenerated and experimentally regenerated scales of *Lepisosteus oculatus* (Holostei) with particular attention to ganoine formation, *Anato Rec.*, **240**: 189–207.

Sire, J.-Y., Geraudie, J., Meunier, F.J. and Zylebeberg, L., 1987, On the origin of ganoine: histological and ultrastructural data on the experimental regeneration of the scales of *Calamoichthys calabricus* (Osteichthyes, Brachyoptergii, Polypteridae), *Am. J. Anat.*, **180**: 391–402.

Slavkin, H.C., Graham, E., Zeichner-David, M. and Hildemann, W., 1983, Enamel-like antigens in hagfish: possible evolutionary significance, *Evolution*, **37**: 404–412.

Smith, M.M., 1978, Enamel in the oral teeth of *Latimeria chalunnae* (Pisces: Actinistia): a scanning electron microscopy study, *J. Zool.*, **185**: 355–369.

Smith, M.M. 1989, Distribution and variation in enamel structure in the oral teeth of sarcopterygians: the significance for the evolution of a protoprismatic enamel, *Hist. Biol.*, **3**: 97–126.

Smith, M.M., 1992, Microstructure and evolution of enamel amongst osteichthyan and early tetrapods. In P. Smith (ed.), *Structure, function and evolution of teeth*, Proc. 8th Int. Symp. Dent. Morphol., Jerusalem, Israel (1989): 73–101.

Smith, M.M. and Hall, B.K., 1990, Developmental and evolutionary origins of vertebrate skeletogenic and odontogenic tissues, *Biol. Rev.*, **65**: 277–374.

Smith, M.M. and Hall, B.K., 1993, A developmental model for evolution of the vertebrate exoskeleton and teeth: the role of cranial and trunk neural crest, *Evol. Biol.*, **27**: 387–448.

Smith, M.M. and Miles, A.E.W., 1969, An autoradiographic investigation with the light microscope of proline-H3 incorporation during tooth development in the crested newt (*Triturus cristatus*), *Archs. Oral Biol.*, **14**: 479–490.

Smith, M.M., Sansom, I.J. and Smith, M.P., 1995, Teeth before armour: the earliest vertebrate mineralised tissues, *Mod. Geol.*, **20**.

Thesleff, I., Mackie, E., Vainio, S. and Chiquet-Ehrismann, R., 1987, Changes in the distribution of tenascin during tooth development, *Development*, **101**: 289–296.

Thesleff, I., Vainio, S. and Jalkanen, M., 1989, Cell-matrix interactions in tooth development, *Int. J. Dev. Biol.*, **33**: 91–97.

Thesleff, I., Vaaktokari, A. and Vainio, S., 1990, Molecular changes during determination and differentiation of the dental mesenchyme cell lineage, *J. Biol. Buccale*, **18**: 179–188.

Thesleff, I., Vaaktokari, A., Kettunen, P. and Åberg, T., 1995, Epithelial–mesenchymal signalling during tooth development, *Conn. Tiss. Res.* (in press).

Vainio, S., Jalkanen, M. and Thesleff, I., 1989, Syndecan and tenascin expression is induced by epithelial–mesenchymal interactions in embryonic tooth mesenchyme, *J. Cell Biol.*, **108**: 1945–1954.

Vainio, S., Karavanova, I., Jowett, A. and Thesleff, I., 1993, Identification of BMP-4 as a signal mediating secondary induction between epithelial and mesenchymal tissues during early tooth development, *Cell*, **75**: 45–58.

Chapter 8

HETEROCHRONY IN DINOSAUR EVOLUTION

John A. Long and Kenneth J. McNamara

INTRODUCTION

Until recently, most of what we know about dinosaurs—their morphology, systematics and phylogeny—was gleaned from often fragmentary, adult material. However, the last decade has seen an upsurge in interest in the developmental history of dinosaurs, arising from new finds of embryonic and juvenile dinosaurs from a number of localities, principally in North America. It is only with descriptions of this preadult material that the ontogenies of many groups of dinosaurs have, to varying degrees, been able to be analysed for trends. The recent publication of two major volumes in which detailed measurements and accounts of many of these ontogenies are presented (Carpenter and Currie, 1990; Carpenter *et al.*, 1994) has enabled us to synthesise much of these data and place them in a phylogenetic context. Furthermore, in cases where stratigraphic and phylogenetic controls are good, it has been possible to analyse patterns of heterochrony in certain dinosaur lineages. We see this chapter as providing a broad overview of the role that heterochrony has played in dinosaur evolution, and anticipate that it may serve to stimulate further research in this area.

Compared with most groups of organisms, very few examples of heterochrony in dinosaurs have been previously described, presumably on account of the paucity of detailed ontogenies. In a review of heterochrony in the fossil record that one of us (McNamara, 1988) carried out the only implied record of heterochrony in dinosaurs was in a review by Hopson (1977) who pointed out that variations in allometries of many cranial and postcranial elements were an important aspect in dinosaur evolution. More specific examples of heterochrony in dinosaurs documented in the literature are exceedingly sparse. Bakker *et al.* (1988) suggested that the adult lachrimal horn in *Nanotyrannus* may have arisen by "neoteny" (i.e., paedomorphosis). Weishampel *et al.* (1993) have suggested that dental evolution in some hadrosaurids may have been driven by progenetic

Evolutionary Change and Heterochrony. Edited by Kenneth J. McNamara © 1995 The Editor and Contributors. Published in 1995 by John Wiley & Sons Ltd.

paedomorphosis. Bonaparte and Vince (1979) have argued that sauropod adults show some paedomorphic characters that occur in juvenile prosauropods, such as *Mussaurus*. However, as Weishampel and Horner (1994) have stated, detailed morphometric analysis is required to verify this.

From the literature one could be forgiven for thinking that heterochrony has played only a very minor role in dinosaur evolution. In our view such an interpretation would be quite misleading. It reflects more the insufficiently detailed analysis of those ontogenies that have been described, combined with a lack of interest amongst most dinosaur researchers with this particular area of evolutionary theory. The only exception to this generalisation is a recent review by Weishampel and Horner (1994) that addresses specifically the impact of heterochrony on dinosaurs, in an overview of the life history syndromes, heterochrony and evolution. In reviewing the wealth of new ontogenetic data emerging from studies of embryonic and postnatal dinosaurs, they point out the potential that such information has to offer for studies of the impact of heterochrony and life history syndromes (i.e., life history strategies, tactics and traits). However, their aim was not to provide detailed descriptions of likely patterns of heterochrony within different groups of dinosaurs.

In this chapter we aim to remedy this to some degree by providing documentation of heterochrony within four major groups of dinosaurs (theropods, hadrosaurids, hypsilophodontids and ceratopsians) to lend support to Weishampel and Horner's view that many evolutionary changes within the Dinosauria are a result of heterochrony. Furthermore, we look at the role that heterochrony may have played in the evolution of birds from theropod dinosaurs.

THEROPODS

Ontogenetic variations

Early studies on growth and variation in Mongolian theropods by Rhozdestvensky (1965) was followed by Russell's (1970) monograph on the Canadian tyrannosaurids in which he provided information on the juvenile material of the tyrannosaurid *Albertosaurus* and reconstructed a possible hatchling theropod dinosaur (Figure 8.1). This reconstruction was based both on his observations of juvenile material as well as on general trends in reptile growth. Madsen (1976) documented many measurements of isolated bones from a large population of *Allosaurus fragilis* specimens, and estimated some extremes of growth for certain bones. His measurements for the individual (non-associated) bones show a size range in postcranial elements from juveniles with femora 245 mm long to adults having femora more than 900 mm long. Similarly, the humeri range in size from 150 mm to 386 mm, although matching the relative sizes of the forelimb bones with those of equivalent growth stages to other postcranial elements is impossible. Thus on the basis of individual elements the allometric coefficients of either the arms or

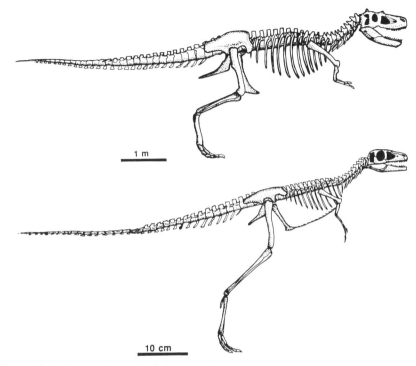

Figure 8.1 Comparison of skeletons of an adult (top) and hatchling (bottom) *Albertosaurus libratus*, scaled to the same size. (Drawing of hatchling after Russell, 1970.)

legs cannot be estimated. Paul (1988) described a 678 mm long juvenile skull that he questioningly attributed to *Albertosaurus libratus*. This is a little more than half the size of the largest known adult skulls (Figure 8.2). There is some doubt as to its species identification as the teeth are as large as in the biggest *Albertosaurus* skulls. However, it does provide some indication of variations in skull proportions that occurred through ontogeny.

Lawson (1976) showed that changes in maxilla proportions occur with growth in three theropods—*Allosaurus*, *Tyrannosaurus* and *Albertosaurus*—although the anomalous growth trend shown by *Albertosaurus* may reflect a possible mixing of different species (one large, one small) rather than a direct allometric growth trend (see comments by Paul, 1988; Bakker *et al.*, 1988). Dr Tony Thulborn (pers. comm.) has statistically analysed the measurements of lower and upper jaw bones of *Allosaurus*, based on Madsen's measurements, and suggests that the population was sexually dimorphic in the shape of the jaw elements and the numbers of teeth in the maxillae and dentaries. Similar sexual dimorphic features were recently identified in *Tyrannosaurus rex* (Carpenter, 1990).

The study of the theropod astragalus by Welles and Long (1974) also demonstrated significant morphological changes with growth. Variations, both

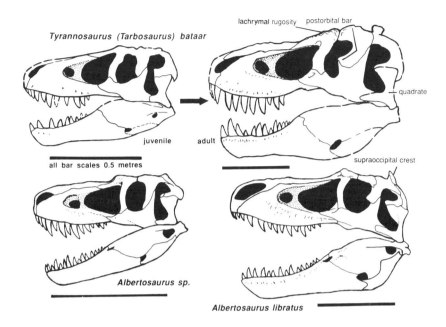

Figure 8.2   Juvenile and adult skulls of *Tyrannosaurus (Tarbosaurus) bataar* (top) and species of *Albertosaurus*.

intraspecific (possibly sexual dimorphic) and ontogenetic, have been documented for small theropods by Raath (1990), and the large theropod *Tyrannosaurus rex* by Carpenter (1990). Growth change in tyrannosaurids has also been documented in *Tyrannosaurus bataar* from the Late Cretaceous of Mongolia by Carpenter (1992), on the basis of a series of skulls at different stages of growth, ranging in length from approximately 750 mm to 1380 mm (Figure 8.2). Many of the ontogenetic changes observed in theropods in the above studies were summarised by Molnar (1990), and these, along with other changes, are listed here:

1. Closure (and some fusion) of cranial sutures, including angular and surangular in the lower jaw (Russell, 1970).
2. Development of supraoccipital alae of the parietals (Russell, 1970).
3. Increase in serration count on teeth (Currie *et al.*, 1990).
4. Skull deepens dorsoventrally with age and muzzle shortens (tyrannosaurids, Carpenter, 1992).
5. Orbits become less rounded and the postorbital bar (where present) develops late in ontogeny (Carpenter, 1992).
6. Metatarsals become more robust and thick (Carpenter 1992).
7. Increase in relative length of presacral column (Russell, 1970).
8. Disproportionate growth of cervical neural spines (Madsen, 1976).
9. Decrease in relative tail length (Russell, 1970).

10. Increase in relative size of limb girdles (Russell, 1970).
11. Decrease in relative length of hind limb (Russell, 1970; Molnar, 1990).
12. Decrease in relative height of astragular ascending process (Welles and Long, 1974).
13. Bone growth (long bones) from rapid fibrolamellar to moderate lamellar-zonal to slow avascular lamellar growth (Varricchio, 1993).

Heterochrony in tyrannosaurid evolution

The largest land predators, the tyrannosaurids, offer the most information on theropod growth and morphological variation, owing to our current knowledge of their ontogeny, in addition to recently published detailed information on their phylogenetic relationships (Bakker *et al.*, 1988; Paul, 1988). Consequently, they serve as an ideal group for examining the role of heterochrony in dinosaur evolution.

Tyrannosaurids differ from other theropods in the possession of several derived features (Carpenter, 1992). These include a number which are probably of heterochronic derivation, being related to ontogenetic changes. Predominant among these, particularly from an adaptionist point of view, is the large head relative to body size. Compared with skulls of juvenile theropods and of adults of smaller species, the relatively larger size of the skull is probably a peramorphic attribute arising from allometric scaling with the large body size. However, even within the skull allometries varied, such that the positive allometric coefficients were probably higher dorsoventrally than anteroposterior coefficients. The result is a change in shape during ontogeny from a relatively long, slender head, to a more massive, less elongate head (Figures 8.1, 8.2). The retention of a more slender skull in the smaller tyrannosaurids *Nanotyrannus* and *Maleevosaurus* is a paedomorphic feature (see below).

The same peramorphic effect of allometric scaling seen in the skull of *Tyrannosaurus* probably underlies the evolution of the massive hind limbs in tyrannosaurids, particularly the larger forms. However, the characteristic very short forelimbs and manus, in which there are only two digits, are paedomorphic features. Extreme dissociated heterochrony involving pronounced peramorphosis of skull and hind limbs, is in contrast to the exteme paedomorphic reduction in the forelimbs and manus. At the familial level, other morphological features in tyrannosaurids that are peramorphic in nature include the exclusion of the frontals from the orbits by lachrymal bones, arising from an increased degree of growth of the lachrymal bone.

Within the clade Tyrannosauridae (Bakker *et al.*, 1988) the main trend is for increase in supraoccipital crests on the head along with lacrymal rugosity development. Advanced forms have a postorbital bar below the orbit (Figure 8.2) and there are various sinuses and cavities developed within the braincase. Other derived features of tyrannosaurids (Carpenter, 1992; Bakker *et al.*, 1988) include the large surangular foramen (in juveniles, and some primitive theropods this is open, not strongly sutured), jugal pierced by large foramen, D-shaped premaxillary teeth sections, supraoccipital wedge with two tabs of bone placed in tandem and the quadratojugal–squamosal suture. However, available evidence indicates that there do not seem to have been any major changes in these characters during ontogeny, based on the limited data from juvenile *Tyrannosaurus bataar* and *Albertosaurus*.

The evolution of tyrannosaurids from other maniraptorans (Holtz, 1994) is one of great growth increase in some features and dissociated heterochrony: peramorphosis of the body size, skull and hind limbs, and reduction (paedomorphosis) of forelimb and manus. Analysis of tyrannosaurid bone structure suggests that, like many other dinosaurs, their growth rate was relatively rapid (de Ricqlès, 1980; Chinsamy, 1992; Varricchio, 1993). This suggests that peramorphic features in tyrannosaurids, even in large forms, were not entirely a function of delayed onset of maturation, but were also caused by acceleration in growth of specific structures. Indeed, the extreme dissociated nature of the heterochronies in tyrannosaurids suggests that acceleration may have been as important, if not more so, than hypermorphosis in generating the peramorphic structures.

The smaller Late Cretaceous tyrannosaurs, such as *Maleevosaurus* (5 m) and *Nanotyrannus* (5 m), are paedomorphic genera, as demonstrated by features such as the more slender snout, wide, rounded orbit, and the lack of a postorbital bar, as well as a small body size. Bakker *et al.* (1988) place *Nanotyrannus* as a primitive sister taxon to the more advanced tyrannosaurids (those with rugose snouts and anterior pneumatic foramina in the basicranium). Bakker *et al.* (1988) suggested "neoteny" as a way of producing the delicate adult lachrymal horn of *Nanotyrannus*, but acknowledged that it most likely arose from a hornless and unswollen ancestral state. This would rather tend to imply peramorphosis, as juvenile tyrannosaurs lack the development of the lachrymal horn ("*Tarbosaurus*"—Carpenter, 1992).

HADROSAURS AND HYPSILOPHODONTIDS

Ontogenetic changes in higher ornithopods

This group is equivalent to the Ornithopoda of Sereno (1986) or Euornithopoda of Weishampel (1990). Our overview of ontogenetic changes is based on *Dryosaurus* (Carpenter *et al.*, 1994), *Maiasaura* and other hadrosaurids (Dodson, 1975; Carpenter *et al.*, 1994; Horner and Currie, 1994; Weishampel, 1981). Weishampel (1981) illustrated the juvenile and adult skulls of several different lambeosaurine hadrosaurids, although most of these were about two-thirds to three-quarters the full adult size. Much of the proportionate changes in skull growth had already occurred by this size, except for the development of their bizarre crests, which are a feature of late stage ontogeny (Figure 8.3). As such, any slight variations in the time of onset of maturation can have pronounced phenotypic effects in the skull crest development.

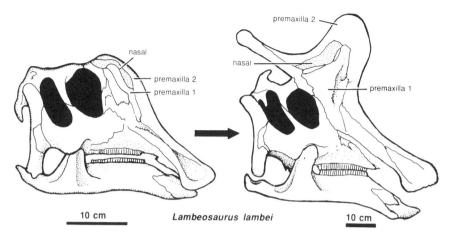

Figure 8.3 Ontogenetic development of the skull of the hadrosaurid *Lambeosaurus lambei* (after Weishampel, 1981).

The discovery of juvenile and embryonic hadrosaurids (*Maiasaura*) and hypsilophodontids (*Orodromeus*) from the Two Medicine Formation of western Montana (Horner, 1984; Horner and Makela, 1979; Horner and Weishampel, 1988) provides preliminary information on the nature of early embryonic development in both these groups. Analysis of the bone structure in these groups (Horner and Currie, 1994) indicates that, like the tyrannosaurids, the young dinosaurs grew at a rapid rate.

Hadrosaurids have provided one of the few examples of heterochrony previously documented in dinosaurs. Weishampel *et al.* (1993) have noted the role of heterochrony (progenetic paedomorphosis) in the development of the miniaturised maxillary dentition and development of a dental battery in *Telmatosaurus* and perhaps all hadrosaurids.

The following general trends in higher ornithopod ontogeny occur:

1. Decrease in size of orbit.
2. Palpebral bones decrease in size relative to orbit.
3. Increase in size (length and width) of premaxillary and nasal bones, eventually expanding to form "bill" in hadrosaurs.
4. Prolongation of snout in large forms.
5. Development of prefrontal brow ridges in some forms (e.g., *Maiasaura*, *Prosaurolophus*).
6. New tooth rows added to both lower and upper jaws.
7. Neural canal decrease in relative size in vertebrae.
8. Neural spines increase in size.

Heterochrony and the position of Tenontosaurus

Tenontosaurus is currently regarded as either a sister taxon of the Hypsilo-phodontidae, placed within the clade Hypsilophodontia (Norman, 1984, 1990; also Dodson, 1980; and others listed in Norman, 1990) or placed as a sister taxon to *Dryosaurus, Camptosaurus*, iguanodontids and hadrosaurids by Sereno (1986). Earlier work by Ostrom (1970) and Galton (1974) placed *Tenontosaurus* with the iguanodontids. Weishampel and Heinrich (1992) place *Tenontosaurus* as the plesiomorphic sister taxon to *Dryosaurus, Camptosaurus* and iguanodontids.

From the observations of ontogenetic changes in *Dryosaurus* (Figure 8.4) it is clear that the development of the long snout in adults is a peramorphic feature arising by positive allometric increase in growth of the nasal and frontals. The major cranial differences between adult hypsilophodontids, *Dryosaurus* and *Tenontosaurus*, are in adults of the latter possessing the longer snout, smaller orbit, straight quadrate bone, long, narrow external antorbital fenestra, absence of premaxillary teeth (present only in hypsilophodontids within this clade), long curved retroarticular process on the mandible, short palpebral bone, and fenestra enclosed entirely by the quadratojugal. Amongst these cranial characters are a few which are apparently paedomorphic (cf., juvenile *Dryosaurus*), such as the short palpebral bones, and relatively larger and elongate antorbital fenestra. However, the most dramatic are peramorphic features of *Tenontosaurus*, such as the relatively straight quadrate bone, the longer snout length, development of the retroarticular process, and the relatively smaller orbits.

The enclosed quadratojugal foramen in hypsilophodontids can be seen in other ornithischian dinosaurs, such as within the monophyletic clade Stegosauria. Here also there are premaxillary teeth present in the most primitive genus (*Huayangosaurus*, Middle Jurassic of China; Galton, 1990) as well as in basal thyreophorans (*Scutellosaurus*, Coombs *et al.*, 1990), but lost independently in both stegosaurids and ankylosaurids. The same paedomorphic loss of teeth within a monophyletic lineage occurs in bird evolution. Thus in *Archaeopteryx* premaxillary teeth are present, but in the late Mesozoic Hesperornithiformes premaxillary teeth are absent, but dentaries and maxillae are toothed (Martin, 1991). Complete loss of dentition occurs in later birds. The presence of premaxillary teeth is only known in small ornithischians (including basal taxa such as *Lesothosaurus*), and the loss of such teeth appears to be a feature of all large members of the group (>4 m). Weishampel and Heinrich (1992) unite *Tenontosaurus* with basal iguanodontians on the basis of 11 shared characters. Of these the following skull features have relevance to growth changes: relative size of nares to orbit; shape of the predentary and its oral margin, also the possible loss of premaxillary teeth. Thus the evidence from general skull growth changes in basal and advanced ornithopods favours the placement of *Tenontosaurus* either as a peramorphic hypsilophodontid which has independently lost the

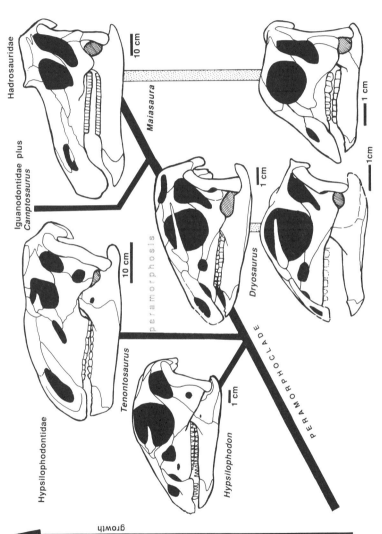

Figure 8.4   Dominant effect of peramorphosis in euornithopod evolution.

premaxillary teeth, paralleling the situation in several ornithischian clades, and has peramorphic growth of many facial characteristics; or as a stem-group member of the iguanodontians, with which it shares many synapomorphies (Weishampel and Heinrich, 1992), none of which seem to be related to growth changes that indicate derivation from any particular level of the basal iguanodontians.

Within the postcranial skeleton *Tenontosaurus* differs principally from hypsilophodontids by a paedomorphic reduction of digit III on the manus (a character shared with higher ornithopods), ossified hypaxial tendons in the tail, straight ischium shaft and position of the obturator process one-third to one-half length of the shaft away from the acetabular margin. These are regarded as hypsilophodontid characters by Norman (1990), although it shares with iguanodontians the deep intercondylar groove on the distal end of the femur, loss of phalanx from manual digit II and the shorter humeral length relative to scapula (Weishampel and Heinrich, 1992).

Most crucial to the phylogenetic arguments are the ontogenetic changes showing which areas of growth are more plastic with respect to rapid change. In the hadrosaur *Maiasaura* the scapula is approximately as long as the humerus in the hatchling (Figure 8.5) becoming proportionately longer than the humerus in maturity. Therefore the longer scapula in higher ornithopods, although based on scant evidence, would appear to be a valid autapomorphic (peramorphic) character, more likely to be a plesiomorphic feature in *Tenontosaurus*, rather than a synapomorphy shared with hypsilophodontids. Other similarities shared with hypsilophodontids appear to be valid characters, only the reduction of digit III on the manus therefore links *Tenontosaurus* with higher ornithopods. The reduction and modification of digits and reduced forearm size is a common trend in the evolution of large theropods, especially tyrannosaurids and other derived maniraptorans like abelisaurids. The increase in growth to reach a large body size (up to 7.5 m) in *Tenontosaurus* could account for reduced size of digits and forearms, these elements being dissociated from the overall peramorphic trends in cranial characters as body size increased, as in tyrannosaurids. However, at this stage the condition in the arms and digits would appear to be an advanced character either linking this taxon with higher ornithopods (as suggested by Weishampel and Heinrich, 1992) or being independently acquired as a convergent feature.

CERATOPSIANS

Growth and variation in ceratopsians, including sexual dimorphism, was first described in detail for ceratopsians by Dodson (1975) although earlier studies had noted changes with growth (Brown and Schlaikjer, 1940; Gray, 1946) and possible sexual dimorphism (Kurzanov, 1972). Although the ontogenetic series for primitive neoceratopsians is now well known (e.g., *Bagaceratops* —Marayanska

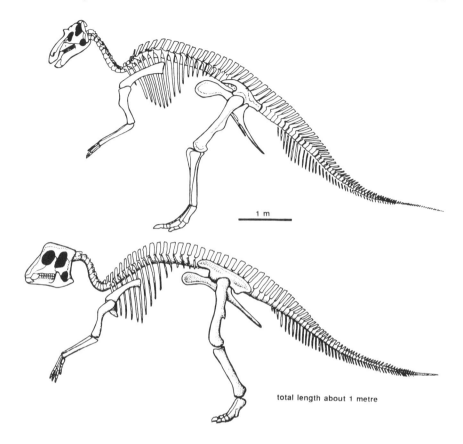

Figure 8.5 Comparison of skeletons of an adult (top) and juvenile (bottom) *Maiasaura peeblesorum*, scaled to the same size.

and Osmolska, 1975; Chapman, 1992; *Protoceratops*—Carpenter *et al.*, 1994), very little is known of growth and variation in ceratopsids, i.e., chasmosaurines and centrosaurines (Dodson and Currie, 1990), apart from recognition of sexual dimorphism (Lehman, 1990). Thus inferences of heterochrony are here based largely on the primitive forms and assumed to follow similar ontogenetic trends apparent in higher neoceratopsians.

Ontogenetic changes in ceratopsians

These are modified from Marayanska and Osmolska (1975), based on *Bagaceratops*, *Leptaceratops* and *Protoceratops*, and from Kurzanov (1990), based on *Breviceratops*. Changes in the growth of postcranial elements have been graphed by Lehman (1990) for *Chasmosaurus*. These observations indicate most growth changes in the scapula, femur, ilium and tibia are allometric. Characteristic ontogenetic changes are:

1. Decrease in orbit diameter.
2. Increase in snout length (slight).
3. Slight increase in frill length, followed by subsequent shortening of the frill.
4. Widening of the frill.
5. Widening of the jugal and quadrate area.
6. Development of the nasal horn (in *Protoceratops*).

Heterochrony in ceratopsians

It is clear from all of these trends, both in size increase from psittacosaurids to primitive and advanced neoceratopsians, and increase in complexity and extent

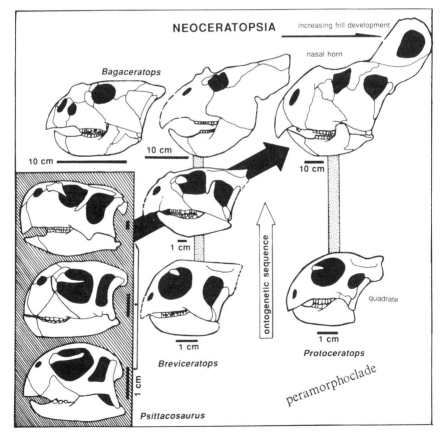

Figure 8.6   Dominant effect of peramorphosis in early neoceratopsian evolution.

of the frill and horn arrangements, that peramorphosis has been a principal factor in ceratopsian evolution (Figure 8.6). The postcranial skeleton is remarkably conservative in the two advanced groups, the Centrosaurinae and the Chasmosaurinae (Dodson and Currie, 1990) with the major evolutionary trends being the flattening of the digits to accommodate the greater weight of the large ceratopsians and increase in size of the olecranon process of the ulna in chasmosaurines. Most postcranial elements, as exemplified by studies on *Chasmosaurus* (Lehman, 1990), grow nearly isometrically without major changes to overall morphology. However, the thickening and broadening of certain limb elements in large ceratopsids (e.g., *Triceratops*) would appear to be a peramorphic development for graviportal posture in accord with dramatic increase in body weight.

THE HETEROCHRONIC EVOLUTION OF BIRDS FROM WITHIN THE THEROPODA

There is a general consensus that theropod dinosaurs are the ancestral group from which birds arose (see Witmer, 1991, for a recent review of the literature), although there is some debate amongst workers as to which particular theropod groups are most closely related to birds (Gauthier, 1986; Witmer, 1991). The most favoured candidate are deinonychosaurian coelurosaurs. Thulborn (1985) proposed that birds are in some respects paedomorphic theropod dinosaurs. He based this interpretation on the concept that feathers may have been present on juvenile theropods, being used as an insulating blanket. However, there is no direct evidence for this. Thulborn has pointed out that there is evidence to suggest that feathers are derivatives of epidermal scales.

There are more concrete ways in which *Archaeopteryx* resembles a juvenile theropod. The most obvious is the shape of the skull and orbits. In young theropods such as *Coelophysis*, the orbit is relatively very large, decreasing in relative size during ontogeny (Figure 8.7). Likewise the braincase is relatively inflated in juvenile theropods, as it is in *Archaeopteryx*. Tooth reduction (see above) is another paedomorphic character. Likewise, the shape of the teeth. A recent discovery in the Gobi Desert in Mongolia (Norell *et al.*, 1994) of a nest of dinosaur eggs in which were also found two embryonic dromaeosaurids (probably *Velociraptor*) skulls, lends support to the role of heterochony in avian evolution. The teeth of these embryonic dinosaurs are simple, peg-like structures, very similar to the teeth present in early, primitive adult birds. As Norell *et al.* (1994) point out, "... ontogenetic modification of tooth morphology is common in archosaurs". Thulborn (1985) has also argued that the relatively long manus, forelimbs and pes are also juvenile theropod characters. However, compared with a juvenile theropod these forelimb elements are even larger, suggesting a peramorphic enlargement has occurred (Figure 8.7).

Martin (1991) has also posed the question of the importance of paedomorphosis in avian evolution, particularly in relation to the determination of sexual

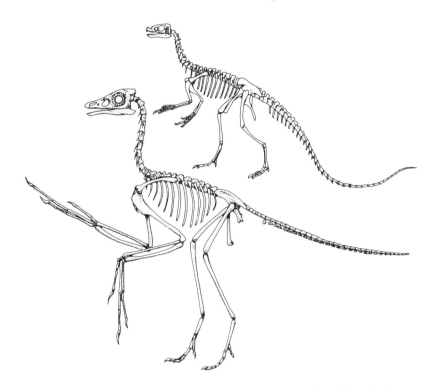

Figure 8.7 Skeletons of juvenile *Coelophysis* (top) and adult *Archaeopteryx* (bottom), showing similarities in overall skeletal and skull structures, but peramorphic development of arm skeletal elements in *Archaeopteryx*. Reproduced from McKinney and McNamara (1991), with permission of Plenum Press.

maturity in birds. Birds have terminal growth. In living birds most growth has been completed when the wings are functional and tarsal bones have fused with the tibia and metatarsals (Martin, 1991). *Archaeopteryx* again displays relatively paedomorphic characters in showing less skeletal fusion than in living birds. As Martin (1991) has pointed out, a list of the unfused or poorly fused elements in *Archaeopteryx* is essentially a catalogue of the juvenile condition found in living birds.

   Martin (1991) has further proposed that study of the ontogenetic changes in modern birds can throw light on the origins of birds, arguing (from a somewhat Haeckelian perspective), that a number of characters, such as pubic reflexion, ilium prolongation and increased fusion, appear in the fossil record in much the same sequence as in ontogeny. Adult Mesozoic birds show characters occurring early in the ontogeny of living birds, such as retention of distinct sutures in the skull. The sternum, interclavicle and uncinate processes of the ribs in *Archaeopteryx* are not ossified. This same condition occurs in the juveniles of living birds. Distinctions occur between *Archaeopteryx* and Cretaceous birds, such as

Apatornis, Ichthyornis and *Hesperornis* in the ilium, ischium and pubis. These are separated by well-defined sutures in *Archaeopteryx*, but are fused in adults of the Cretaceous birds (Martin, 1991). What such a pattern is reflecting is an overall trend of peramorphosis within birds, in contrast to the original predominantly paedomorphic origination of birds from theropods. However, tooth reduction in Hesperorniformes, and subsequent loss in later birds, is a paedomorphic trend.

CONCLUSIONS

Like other groups of organisms, heterochrony appears to have been a major factor in the evolution of dinosaurs. Many of the features that we have documented here in this small selection indicate that peramorphosis was an important factor in the evolution of the group and may well have been more common than paedomorphosis. The high growth rate of a number of dinosaur groups, as indicated by studies of bone structure, indicates that acceleration was an important heterochronic process. The large body size attained in many clades suggests that either hypermorphosis was also a contributory factor, or that, as in the case of some large terrestrial mammals (see Chapter 4), growth may have been indeterminate. The late development in ontogeny of structures such as skull crests in hadrosaurids suggests that like some living reptiles, growth rates of particular structures may have actually increased late in ontogeny.

Examples of paedomorphic genera are less obvious. However, genera that are small for their clade, such as *Nanotyrannus*, retain typically juvenile characters as adults, and were specialised paedomorphic forms. Future research aimed at analysing bone structure of such a paedomorphic taxon should provide some indication of the likely process that led to the paedomorphosis. If bone growth rate appears to be the same as in other tyrannosaurids, then a strong case could be made for progenesis. However, if growth rate appears to have been slower, then the genus is likely to be neotenic.

As with many groups of organisms, the success of the Dinosauria lies in the dissociated nature of the heterochrony. Moreover, the evolution of birds from dinosaurs likewise exemplifies such dissociated heterochrony, but with an opposite polarity for the same skeletal elements. Thus, in tyrannosaurid dinosaurs the large body size, skull and hind limbs are the product of peramorphic processes, whereas reduction in the manus and digits is paedomorphic. On the contrary, the evolutionary novelties that lead to the success of the birds include the paedomorphic retention of small body size, unfused bones and paedomorphic skull characters, i.e., the large orbit, and subsequent tooth loss. In contrast to this is the peramorphic enlargement of the forelimbs, the opposite situation to that found in tyrannosaurid dinosaurs.

The discovery by Varricchio (1993) that *Troodon* passed through three ontogenetic growth phases, as demonstrated by changes in bone microstructure, and had determinate growth, provides the potential for determining relative growth rates and maturation times in different dinosaur taxa. The presence of lines of arrested growth, which Varricchio (1993) considers may be a reflection of the seasonal climate in which the dinosaurs lived, may give us a handle on actual growth rates and time of onset of maturation. From his data, Varricchio (1993) concluded that the species of *Troodon* that he was studying may have reached maturity after three to five years, at a body weight of about 50 kg. Similar analysis of bone microstructure in other dinosaurs may thus allow us to suggest the heterochronic mechanisms that were involved in the evolution of dinosaurs.

ACKNOWLEDGMENTS

We thank Dr David Weishampel (Johns Hopkins University, Baltimore) and Dr Tom Rich (Museum of Victoria, Melbourne) for their helpful comments on the manuscript. JAL also acknowledges the support of Indiana and Purdu Universities, Indianapolis, for financial support to attend the 1994 Dino Fest meeting and to discuss the subject of dinosaur heterochrony with various specialists at that meeting.

REFERENCES

Bakker, R.T., Williams, M. and Currie, P., 1988, *Nanotyrannus*, a new genus of pygmy tyrannosaur, from the latest Cretaceous of Montana, *Hunteria*, 1(5): 1–30.
Bonaparte, J.F. and Vince, M., 1979, El hallazgo del primer nido de Dinosaurios Triásicos (Saurischia, Prosauropoda), Triásico Superior de Patagonia, Argentina, *Ameghiniana*, 16: 173–182.
Brown, B. and Schlaikjer, E.M., 1940, The structure and relationships of *Protoceratops*, *Ann. N.Y. Acad. Sci.*, 40: 133–266.
Carpenter, K., 1990, Variation in *Tyrannosaurus rex*. In K. Carpenter and P.J. Currie (eds), *Dinosaur systematics. Approaches and perspectives*, Cambridge University Press, Cambridge: 141–145.
Carpenter, K., 1992, Tyrannosaurids (Dinosauria) of Asia and North America. In N. Mateer and Chen Pei-ji (eds), *Aspects of nonmarine Cretaceous geology*, China Ocean Press, Beijing: 250–268.
Carpenter, K.J. and Currie, P.J. (eds), 1990, *Dinosaur systematics. Approaches and perspectives*, Cambridge University Press, Cambridge.
Carpenter, K., Hirsch, K.F. and Horner, J. (eds), 1994, *Dinosaur eggs and babies*, Cambridge University Press, Cambridge.
Chinsamy, A., 1992, Ontogenetic growth of the dinosaurs *Massospondylus carinatus* and *Syntarsus rhodesiensis*, *J. Vert. Pal.*, 12: 23A.
Coombs, W. P., Weishampel, D.B. and Witmer, L.M., 1990, Basal Thyreophora. In D. Weishampel, P. Dodson and H. Osmólska (eds), *The Dinosauria*, University of California Press, Berkeley: 427–434.
Currie, P.J., Rigby, J.K. and Sloan, R.E., 1990, Theropod teeth from the Judith River Formation of southern Alberta, Canada. In K. Carpenter and P.J. Currie (eds), *Dinosaur systematics. Approaches and perspectives*, Cambridge University Press, Cambridge: 107–125.
Dodson, P., 1975, Taxonomic implications of relative growth in lambeosaurine hadrosaurs, *Syst. Zool.*, 24: 37–54.

Dodson, P., 1976, Quantitative aspects of relative growth and sexual dimorphism in *Protoceratops*, *J. Paleont.*, **50**: 929–940.

Dodson, P., 1980, Comparative osteology of the American ornithopods *Camptosaurus* and *Tenontosaurus*, *Mem. Geol. Soc. France*, **139**: 81–85.

Dodson, P. and Currie, P.J., 1990, Neoceratopsia. In D. Weishampel, P. Dodson and H. Osmólska (eds), *The Dinosauria*, University of California Press, Berkeley: 593–618.

Galton, P.M., 1974, Notes on *Thescelosaurus*, a conservative ornithopod dinosaur from the Upper Cretaceous of North America, with comments on ornithopod classification, *J. Paleont.*, 48: 1048–1067.

Galton, P.M., 1990, Stegosauria. In D. Weishampel, P. Dodson and H. Osmólska (eds), *The Dinosauria*, University of California Press, Berkeley: 435–455.

Gauthier, J., 1986, Saurischian monophyly and the origin of birds. In K. Padian (ed.), *The origin of birds and the evolution of flight*, *Mem. Calif. Acad. Sci.*, **8**: 1–55.

Gray, S.W., 1946, Relative growth in a phylogenetic series and an ontogenetic series of one of its members, *Am. J. Sci.*, **244**: 792–807.

Holz, T.R., 1994, The phylogenetic position of the Tyrannosauridae: implications for theropod systematics, *J. Paleont.*, **68**: 1100–1117.

Hopson, J.A., 1977, Relative brain size and behaviour in archosaurian reptiles, *Ann. Rev. Ecol. Syst.*, **8**: 429–448.

Horner, J.R., 1984, The nesting behaviour of dinosaurs, *Sci. Am.*, **250**: 130–137.

Horner, J.R. and Currie, P.J., 1994, Embryonic and neonatal morphology and ontogeny of a new species of *Hypacrosaurus* (Ornithischia, Lambeosauridae) from Montana and Alberta. In K. Carpenter, K.F. Hirsch and J. Horner (eds), *Dinosaur eggs and babies*, Cambridge University Press, Cambridge: 312–336.

Horner, J.R. and Makela, R., 1979, Nest of juveniles provides evidence of family structure among dinosaurs, *Nature*, **282**: 296–298.

Horner, J.R. and Weishampel, D.B., 1988, A comparative embryological study of two ornithischian dinosaurs, *Nature*, **332**: 256–257.

Kurzanov S.M., 1972, Sexual dimorphism in protoceratopsids, *Pal. Jl*, **1972**: 91–97.

Kurzanov, S., 1990, A new Late Cretaceous protoceratopsid genus from Mongolia, *Pal. Jl*, **4**: 85–91.

Lawson, D.A., 1976, *Tyrannosaurus* and *Torosaurus*, Maestrichtian dinosaurs from Trans-Pecos, Texas, *J. Paleont.*, **50**: 158–164.

Lehman, T.H., 1990, The ceratopsian subfamily Chasmosaurinae: sexual dimorphism and systematics. In K. Carpenter and P.J. Currie (eds), *Dinosaur systematics. Approaches and perspectives*, Cambridge University Press, Cambridge: 211–229.

Madsen, J.H., 1976, *Allosaurus fragilis*: a revised osteology, *Bull. Utah Geol. Min. Surv.*, **109**: 1–163.

Martin, L.D., 1991, Mesozoic birds and the origin of birds. In H.-P. Schultze and L. Trueb (eds), *Origins of the higher groups of tetrapods*, Cornell University, Ithaca: 485–540.

Maryánska, T. and Osmólska, H., 1975, Protoceratopsidae (Dinosauria) of Asia, *Pal. Polonica*, **33**: 45–102.

McKinney, M.L. and McNamara, K.J., 1991, *Heterochrony: the evolution of ontogeny*, Plenum, New York.

McNamara, K.J., 1988, The abundance of heterochrony in the fossil record. In M.L. McKinney (ed.), *Heterochrony in evolution: a multidisciplinary approach*, Plenum, New York: 287–325.

Molnar, R.E., 1990, Variation in theory and in theropods. In K. Carpenter and P.J. Currie (eds), *Dinosaur systematics. Approaches and perspectives*, Cambridge University Press, Cambridge: 71–79.

Norell, M.A., Clark, J.M., Demberelyin, D., Rhinchen, B., Chiappe, L.M., Davidson, A.R., McKenna, M.C., Altangerel, P. and Novacek, M.J., 1994, A theropod dinosaur

embryo and the affinities of the Flaming Cliffs dinosaur eggs, *Science*, **266**: 779–782.
Norman, D., 1984, A systematic reappraisal of the reptile order Ornithischia. In W.E Reif
    and F. Westphal (eds), *Third symposium on Mesozoic terrestrial ecosystems, short
    papers*, Attempto Verlag, Tubingen: 157–162.
Norman, D., 1990, A review of *Vectisaurus valdensis*, with comments on the family
    Iguanodontidae. In K. Carpenter and P.J. Currie (eds), *Dinosaur systematics.
    Approaches and perspectives*, Cambridge University Press, Cambridge: 147–161.
Ostrom, J., 1970, Stratigraphy and paleontology of the Cloverley Formation (lower
    Cretaceous) of the Bighorn Basin area, Wyoming and Montana, *Peabody Mus., Nat.
    Hist. Bull.*, **35**: 1–234.
Paul, G.S., 1988, *Predatory dinosaurs of the world*, Simon and Schuster, New York.
Raath, M.A., 1990, Morphological variation in small theropods and its meaning in
    systematics: evidence from *Syntarsus rhodesiensis*. In K. Carpenter and P.J. Currie
    (eds), *Dinosaur systematics. Approaches and perspectives*, Cambridge University Press,
    Cambridge: 91–105.
Ricqlès, A. de, 1980, Tissue structures of dinosaur bone: functional significance and
    possible relation to dinosaur physiology. In R.D.K. Thomas and E.C. Olsen (eds), *A
    cold look at warm-blooded dinosaurs*, Westview, Boulder: 103–140.
Rozhdestvensky, A.K., 1965, Growth changes and some problems of systematics of Asian
    dinosaurs, *Pal. Zh.*, **1965**: 95–109 (In Russian).
Russell, D.A., 1970, Tyrannosaurs from the Late Cretaceous of western Canada, *Nat.
    Mus. Nat. Sci. Publ. Pal.*, **1**: 1–34.
Sereno, P.C., 1986, Phylogeny of the bird-hipped dinosaurs (order Ornithischia), *Nat.
    Geog. Res.*, **2**: 234–256.
Thulborn, R.A., 1985, Birds as neotenous dinosaurs, *Rec. N. Z. Geol. Surv.*, **9**: 90–92.
Varricchio, D.J., 1993, Bone microstructure of the Upper Cretaceous theropod *Troodon
    formosus*, *J. Vert. Paleont.*, **13**: 99–104.
Weishampel, D.B., 1981, The nasal cavity of lambeosaurine hadrosaurids (Reptilia:
    Ornithischia): comparative anatomy and homologies, *J. Palaeont.*, **55**: 1046–1058.
Weishampel, D.B., 1990, Hadrosauridae, In D. Weishampel, P. Dodson and H.
    Osmólska (eds), *The Dinosauria*, University of California Press, Berkeley: 534–561.
Weishampel, D.B. and Heinrich, R.E., 1992, Systematics of the Hypsilophodontidae and
    basal Iguanodontia (Dinosauria: Ornithopoda), *Hist. Biol.*, **6**: 159–184.
Weishampel, D.B. and Horner, J.R., 1994, Life history syndromes, heterochrony, and the
    evolution of Dinosauria. In K. Carpenter, K.F. Hirsch and J.R. Horner (eds), *Dinosaur
    eggs and babies*, Cambridge University Press, Cambridge: 227–243.
Weishampel, D.B., Norman, D.B. and Grigorescu, D., 1993, *Telmatosaurus transsylvanicus*
    from the Late Cretaceous of Romania: the most basal hadrosaurid dinosaur,
    Palaeontology, **36**: 361–385.
Welles, S.P. and Long, R.A., 1974, The tarsus of theropod dinosaurs, *Ann. S. Afr. Mus.*,
    64: 191–218.
Witmer, L.M., 1991, Perspectives on avian origins, In H.-P. Schultze and L. Trueb (eds),
    Origins of the higher groups of tetrapods, Cornell University, Ithaca: 427–466.

Chapter 9

HETEROCHRONY AND THE EVOLUTION OF AVIAN FLIGHTLESSNESS

Bradley C. Livezey

INTRODUCTION

In his landmark work on heterochrony, Gould (1977, p. 346) noted the phenomenon of delayed breeding in tropical seabirds, polygamous New World blackbirds (Icteridae), and grouse (Tetraonidae), and concluded:

> These cases do not involve neoteny: birds, when adult, do not retain juvenile features (so far as the literature records), and there is no delay of somatic development relative to reproduction.

The inclusiveness of this statement is doubly remarkable in light of the historical thoroughness that otherwise typified Gould's (1977) review. On the contrary, however, there is a long tradition of citing "neoteny" in the ornithological literature, one intertwined with the related concept of recapitulation.

Perhaps the earliest diagnosis of avian heterochrony, and one of the most compelling, was that by Strickland and Melville (1848, p. 4) concerning the extinct, flightless dodo (Columbiformes: *Raphus cucullatus*), which was described as having:

> ... wings too short and feeble for flight, the plumage loose and decomposed, and the general aspect suggestive of gigantic immaturity.

These authors continued (1848, pp. 33–4):

> We cannot form a better idea of it than by imagining a young Duck or Gosling enlarged to the dimensions of a Swan. It affords one of those cases ... where a species, or a part of the organs in a species, remains permanently in an underdeveloped or infantine state. Such a condition has reference to peculiarities in the mode of life of the animal, which render certain organs unnecessary, and they therefore are retained through life in an imperfect state, instead of attaining that fully developed condition which marks the mature age of the generality of animals.

Evolutionary Change and Heterochrony. Edited by Kenneth J. McNamara © 1995 The Editor and Contributors. Published in 1995 by John Wiley & Sons Ltd.

. . . And lastly, . . . the Dodo is (or rather was) *a permanent nestling*, clothed with down instead of feathers, and with the wings and tail so short and feeble, as to be utterly unsubservient to flight [emphasis in original].

Unfortunately, like the dodo, ornithological study of heterochrony has remained in a protracted infancy, retarded in part by a poor appreciation of the diversity of processes involved in the phenomenon, as well as a confusion between pattern and process. Also, study of the evolutionary role of ontogeny in avian flightlessness suffered collateral damage from contemporary scientific debates concerning continental drift, homology, recapitulation and morphological convergence. With the exception of recent investigations of anatomical heterochrony in a few flighted species of birds (Fry, 1983; Dawson *et al.*, 1994) and the study of heterochrony of plumage and behaviour (Lawton and Lawton, 1986; McDonald and Smith, 1990, 1994), ornithological interest in heterochrony continues to centre on flightless species. In this paper I review the study of the ontogenetic basis of avian flightlessness, with an emphasis on the history of study and most promising directions for future research.

DIVERSITY OF FLIGHTLESS BIRDS

Flightless birds of the Cretaceous

Fossil remains of large, flightless, foot-propelled diving birds—the Hesperornithiformes—were first described in detail by Marsh (1880). Two families are recognized (Table 9.1). Members of the group currently are considered to represent one of the earliest branches in the class Aves (Gingerich, 1973; Cracraft, 1986). A flighted ancestor generally is assumed for the Hesperornithiformes, but this is not essential to the currently inferred phylogenetic position of the order (Cracraft, 1986). Members of the order are highly specialized for foot-propelled diving (Marsh, 1880; Martin and Tate, 1976). Available skeletal elements, including the comparatively distal elements described by Martin and Tate (1976), substantiate the radically shortened pectoral limbs of the Hesperornithiformes.

Lowe (1935, p. 410) considered *Hesperornis* to be:

. . . as reptilian as it was possible to be without losing its claim to be avian . . . an aquatic palæognath, just as the Ostriches (Struthiones) were cursorial palæognathes . . . there is little reason to think that either had volant ancestors . . .

Lowe (1935, p. 410) also likened *Hesperornis* to a "swimming ostrich" and later (1942, p. 17) included *Hesperornis* among the "aquatic or swimming palæognaths". Preliminary consideration of the probable ontogenetic mechanism(s) involved in producing the uniquely derived body form of this group indicates pectoral paedomorphosis combined with significant non-pectoral peramorphosis, but absence of developmental data and a flighted ancestral morphotype precludes detailed inferences.

"Ratites" or flightless paleognathous birds

Undoubtedly the most celebrated flightless birds are the "ratites" or flightless Paleognathiformes. Named for the flat, raft-like conformation of their sterna (Feduccia, 1980), "ratites" are relatively large, flightless land birds and include the ostriches, kiwis, and extinct moas. Prior to the recognition of continental drift, many ornithologists reasoned that these flightless giants of the southern continents were necessarily polyphyletic (e.g., Fürbringer, 1888; Mayr and Amadon, 1951; Storer, 1971), and evolutionary interpretations of flightlessness were made accordingly. Opinions regarding the phylogenetic relationships among ratites remain diverse, but monophyly of the group remains the favoured hypothesis (e.g., Bock, 1963; Parkes and Clark, 1966; Cracraft, 1974, 1986; Prager *et al.*, 1976; Sibley and Ahlquist, 1990; Stapel *et al.*, 1984; Bledsoe, 1988). Eight taxonomic families are recognized (Mayr and Amadon, 1951), distributed throughout the southern hemisphere (Table 9.1). Although varying in cursorial abilities, all ratites are (were) strictly terrestrial.

Often popularly considered to typify the flightless avian morphotype, ratites are diverse and combine profoundly primitive features (indicating an early divergence from most other modern birds) with a variety of extremely derived, often unique characters (only some of which are related to flightlessness). There is considerable variation in body size among the families of ratites; kiwis (Apterygidae) average 1–3 kg in total body mass, whereas the other extant ratites have mean body masses of 30–100 kg. Variation among ratites in osteology (especially the paleognathous palate) and the integument, widely interpreted as the result of divergence subsequent to common ancestry, has been cited as evidence of the polyphyly of ratites by several systematists (Fürbringer, 1888; Pycraft, 1900; McDowell, 1948; Houde, 1988; Bock and Bühler, 1990).

Flightless "carinate" (neognathous) birds

Neognathes, traditionally referred to as "carinates" because of the presence of a keel or carina on the sternum (Newton, 1896), include flightless members from a number of taxonomic orders, most of which show significant reduction of this purportedly diagnostic feature (Table 9.1). Unlike ratites, the higher-order phylogenetic relationships of flightless neognaths (possibly excluding the penguins) have not been a significant source of confusion, and most systematic issues have concerned the lesser determination of the closest flighted confamilials of each flightless species for purposes of comparison. The Gruiformes (cranes, rails, and allies) include the greatest number and diversity of flightless members; the rails (Rallidae) alone account for roughly one-third of the known flightless neognaths (Table 9.1). Penguins, the only modern neognathous order in which all species are flightless, are second-highest in number of flightless species

Table 9.1 Taxonomic diversity and geographical distributions of adequately documented flightless species of birds, after Raikow (1985). Groups preceded by daggers are extinct.

Taxonomic group			Flightless members			No. flightless species	
Order	Family	Subgroup(s)	Common name(s)	Period of occurrence	Distribution	(Sub)fossil	Modern
†Hesperornithiformes	Hesperornithidae	—	—	Cretaceous	Europe, N. and S. America	4 +	—
	Baptornithidae	—	—	Cretaceous	N. (possibly S.) America	2 +	—
Paleognathiformes	Casuariidae	—	Cassowaries	Pleistocene–Recent	Australia, New Guinea	0	3
	Dromaiidae	—	Emus	Pleistocene–Recent	Australia	1	2
	Apterygidae	—	Kiwis	Pleistocene–Recent	New Zealand	0	3
	†Dinornithidae	—	Moas	Miocene–Recent	New Zealand	11	0
	†Aepyornithidae	—	Elephant birds	Pleistocene–Recent	Madagascar (Africa, Europe)	2–9	—
	Struthionidae	—	Ostriches	Pliocene–Recent	Eurasia, Africa	5 +	1–2
	Rheidae	—	Rheas	Eocene–Recent	S. America	1	2
	†Dromornithidae[a]	—	Mihirung birds	Miocene–Recent (?)	Australia	8	—
Galliformes	Megapodiidae	†Sylviornis[b]	—	Recent	New Caledonia	1	0
Anseriformes	†Cnemiornithidae	Cnemiornis	New Zealand geese	Pleistocene–Recent	New Zealand	2	—
	Anatidae	Anser, Branta	Typical geese	Pleistocene	Hawaii	2	0
		Three genera[c]	†Moa-nalos	Pleistocene	Hawaii	4	0
		Tachyeres	Steamer-ducks	Recent	South America	0	3
		Anas	Dabbling ducks	Recent	New Zealand, Amsterdam I.	1 +	1 +
		†Chendytes	"Diving geese"	Pleistocene	Pacific N. America	2	—
		Mergus	Mergansers	Recent	New Zealand	1	1
Sphenisciformes	Spheniscidae		Penguins	Eocene–Recent	Southern hemisphere	18 +	18
Podicipediformes	Podicipedidae	Involves three genera[d]	Grebes	Recent	Central and S. America	0	3
Pelecaniformes	Phalacrocoracidae	Compsohalieus	Marine cormorants	Recent	Galápagos Islands	0	1
	†Plotopteridae	—	—	Oligocene/Miocene	N. Pacific	3	—
Ciconiiformes	Threskiornithidae	†Apteribis	Ibises	Pleistocene	Hawaii	2	—
		†Xenicibis	Ibises	Pleistocene	West Indies	1	—
†Gastornithiformes[e]	Gastornithidae	Gastornis	—	Paleocene	Europe, Asia	3	—
	Diatrymidae	Diatryma	—	Eocene	Europe, N. America	4 +	—

Order	Family	Genera		Temporal range	Distribution		
Gruiformes	Mesitornithidae[f]	Mesitornis, Monias	Roa-talos	Pleistocene–Recent	Madagascar	0	3
	†Phorusrhacidae	—	—	Eocene(?)–Pliocene	N. and S. America, Europe	10+	—
	†Bathornithidae	Paracrax, others(?)	—	Eocene–Miocene	N. America	3+	—
	Rhynochetidae[f]	Rhynochetos	Kagus	Pleistocene–Recent	New Caledonia	1	1
	†Apterornithidae[f]	Apterornis	Adzebills	Pleistocene–Recent	New Zealand	2	0
	Gruidae	Grus	Cranes	Pleistocene	Cuba	1	0
	Rallidae	Involves 16+ genera[g]	Rails	Pleistocene–Recent	Oceanic islands worldwide	25+	30+
Charadriiformes	Alcidae	†Mancallinae	Lucas auks	Miocene–Pleistocene	Pacific N. America	8	0
		†Pinguinus	Great auks	Pliocene	Coastal N. Atlantic	1	1
Strigiformes	Strigidae	†Ornimegalonyx	Cave owls	Pleistocene	Cuba	1+	0
Psittaciformes	Psittacidae	Strigops habroptilus	Kakapo or owl parrot	Recent	New Zealand	0	1
Columbiformes	†Raphidae	Raphus, Pezophaps	Dodo and solitaire	Recent	Mascarene Islands	0	2+
Passeriformes	Acanthisittidae	Several genera	New Zealand wrens	Pleistocene–Recent	New Zealand	3	1
	Atrichornithidae[f]	Atrichornis	Scrub-birds	Recent	New Zealand	0	2
	Menuridae[f]	Menura	Lyretails	Recent	Australia	0	2

[a] Inclusion among Paleognathiformes provisional.
[b] Classification tentative.
[c] Relationships of the three extinct genera—*Thambetochen* (2 spp.), *Ptaiochen* (1 sp.) and *Chelychelynechen* (1 sp.)—within the Anseriformes remain controversial.
[d] *Rollandia microptera*, *Podilymbus gigas* and *Podiceps taczanowskii*.
[e] Composition and position of this order remain controversial.
[f] Actually weakly flighted.
[g] Flightless species (many extinct) in a number of genera, including *Porphyrio* (swamphens), †*Nesotrochis*, †*Aphanapteryx*, †*Diaphorapteryx*, *Habroptila*, *Tricholimnas*, *Rallus* (typical rails), *Gallirallus* (wekas), *Atlantisia*, *Porzana* (crakes), *Amaurornis* (bush-hens), *Tribonyx* (water-hens) and *Fulica* (coots).

(Table 9.1; Livezey, 1989a). Flightlessness has evolved in at least three genera of grebes (Livezey, 1989b), nine different lineages of waterfowl (Livezey and Humphrey, 1986; Livezey, 1989c, 1990, 1993a), and several genera of auks (Livezey, 1988). Most other instances of flightlessness among carinates are limited to comparatively few species (Table 9.1), but include members of such diverse groups as the cormorants (Livezey, 1992a), parrots (Livezey, 1992b), and dodos (Livezey, 1993b).

Flightless carinates include a number of terrestrial species, mostly limited to islands, including the dodo, solitaire (*Pezophaps solitaria*), several ibises, and most flightless rails. Flightless carinates inhabiting marine coastlines include(d) surface-feeding waterfowl like the Auckland Islands teal (*Anas aucklandica*), foot-propelled diving species like the steamer-ducks (*Tachyeres* spp.) and Galápagos cormorant (*Compsohalieus harrisi*), and pelagic, wing-propelled diving species like the great auk (*Pinguinus impennis*) and the penguins. Flightless grebes, also foot-propelled diving birds, inhabit(ed) continental, freshwater lakes. Several flightless birds are capable of arboreal climbing, including the New Guinea flightless rail (*Amaurornis ineptus*) and a nocturnal parrot, the kakapo (*Strigops habroptilus*).

HETEROCHRONY AND THE EVOLUTION OF RATITES

Historical perspective

Degeneration and ontogeny
Originally grouped taxonomically on the basis of their truly "keel-less" sterna (Merrem, 1813) and short wings (Lesson, 1831), ratites are unique among birds in a variety of morphological characters, including variably "loose" plumage structure, synostosis of the scapula and coracoid, substantially reduced pectoral appendages, and powerfully developed pelvic limbs (e.g., Beddard, 1898; Pycraft, 1900; Glutz von Blotzheim, 1958; McGowan, 1982; Raikow, 1985). Most early naturalists, including Darwin (1859), attributed the diminutive pectoral limb of ratites to "disuse" or "degeneration". Other anatomists also interpreted the "degenerate" characters of ratites as the products of truncated development (T.J. Parker, 1882; Newton, 1896; Duerden, 1920; Gregory, 1935).

Eventually, the idea that adult ratites retained "juvenile" anatomical characters gained support, and the hypothesis was extended beyond the pectoral appendage to the cranium, pelvic girdle, and integument (de Beer, 1930, 1956, 1975; McDowell, 1948; Webb, 1957). Thus commenced an ornithological tradition of equating "neoteny" with the "process of retention of juvenile characters" (James and Olson, 1983, p. 31), and this term is frequently considered to include any process or outcome of heterochrony in birds (Feduccia, 1980; James and Olson, 1983; Olson, 1985). Under currently accepted nomenclature, however, most ornithological instances represent the *result* "paedomorphosis"; the underlying heterochronic *mechanism*, one plausible candidate being neoteny, generally remains undiagnosed. A growing body of evidence led to the traditional orthodoxy expousing the "neoteny" of ratites and the current consensus that the ratites evolved from flighted ancestors, thus establishing that flightlessness among Paleognathiformes is a derived condition and probably involves heterochrony.

Lowe's quandary, or "ontogeny vs. phylogeny"
An especially illuminating ornithological example of the intimate conceptual relationship between recapitulation and heterochrony is provided by the writings of Percy Roycroft Lowe, curator in the Bird Room at the British Museum (Natural History) during 1919–1935. Although vestiges of recapitulationist theory lingered in the ornithological writings of some prominent ornithologists through the turn of the century and many considered the ratites to be anatomically "archaic" or (in part) "reptilian", the relevance of recapitulation to the evolution of ratites remained controversial (W.K. Parker, 1888a; Newton, 1896; Beddard, 1898; Wiglesworth, 1900; Heilmann, 1927). Lowe (1926) initially was provoked by the Lamarckian interpretations by Duerden (1920) of the anatomy of the ostrich (*Struthio camelus*), in which a number of characters also were described as "degenerative". As evident from his earliest studies of avian anatomy, Lowe was a die-hard recapitulationist (*sensu* Haeckel, 1866). Accordingly, Lowe (1928a, p. 187) instead interpreted the retention of cranial sutures in ratites as:

> . . . a belated manifestation of a reptilian character in the "Ratite" skull.

Lowe (1928a, p. 245) recognized that:

> . . . the adults of the existing Struthiones are clothed in prepennal down and have not reached a much more advanced stage of development than the downy chick of a fowl.

However, he concluded (Lowe, 1928a, p. 244) that:

> . . . the Struthiones represent a perfectly natural group descended from some common ancestor which left the main avian stem before flight had been attained.

This interpretation of the "primary" nature of the flightlessness of ratites, combined with ancillary anatomical work on other avian groups, led Lowe (1933, p. 533) to define three subclasses of Recent birds, all of which he thought had ". . . specialized independently from a common generalized ancestor". These were:

> (1) a true aquatic, represented solely by the Penguin; (2) a cursorial, represented by such forms as the "Ostriches"; (3) a flying, represented by the Carinate division of birds.

Lowe (1935, p. 411) clarified his recapitulationist interpretation of the evolution of ratites, one combined with an almost Aristotelean reverence for avian flight:

. . . flight in birds was only a comparatively recent crowning feat in their evolution, and that for infinitely the greater part of the total time consumed since their first appearance birds may be regarded as having been flightless. . . . The flying, or the highest type of bird . . . seem[s] to have appeared . . . toward the end of the Cretaceous . . . But in some groups of birds, such as the Struthionids and the Penguins, may . . . have missed the evolutionary bus.

Lowe (1935, p. 412) concluded:

In the Struthiones the entire make-up of the adult is chick-like and reflects the beginning of the story of avian evolution instead of the end.

Lowe (1935, p. 414) considered the scapulocoracoid of ratites to be of "definitely reptilian" form.

Lowe (1942, p. 6) continued to make important anatomical observations on ratites, and cited the ". . . elementary simplicity of a high degree" of a wing muscle, the "Flexor carpi ulnaris" [M. expansor secundariorum] of *Rhea*, but interpreting the genus as:

. . . a simple, primitive, and blind-alley phase of a process of evolution which in other birds was eventually to end in the more complicated flying technique seen in the more highly placed Neognathæ.

Lowe (1942, pp. 6–7) also explicitly challenged the hypothesis of the "neoteny" of ratites:

. . . if their morphology could be properly ascribed to a process of Neoteny, . . . [this muscle] would have presented . . . the normal youthful conditions seen in the just hatched chick of an ordinary flying bird . . . and would have comparable in morphological details with those of the adult bird. But in *Rhea*, . . . the muscle bears little or no resemblance to that seen in the normal Neognathous chick, flying or non-flying . . . There is no indication in it of a possibility of flight in the ancestor or . . . owner of the wing, when the adult stage had been reached.

Lowe (1942, p. 9) noted that in *Rhea*:

. . . the stage of development in the wing . . . *considerably antedates in point of evolution any mere juvenile phase in the Neognathæ* [emphasis in original].

This view of primitive flightlessness of ratites led Lowe (1944: p. 518) to conclude that the feathers of *Archaeopteryx* were not strictly homologous with those of modern Aves. As for the theory of continental drift and the biogeography of ratites, Lowe (1944, pp. 521–2) preferred the scenario of the wide distribution and relictual persistence of primitively flightless "struthious stock" (either prior or subsequent to movement of continental plates) to the alternative of derived avian flightlessness, which he believed required that:

. . . at some time or other every family, genus or species, no matter how wide their geographical separation, all, without exception, *independently* lost the power of flight and became degenerate.

Contemporary opposition to Lowe's theory regarding the ratites was strong, although many of the "primitive" characters cited by Lowe were used by his critics to support the alternative hypothesis of heterochrony (e.g., McDowell, 1948). Murphy (*in* Gregory, 1935, p. 12) wrote:

> Dr Lowe regards both of these characteristics [absence of apterylae and barbicels] as ancestral and primitive . . . that, in the case of the ostrich-like birds, feather evolution has not proceeded much beyond an embryonic or "early avian" stage . . . The result may rather have been attained through the actual dropping out or cutting off of the final stage of feather development somewhere along the line of the ostrich's phylogeny.

Tucker (1938, p. 224) submitted similar arguments:

> Doubtless these ancestral forms were a good deal more primitive than present-day birds, and some of the Ratite characters, like the palaeognathous palate, may be a direct legacy from them. But . . . the conclusion seems that . . . [they] . . . are not in fact primitive in the phylogenetic sense, but are literally embryonic characters carried over into the adult, constituting a striking instance of neoteny or retarded development.

The hypothesis of the "primary flightlessness" of ratites championed by Lowe won support from several other anatomists (Holmgren, 1955; Glutz von Blotzheim, 1958). Although Lowe's hypothesis remained a minority view, it continued to have some influence on subsequent opinion. McGowan (1982, 1986) reconsidered the possible primitiveness of flightlessness in ratites based on an assessment of the evolutionary changes that would be necessary to transform a fully flighted bird into one possessing ratite-like characteristics. Also, McGowan (1984, p. 735) cited Lowe (1928a, 1935) in support of the conclusion that:

> Indeed, most of the features that distinguish ratites from carinates are now considered to be primitive rather than derived.

Toward synthesis

Anatomical and embryological research on ratites (e.g., Owen, 1842; T.J. Parker, 1892; Lutz, 1942; Frank, 1954; Lang, 1956; Webb, 1957; Müller, 1963) complemented studies of development in other birds (e.g., W.K. Parker, 1888b; Steiner, 1922; Fell, 1939; Montagna, 1945; Klima, 1962; Sullivan, 1962), and led to broad support for the importance of heterochrony in the ratites. de Beer (1956, 1975) became a leading proponent of this hypothesis, citing (1956, p. 65):

> . . . the presence in the Ratites of nestling-down, permanent sutures between the bones of the skull, and the dromaeognathous [paleognathous] structure of the palate . . . [as] demonstrably the result of neoteny or the secondary retention of features which were juvenile in the ancestors of the Ratites.

de Beer (1956, p. 66-7) concluded:

> For those, if there be any, who still believe in the theory of recapitulation, it would no doubt be tempting to say that the neognathous palate "recapitulates" in its development the condition of the paleognathous palate which would therefore be

ancestral. But in view of the overwhelming evidence that the Ratites are secondarily descended from flying birds . . . [and] the fact that the Ratites already show neoteny in two other features . . . it is impossible to believe that in their palates the Ratites are primitive.

Cracraft (1974, 1986) presented phylogenetic analyses of the ratites, in which the evolution of flightlessness was inferred (at least on the grounds of evolutionary parsimony) to have occurred only once (in the common ancestor of the group). Although differing in some topological findings, a phylogenetic analysis by Bledsoe (1988) also supported the monophyly of the ratites, and patterns of molecular (phenetic) similarity are consistent with this hypothesis (Prager *et al.*, 1976; Sibley and Ahlquist, 1990; Stapel *et al.*, 1984). Several other authors, however, questioned the monophyly of the ratites (Houde and Olson, 1981; McGowan, 1984; Feduccia, 1980, 1985; Olson, 1985; Bock and Bühler, 1990), raising doubts concerning the homology of the anatomical characters associated with flightlessness in the included families. None of the latter critics, however, proposed explicit phylogenetic alternatives to the hypothesis of monophyly.

Significant disagreement concerning the polarities and ontogeny of the distinctive characters of ratites remains. Unfortunately, some of the debate derives less from substantive differences in inference than from confusion concerning the concept of homology. Cracraft (1981, p. 689) correctly emphasized that:

> . . . an argument of neoteny is a *prima facie* admission that the similarities in question are derived and not primitive, that is, the adult condition of the ancestor is "replaced" in the descendant by the juvenile condition.

This does not imply that any two similar "neotenic" states are necessarily synapomorphic, i.e., *shared* derived characters, but that both are derived and *possibly* homologous. In a discussion of the implications of recapitulation for the anatomy of ratites, however, Feduccia (1980, p. 135) confused truly primitive characters with derived, pseudo-primitive characters produced secondarily through heterochrony. Similarly, arguments by Olson (1985, p. 100) concerning "neotenic" characters of ratites indicate confusion on both the issues of homology and polarity.

Even where the derived nature of "neotenic" characters is appreciated, there is a tradition of assuming that such characters evolve "willy-nilly" in flightless birds, nullifying their utility in phylogenetic inference (e.g., Olson, 1973a,b, 1982). Feduccia (1980, p. 135) expressed an optimism concerning the evolutionary power of heterochrony by suggesting that the neognathous order ". . . Gruiformes would be the best candidates for some of the various forebears. . . ." of the "neotenic forms" of the ratites. Olson (1983) offered an extreme view of the potential of heterochrony when he implied that "neoteny" might have produced the kiwis and moas independently from different carinate groups (ibises and geese, respectively).

Impugning the phylogenetic informativeness of "neotenic" characters through overstatement is also typical. In a critique of the hypothesis of ratite monophyly, Feduccia (1985, p. 186) asserted:

> Following Cracraft's use of neotenic characters as shared, derived features, we might logically include all flightless rails in the same genus as they all possess reduced wings, a keel-less sternum and an open ilioischiatic fenestra. . . .

Setting aside the inaccuracies of the anatomical generalizations (adult flightless rails are neither strictly acarinate nor possess open ilioischiatic fenestrae) and the unsupported suggestion that monophyly would necessarily imply membership in the same *genus*, this statement implies that either one must consider *all* "neotenic" characters as homologous or consider them *all* to be convergent. The truth, however, probably lies somewhere between these two extremes. Even the harshest critics of "neotenic" characters in phylogenetic reconstruction would admit the homology of the many flightlessness-related characters of subgroups of ratites, e.g., those uniting the kiwis (T.J. Parker, 1892; Pycraft, 1900; McGowan, 1982).

For those characters generally acknowledged to have been the result of heterochrony, imprecision prevails regarding the mechanisms and organ systems involved. Related to the tradition of synonymizing paedomorphosis (perhaps heterochrony) with neoteny in regard to paleognathous birds, differential diagnoses among the three ontogenetic mechanisms producing paedomorphosis (neoteny, postdisplacement and progenesis) and the three counter-mechanisms resulting in peramorphosis (acceleration, predisplacement and hypermorphosis) are all but lacking in the ornithological literature. The diminutive pectoral apparatus and delay in closure of cranial sutures of ratites may, in fact, involve neoteny, especially in light of the retarded progress of the latter until several years after hatching (Elzanowski, 1988; Beale, 1991). Also, the obtuse scapulo-coracoid angle shared by adult ratites and embryonic carinates also suggests neoteny *sensu stricto* (Nauck, 1930; Kaelin, 1941), but postdisplacement has not been empirically excluded. The radically derived anatomy of ratites, however, defies the proposal of any single heterochronic mechanism as completely explanatory (Table 9.2). Levinton (1988) rightly emphasized the potentially complex interactions of diverse heterochronic perturbations in different organ systems within lineages, and it is evident that most or all ratites manifest (at least) peramorphosis of the pelvic limb as well as paedomorphosis of the pectoral limb (McKinney and McNamara, 1991). Elzanowski (1988) critiqued many supposed examples of "neoteny" in ratites, including several features of the skull, pectoral limb, pelvis, and integument, and proposed that some of these instead may be the result of peramorphosis or may simply be plesiomorphous. Critical data for the diagnosis of the ontogenetic basis of the unique morphology of ratites are lacking; however, it is clear that several developmental mechanisms are involved, producing a complex constellation of variably localized heterochronic changes in skull, pectoral and pelvic appendages, and integument (Table 9.2).

Table 9.2 Changes in body size, plumage, skull, appendages, and underlying heterochronic mechanisms hypothesized in selected flightless birds.

Species	Body size	Plumage	Skull	Girdle and appendage		Hypothesized heterochronic mechanism(s)
				Pectoral	Pelvic	
Ratites	Increase[a]	Paedomorphosis, peramorphosis	Paedomorphosis, peramorphosis	Paedomorphosis, peramorphosis	Paedomorphosis, peramorphosis	Neoteny of feather structure, hypermorphosis of pterylosis; possible localized paedomorphosis of palate and cranial sutures, possible generalized cephalic hypermorphosis; neoteny or hypermorphosis of pectoral girdle, neoteny of pectoral limb; neoteny and/or hypermorphosis of pelvic girdle, hypermorphosis of pelvic limb
New Zealand geese	Increase	—[b]	Peramorphosis	Paedomorphosis	Peramorphosis	—[b]
Flightless steamer-ducks	Increase	No change	No change	Paedomorphosis	No change	Pectoral postdisplacement[c]
Auckland Islands teal	Decrease	Paedomorphosis	No change	Paedomorphosis	No change	Generalized progenesis
Auckland Islands merganser	No change	Paedomorphosis	Peramorphosis	Paedomorphosis	No change	Neoteny of plumage pattern and pectoral apparatus
Titicaca grebe	Increase	No change	Peramorphosis	Paedomorphosis	Peramorphosis	Pectoral postdisplacement,[c] cranial and pelvic hypermorphosis
Galápagos cormorant	Increase	Paedomorphosis	Peramorphosis	Paedomorphosis	Peramorphosis	Neoteny of plumage pattern and pectoral apparatus, cranial and pelvic hypermorphosis
Adzebills	Increase	—[b]	Peramorphosis	Paedomorphosis	Peramorphosis	—[b]
Weka[d]	Increase	Negligible change	Negligible change	Paedomorphosis	Peramorphosis	Pectoral neoteny, pelvic hypermorphosis
Inaccessible Island rail[d]	Decrease	Paedomorphosis	No change	Paedomorphosis	Peramorphosis	Plumage and pectoral progenesis, pelvic hypermorphosis
Kakapo	Increase	Indeterminate	No change	Paedomorphosis	Peramorphosis	Pectoral neoteny, pelvic hypermorphosis
Dodo	Increase	Paedomorphosis	Peramorphosis	Paedomorphosis	Peramorphosis	Plumage and pectoral neoteny, cranial and pelvic hypermorphosis

[a] Evolutionary size increases inferred for most ratites may not apply to kiwis; small size of Apterygidae may be primitive or represent a derived reversal.
[b] Subfossil remains only, characters of integument unknown, heterochronic mechanisms not diagnosable.
[c] Inference tentative, based on static morphometric analysis.
[d] Currently under study, inferences preliminary.

HETEROCHRONY AND THE EVOLUTION OF FLIGHTLESS CARINATES

Historical perspective

Penguins and Lowe's "primary" flightlessness
The anatomical peculiarities of penguins were widely appreciated by the early twentieth century, but early research on the ontogeny of penguins produced diverse interpretations. For example, Wray (1887, p. 353) found that:

> The embryo of the Penguin shows in its wings no signs of being a degeneration or modification of the specialized flight-wing of other Carinates.

In contrast, Pycraft (1907, p. 18) concluded:

> The wing of the embryo penguin . . . will be found to differ remarkably from that of the adult, and entirely confirms the contention that the paddle of the modern Penguin has been derived from a functional flying wing . . . It agrees in all essentials with that of the adult flying bird more closely than at any other later stage of development.

True to his recapitulationist perspective, Lowe (1933, p. 522) interpreted the ontogeny of penguins differently:

> We find nothing . . . resembling an approach to the morphological details to be noted in a similar series proper to the embryo of a flying bird.

Lowe (1933, p. 534) concluded:

> . . . while the Penguin seems to have specialized *directly* from the primitive non-flying generalized ancestor, all other swimming and diving birds are neither more nor less than flying Carinate types, and are, in fact, what one might term only pseudo-aquatic [emphasis in original].

As with Lowe's interpretation of the ratites, criticism of his analysis of flightlessness in the penguins was forthcoming. Murphy (*in* Gregory, 1935, pp. 14–15) wrote:

> Dr Lowe . . . sees through his study of the penguin feather the outcome, not of a process of degeneration, but one of failure to develop. . . . The arrangement of the wing feathers in the penguins may be said to be larval in character, with specialities superimposed . . . Many penguin characters, including those of feathers, are doubtless to a certain extent larval, but this by no means precludes the likelihood that in earlier stages this same feather structure may have been succeeded by others which have since dropped away.

Wiman and Hessland (1942) and Simpson (1946) rejected Lowe's (1933) recapitulationist view of the flightlessness of penguins, arguing instead that the Sphenisciformes were derived from an ancestor capable of aerial flight. Also, Simpson (1946) described a partial reversal in tarsometatarsal fusion in modern penguins (compared with Miocene penguins), a shift that McDowell (1948, p. 540) credited to "neoteny".

Other flightless carinates and Lowe's "secondary" flightlessness
Between the 19th century series of publications treating the anatomy of the
extinct dodo and solitaire (e.g., Strickland and Melville, 1848; quoted above) and
the early part of the twentieth century, possible ontogenetic mechanisms of
flightless carinates exclusive of the penguins received relatively little attention.
Typical of the period were the references by Newton (1896) to "disuse" leading to
an "aborted condition" of carina sterni of the flightless parrot, the kakapo.
Similar observations were made for the takahe (Rallidae: *Porphyrio mantelli*) by
T.J. Parker (1882) and the Galápagos cormorant by Gadow (1902). A notable
exception was the study of the diminutive, flightless Inaccessible Island rail
(Rallidae: *Atlantisia rogersi*) by Lowe (1928b). Lowe (1928b, pp. 105–6) explicitly
considered ontogenetic mechanisms, and wrote:

> . . . the almost wholly black coloration characteristic of the chicks of the Ralline
> family is apparently retained for a much longer time than is usual in immature
> examples, would seem to suggest that *Atlantisia* is a generalised and so presumably
> a near representative of some more primitive type. In the most adult example . . .
> there is still a very evident air of immaturity . . . borne out by the discovery that the
> remiges or wing-feathers, as regards the development of the rami and radii, exhibit
> microscopically a stage of evolution which has not advanced beyond that seen in the
> body or contour feathers of such a volant form as *Rallus aquaticus*.

Lowe (1928b, pp. 108–9) continued:

> The view might be taken that such a condition is an indication of an arrest or a
> retardation of development . . . At any rate, the almost invariable explanation of the
> morphological changes present in flightless rails or other birds is that they are
> simple manifestations of degeneration or retrogression resulting from mere disuse of
> the wing . . . merely secondary changes practically akin to atrophy.

Lowe (1928b) reserved "secondary" flightlessness among Rallidae to those in
which he found no qualitative anatomical changes, especially in feather structure
(e.g., *Dryolimnas*, *Porphyriornis* and *Porzana*), and noted the relationship
between flightlessness and the generally slow pectoral development of rails. Thus
Lowe (1928b) resurrected Fürbringer's (1888) distinction between flightless birds
having no flighted ancestors (*Prot-Apterornithes*) and those having secondarily
lost flight (*Deuter-Apterornithes*). Fürbringer (1888), however, had reserved the
former for ancient, largely hypothetical proto-birds (*Protorthornithes*) and
included the ratites (which he believed to be polyphyletic) in the latter group
(Newton, 1896).
    Subsequently, Lowe (1934) included the flightless steamer-ducks in the
"secondary" group, inferring (p. 482):

> . . . that the rate of growth of the wing of the non-flying embryo has either been
> already relatively retarded before hatching or the rate of general body growth
> relatively accelerated.

Lowe (1935, pp. 411–12) contrasted the condition of flightless carinates with that
of ratites:

. . . when a carinate flying bird becomes flightless it still remains . . . essentially similar in anatomical details to the generality of flying birds. Even its wings do not undergo essential change, except as regards a simple atrophy relative to the rest of the body. A diminution or want of development in size may, of course, occur in the sternum, fore-arm, and muscles, but in such flightless birds such as the ocydromine Rails of New Zealand the various component parts of the remigial and body-feathers—barbs, barbules, and barbicels—all reflect the essential structure and disposition of the normal carinate flying and body feathers, except that their development may have been inhibited, and so caused to lag behind in a juvenile phase.

Lowe (1935, pp. 424–8) extended "secondary flightlessness" to some rallids (now evidently including *Gallirallus, Diaphorapteryx* and *Atlantisia*), the dodo and solitaire, and the Galápagos cormorant, but his discussion of "temporary" and "permanent" retardation of growth indicates a misunderstanding of the physiological causes of paedomorphosis. Lowe (1935, p. 430) reaffirmed that although neoteny may have contributed to the pectoral retardation of flightless steamer-ducks and rails:

. . . the inhibiting factor has not converted them from a neognathous to a palæognathous condition. They present nothing approaching to the larval make-up of the Ostrich.

In one of his last papers on avian flightlessness, Lowe (1942, p. 7) revealed a reluctance to credit even "secondary flightlessness" to neoteny (narrowly conceptualized), writing:

It may be that some would explain these cases of secondary flightlessness in Neognathous birds by the phenomenon of Neoteny, but by whatever name we may choose to designate them, I am driven to believe that they will ultimately be proved to be due either to the action of some *permanent* inhibitory growth-factor or hormone, such as acts *temporarily* as a normal occurrence in the chicks of Ducks and Rails, or, on the other hand, to the absence of some normal stimulating growth-factor [emphasis in original].

Toward synthesis

Descriptive support for a "neotenic" basis of flightlessness in rails and other flightless land birds continued (e.g., Olson, 1973a,b; Feduccia, 1980; James and Olson, 1983), and resemblances between adult flightless birds and subadult flighted species provided new insights into the ontogeny of flightlessness. These correspondences include: reduction of the carina and incomplete development of the caudal margin of the sternum (T.J. Parker, 1882; Lowe, 1928b, 1934; Fell, 1939; Klima, 1962); obtuse scapulocoracoid angles (Nauck, 1930; Kaelin, 1941); and disproportionate reduction of the distal wing elements relative to proximal elements (Gadow, 1902; Steiner, 1922; Montagna, 1945; Sullivan, 1962; Levinton, 1988).

Contrary to the prominence accorded penguins in early studies of the ontogeny of flightlessness, the possible role of heterochrony in the evolution of flightlessness in wing-propelled diving birds (by any name) has received minimal attention in recent decades. Some of this disinterest reflected an assumption that "neoteny" necessarily produces an underdeveloped wing. Olson (1977, p. 690), in a discussion of flightless auks, stated:

> The great modifications seen in the wing . . . are not the result of neoteny, as seen in many other flightless birds . . . . Instead, these modifications represent highly derived specializations for wing-propelled diving.

However, it is now appreciated that neoteny and other heterochronic mechanisms leading to paedomorphosis are not limited in effect to simple underdevelopment, but can generate important innovations (McKinney and McNamara, 1991). Some of the morphological specializations of flightless wing-propelled diving birds that have obscured possible paedomorphosis (e.g., proportions within the wing skeleton) may themselves be the result of heterochrony (Livezey 1988, 1989a).

An overly narrow view of "neoteny" also influenced expectations concerning the timing and generality of the resultant morphological effects and this led to speculations concerning likely candidates for "neotenic" flightlessness based on general developmental differences among taxonomic groups. Feduccia (1980, p. 112) assumed that "neoteny" necessarily affects all organ systems equally. The possibility of localized paedomorphosis of the pectoral apparatus seems not to have been considered, although this appears to characterize several other flightless carinates (Livezey and Humphrey, 1986; Livezey, 1992a,b, 1993a). Moreover, the discovery by Poplin and Mourer-Chauviré (1985) of a large, extinct flightless megapode from New Caledonia indicates that paedomorphosis targeting specific organ systems also occurs in the Galliformes.

Further study of flightless carinates has indicated a substantial diversity of underlying heterochronic change, although diagnostic examination of ontogeny has yet to be accomplished for any taxonomic group (Table 9.2). In addition to neoteny, preliminary assessments indicate that pectoral paedomorphosis in several groups evidently may involve postdisplacement or progenesis (Livezey and Humphrey, 1986; Livezey, 1989b,c, 1990, 1992a,b, 1993a,b,c, 1994). Most instances of avian flightlessness are associated with increased body size (Table 9.2), and this in part reflects peramorphosis of regions other than the pectoral limb. Peramorphosis of the pelvic limb is especially conspicuous in some groups (McKinney and McNamara, 1991; Livezey, 1992a,b, 1993a,b, 1994), associated with a "compensatory" shift in appendicular specialization recognized by Gadow (1902) for the Galápagos cormorant. Andors (1988) proposed that the primary heterochronic mechanism that produced the extraordinary body form of the giant, flightless "terror cranes" (Diatrymidae) may have been hypermorphosis. Probably the most extreme pectoral paedomorphosis among carinates is that of the extinct gruiform *Apterornis* (Livezey, 1994). A diversity of heterochronic mechanisms and evolutionary trends in body size and sexual dimorphism is evident in flightless rails (Table 9.2; Livezey, in prep.).

Increased sexual size dimorphism also is associated with loss of flight in a number of species, and is especially marked in those lineages in which non-pectoral peramorphosis is indicated (Livezey and Humphrey, 1986; Livezey, 1989b, 1990, 1992a,b, 1993b). Furthermore, in the flying steamer-duck (Anatidae: *Tachyeres patachonicus*), a species with negative allometry between wing area and body mass, sexual differences in body size produce flightlessness in a minority of males in some marine populations, whereas the smaller females remain capable of flight (Humphrey and Livezey, 1982).

HETEROCHRONY AND AVIAN FLIGHTLESSNESS: FUTURE DIRECTIONS

Morphometric and embryological approaches

A revolution in the mathematical description of relative (allometric) growth was begun by Huxley (1932), and the quantitative nature of the ontogenetic processes involved in heterochrony was recognized soon thereafter (de Beer, 1940). The mechanisms underlying heterochrony and its morphological products are well summarized graphically (Tissot, 1988; McKinney and McNamara, 1991). Most studies of avian heterochrony, however, remain descriptive, non-quantitative, and based on static analyses of adults. A number of bivariate and multivariate approaches can provide useful numerical summaries of ontogenetic change and its relationship to interspecific differences among adults (Livezey, 1989b, 1990, 1992a, 1993c). Critical ornithological studies based on known-age developmental series of flighted and flightless relatives, including pre-hatch embryos and detailed records of ossification, muscle primordia, and pterylosis, have yet to be performed. Such research would be particularly powerful for an analysis of heterochrony in the ratites, for which the ontogeny of ostriches (currently the subject of an expanding commercial aviculture) would be most appropriate. Where possible, the physiological and behavioral corollaries of ontogeny should be examined as well.

James and Olson (1983, p. 31) envisioned a streamlined ontogeny of avian flightlessness:

> The genetic mechanism for the evolution of a flightless bird from a flying one is actually quite simple. All birds are flightless when they are small chicks, and the young of flying birds have the same features that characterize the adults of flightless birds . . . Merely by retaining the skeletomuscular structure of infancy into adulthood—probably by the alteration of a few regulatory genes—almost any bird species could become flightless.

The supposed role of regulatory genes in producing paedomorphosis, however, remains largely hypothetical and probably is overly simplistic. An understanding of the genetic bases of heterochrony in birds is crucial to an assessment of the evolutionary "ease" with which the resultant conditions (e.g., paedomorphic flightlessness) arise.

Recent work by Dawson *et al.* (1994) revealed that thyroidectomized starlings (Sturnidae: *Sturnus vulgaris*) attained sexual maturity but displayed several "juvenile" characters judged to be comparable to those found in *adult* ratites, including long contour feathers lacking barbicels, aborted palatal and rostral development, and incomplete closure of cranial sutures (although presence of the latter in adult ratites was contested by Elzanowski (1988)). Decades earlier, Edinger (1942) suggested that the giantism and plumage structure of ratites might be related to hyperactivity of the pituitary. Although the speculations by Edinger (1942) and Dawson *et al.* (1994) concerning the role of endocrinology in the evolution of modern ratites are less than convincing, these reports indicate the considerable promise that experimental embryology may hold for the elucidation of the ontogenetic processes underlying avian heterochrony. Combined with a new appreciation for the physiological implications of avian flightlessness (Calder and Dawson, 1978; Vleck *et al.*, 1980; Livezey, 1992b, 1993b), the extension of the study of heterochrony beyond simple anatomical comparison and into embryological and physiological dimensions is the next logical step (McNab, 1994).

Phylogenetic perspective on avian heterochrony

An understanding of phylogenetic relationships is critical for comparative studies and the detection of evolutionary change. From the ontogenetic standpoint, avian transitions to flightlessness provide multiple (hierarchically nested) evolutionary "replicates", in that loss of flight results from a variety of subtly and (in some cases) distinctly different perturbations in ontogeny. Late development of the pectoral appendage in birds makes paedomorphosis the most obvious ontogenetic inference in flightless birds, but as more detailed studies of avian flightlessness are made it is likely that several heterochronic mechanisms will be documented. For example, the direction of changes in body size and plumage dichromatism that accompanied paedomorphosis in the flightless Auckland Islands teal (Livezey, 1990, 1991) and Auckland Islands merganser (Livezey 1989c) were determinable only within the context of corroborated phylogenetic hypotheses (Livezey, 1990, 1991). Even if a phylogenetic hypothesis remains controversial, it permits the placement of the associated evolutionary inferences within an explicit historical context.

    Unfortunately, progress in the study of heterochrony in flightless birds has been hampered significantly by inadequately known phylogenetic relationships within the taxonomic groups concerned. Some of the difficulties arise from poor understanding of phylogenetic methodology. For example, McGowan (1984, p. 734) reported that carinates, unlike theropod dinosaurs and ratites, possess a (presumably derived) pretibial bone, and therefore concluded that "... they [carinates] could not have given rise to the ratites ..." If correct, this conclusion has important implications for the evolution of flightlessness and inferences concerning heterochrony in the group. Phylogenetic inferences based on single characters carry little weight, however, and can only be assessed credibly with other characters under the criterion of global parsimony of evolutionary change (wherein a reversal or loss of a specific character is a possibility). Also, there is a misleading ornithological tradition of treating diversity of form, regardless of the polarities of the characters involved, as evidence of distant relationship or polyphyly (Feduccia, 1980; Olson, 1985; Bock and Bühler, 1990), whereas such differences may simply reflect phylogenetically uninformative, autapomorphic differentiation (Cracraft, 1974). Some anatomical diversity in ratites (e.g., within the pelvis) may be the result of minor perturbations of shared ontogenetic trajectories within lineages (Elzanowski, 1988).

The dearth of explicit phylogenetic hypotheses has led to misunderstandings concerning evolutionary patterns and rates of change. For example, McGowan (1986, pp. 342–3) noted that pectoral changes evident in a flightless carinate, the weka (Rallidae: *Gallirallus australis*), were comparatively minor compared with those of ratites, and stated on questionable biogeographical grounds that: ". . . the Weka could have a genealogy dating back as long as that of ratites." Although the meaning of this statement is unclear, it would be precluded by a phylogenetic hypothesis for rails within which the capacity for flight could be mapped. Assumption of the polyphyly of ratites and the multiple losses of flight in rails has led to a widespread belief that "neoteny" can produce flightlessness rapidly (Olson, 1973a,b, 1982, 1985; James and Olson, 1983), although only the estimated ages of oceanic islands inhabited by endemic flightless birds are germane to questions of absolute rates of change, and these only provide upper limits.

Interpretations of individual characters, taken together, form the basis for understanding morphological evolution and phylogeny, and each character must be considered on its own merits. In this context, the assumption by de Beer (1956) that the palate of ratites is "neotenic" simply because there is evidence of neoteny in cranial sutures and feather structure is questionable, given the potential for different ontogenetic perturbations among different anatomical structures (McKinney and McNamara, 1991). Also, most ornithological citations of recapitulation remain vague. For example, Houde and Olson (1981, p. 1237) cited ". . . the existence of at least some of the features of the paleognathous palate in the early ontogeny of some neognathous birds" in support of the primitiveness of the paleognathous palate in ratites. The theory of recapitulation suffered, in part, from its early, uncritical popularity (Churchill, 1980), and the importance of the relationship between ontogenetic and phylogenetic transform-ation to evolutionary study is far from resolved (Løvtrup, 1978, 1989; Rieppel, 1990; Patterson, 1983; Mayr, 1994). There is evidence for limited recapitulation in ontogenetic sequences showing terminal additions (McKinney and McNamara, 1991) and this may facilitate the determination of character polarities (Nelson, 1978; de Quieroz, 1985).

The history of study of avian flightlessness, especially that of the ratites, underscores the reciprocal relationship between the study of ontogeny and phylogenetic reconstruction. Heterochrony is by definition an evolutionary problem and therefore is interpretable only within an historical context. A phylogenetic approach to heterochrony also reveals the hierarchy of ontogenetic evolution, which for Aves may range from the class-level neoteny hypothesized by Thulborn (1985) to heterochrony of morphology, physiology, and behaviour through progressively exclusive taxonomic groups. A phylogenetic approach may indicate that the "intermediate" metabolic rates of ratites (Calder and Dawson, 1978; Vleck *et al.*, 1980) are explainable by the basal phylogenetic position of the group (see Duncker, 1989). This possible plesiomorphy contrasts with the partial, possibly heterochronically produced reversals in metabolic parameters suggested in some flightless carinates, e.g., the kakapo and dodo (Livezey, 1992b, 1993b). Moreover, it may be that the best corroborated phylogenetic hypothesis would vindicate to varying extents each of the seemingly competing (but not mutually exclusive) theories for the evolution of ratites and their flightlessness, including primitiveness or intermediacy of some characters, reversals through paedomorphosis, and genuine novelties (perhaps shared by entire clades) associated with heterochrony.

ACKNOWLEDGMENT

This study was supported in part by National Science Foundation grant BSR-9396249.

REFERENCES

Andors, A.V., 1988, *Giant groundbirds of North America (Aves, Diatrymidae)*, Ph.D. Diss., Columbia University, New York.
Beale, G., 1991, The maturation of the skeleton of a kiwi (*Apteryx australis mantelli*)—a ten year radiological study, *J. Royal Soc. N.Z.*, **21**: 219–220.
Beddard, F.E., 1898, *The structure and classification of birds*, Longmans and Green, London.
Bledsoe, A.H., 1988, A phylogenetic analysis of postcranial skeletal characters of the ratite birds, *Ann. Carnegie Mus.*, **57**: 73–90.
Bock, W.J., 1963, The cranial evidence for ratite affinities. In C.G. Sibley (ed.), *Proceedings XIII International Ornithological Congress*, American Ornithologists' Union, Washington, D.C.: 39–54.
Bock, W.J. and Bühler, P., 1990, The evolution and biogeographical history of the palaeognathous birds. In R. van den Elzen, K.-L. Schuchmann and K. Schmidt-Koenig (eds), *Current topics in avian biology*, Deutschen Ornithologen-Gesellschaft, Bonn: 31–36.
Calder, W.A., III and Dawson, T.J., 1978, Resting metabolic rates of ratite birds: the kiwi and the emu, *Comp. Biochem. Physiol.* (A), **60**: 479–481.
Churchill, F.B., 1980, The modern evolutionary synthesis and the biogenetic law. In E. Mayr and W.B. Provine (eds), *The evolutionary synthesis: perspectives on the unification of biology*, Harvard University Press, Cambridge: 112–122.

Cracraft, J., 1974, Phylogeny and evolution of the ratite birds, *Ibis*, **116**: 494–521.

Cracraft, J., 1981, Toward a phylogenetic classification of the Recent birds of the world (Class Aves), *Auk*, **98**: 681–714.

Cracraft, J., 1986, The origin and early diversification of birds, *Paleobiology*, **12**: 383–399.

Darwin, C., 1859, *On the origin of species by means of natural selection, or the preservation of favoured races in the struggle for life*, John Murray, London.

Dawson, A., McNaughton, F.J., Goldsmith, A.R. and Degen, A.A., 1994, Ratite-like neoteny induced by neonatal thyroidectomy of European starlings, *Sturnus vulgaris*, *J. Zool. London*, **232**: 633–639.

de Beer, G., 1930, *Embryology and evolution*, Oxford University Press, Oxford.

de Beer, G., 1940, Embryology and taxonomy. In J.S. Huxley (ed.), *The new systematics*, Clarendon Press, Oxford: 365–393.

de Beer, G., 1956, The evolution of ratites, *Bull. Brit. Mus. (Natur. Hist.)*, **4**: 59–70.

de Beer, G., 1975, The evolution of flying and flightless birds, *Oxford Biol. Reader*, **68**:1–16.

de Queiroz, K., 1985, The ontogenetic method for determining character polarity and its relevance to phylogenetic systematics, *Syst. Zool.*, **34**: 280–290.

Duerden, J.E., 1920, Methods of degeneration in the ostrich, *J. Genet.*, **9**: 131–193.

Duncker, H.-R., 1989, Structural and functional integration across the reptile-bird transition: locomotor and respiratory systems. In D.B. Wake and G. Roth (eds), *Complex organismal functions: integration and evolution in vertebrates*, J. Wiley, Chichester: 147–169.

Edinger, E.H., 1942, The pituitary body in giant animals, fossil and living: a survey and a suggestion, *Quart. Rev. Biol.*, **17**: 31–45.

Elzanowski, A., 1988, Ontogeny and evolution of the ratites. In H. Ouellet (cd.), *Acta XIX Congressus Internationalis Ornithologici*, University of Ottawa Press, Ottawa: 2037–2046.

Feduccia, A., 1980, *The age of birds*, Harvard University Press, Cambridge.

Feduccia, A., 1985, The morphological evidence for ratite monophyly: fact or fiction. In V.D. Ilyichev and V.M. Gavrilov (eds), *Acta XVIII Congressus Internationalis Ornithologici*, Academy of Sciences, Moscow: 184–190.

Fell, H.B., 1939, The origin and developmental mechanics of the avian sternum, *Philos. Trans. Royal Soc. London (Ser. B)*, **229**: 407–464.

Frank, G.H., 1954, The development of the chondrocranium of the ostrich, *Ann. Univ. Stellenbosch*, **30**: 179–248.

Fry, C.H., 1983, The jacanid radius and *Microparra*. A neotenic genus, *Gerfaut*, **73**: 173–184.

Fürbringer, M., 1888, *Untersuchungen zur Morphologie und Systematik der Vögel, zugleich ein Beitrag zur Anatomie der Stütz—und Bewegungsorgane*, 2 Volumes, T.J. Van Holkema, Amsterdam.

Gadow, H., 1902, The wings and the skeleton of *Phalacrocorax harrisi*, *Novit. Zool.*, **9**: 169–176.

Gingerich, P.D., 1973, Skull of *Hesperornis* and early evolution of birds, *Nature*, **243**: 70–73.

Glutz von Blotzheim, U., 1958, Zur Morphologie und Ontogenese von Schultergürtel, Sternum und Becken von Struthio, Rhea und Dromiceius, *Rev. Suisse Zool.*, **65**: 609–772.

Gould, S.J., 1977, *Ontogeny and phylogeny*, Belknap Press, Cambridge.

Gregory, W.K., 1935, Remarks on the origins of the ratites and penguins [with discussion by R. C. Murphy], *Proc. Linnean Soc. New York*, **45–46**: 1–18.

Haeckel, E., 1866, *Generelle Morphologie der Organismen: Allgemeine Grundzüge der organischen Formen-Wissenschaft, mechanisch begründet durch die von Charles Darwin reformirte Descendenz-Theorie*, Reimer, Berlin.

Heilmann, G., 1927, *The origin of birds*, D. Appleton, New York.

Holmgren, N., 1955, Studies on the phylogeny of birds, *Acta Zool.*, **36**: 244–328.

Houde, P.W., 1988, *Paleognathous birds from the early Tertiary of the northern hemisphere*, Nuttall Ornithological Club, Cambridge.

Houde, P.W. and Olson, S.L., 1981, Paleognathous carinate birds from the early Tertiary of North America, *Science*, **214**: 1236–1237.

Humphrey, P.S. and Livezey, B.C., 1982, Flightlessness in flying steamer-ducks, *Auk*, **99**: 368–372.

Huxley, J.S., 1932, *Problems of relative growth*, Lincoln MacVeagh, New York.

James, H.F. and Olson, S.L., 1983, Flightless birds, *Nat. Hist.*, **92**: 30–40.

Kaelin, J., 1941, Über den Coracoscapularwinkel und die Beziehungen der Rumpfform zum Lokomotionstypus bei den Vögeln, *Rev. Suisse Zool.*, **48**: 553–557.

Klima, M., 1962, The morphogenesis of the avian sternum, *Pr. Brn. Zákl. Csl. Akad. Ved*, **34**: 151–194.

Lang, C., 1956, Das Cranium der Ratiten mit besonderer Berücksichtigung von *Struthio camelus*, *Z. Wiss. Zool.*, **159**: 165–224.

Lawton, M.F. and Lawton, R.O., 1986, Heterochrony, deferred breeding, and avian sociality. In R.F. Johnston (ed.), *Current ornithology*, Volume 3, Plenum, New York: 187–221.

Lesson, R.-P., 1831, *Traite d'ornithologie*, Volume 1, Levrault, Paris.

Levinton, J., 1988, *Genetics, paleontology, and macroevolution*, University Press, Cambridge.

Livezey, B.C., 1988, Morphometrics of flightlessness in the Alcidae, *Auk*, **105**: 681–698.

Livezey, B.C., 1989a, Morphometric patterns in Recent and fossil penguins (Aves, Sphenisciformes), *J. Zool. London*, **219**: 269–307.

Livezey, B.C., 1989b, Flightlessness in grebes (Aves, Podicipedidae): its independent evolution in three genera, *Evolution*, **43**: 29–54.

Livezey, B.C., 1989c, Phylogenetic relationships and incipient flightlessness of the extinct Auckland Islands merganser, *Wilson Bull.*, **101**: 410–435.

Livezey, B.C., 1990, Evolutionary morphology of flightlessness in the Auckland Islands teal, *Condor*, **92**: 639–673.

Livezey, B.C., 1991, A phylogenetic analysis and classification of Recent dabbling ducks (Tribe Anatini) based on comparative morphology, *Auk*, **108**: 471–507.

Livezey, B.C., 1992a, Flightlessness in the Galápagos cormorant (*Compsohalieus* [*Nannopterum*] *harrisi*): heterochrony, giantism, and specialization, *Zool. J. Linnean Soc.*, **105**: 155–224.

Livezey, B.C., 1992b, Morphological corollaries and ecological implications of flightlessness in the kakapo (Psittaciformes: *Strigops habroptilus*), *J. Morphol.*, **213**: 105–145.

Livezey, B.C., 1993a, Morphology of flightlessness in *Chendytes*, fossil seaducks (Anatidae: Mergini) of coastal California, *J. Vert. Paleontol.*, **13**: 185–199.

Livezey, B.C., 1993b, An ecomorphological review of the dodo (*Raphus cucullatus*) and solitaire (*Pezophaps solitaria*), flightless Columbiformes of the Mascarene Islands, *J. Zool. London*, **230**: 247–292.

Livezey, B.C., 1993c, Comparative morphometrics of *Anas* ducks, with particular reference to the Hawaiian duck *Anas wyvilliana*, Laysan duck *A. laysanensis*, and Eaton's pintail *A. eatoni*, *Wildfowl*, **44**: 74–99.

Livezey, B.C., 1994, The carpometacarpus of *Apterornis*, *Notornis*, **41**: 51–60.

Livezey, B.C. and Humphrey, P.S., 1986, Flightlessness in steamer-ducks (Anatidae: *Tachyeres*): its morphological bases and probable evolution, *Evolution*, **40**: 540–558.

Løvtrup, S., 1978, On von Baerian and Haeckelian recapitulation, *Syst. Zool.*, **27**: 348–352.

Løvtrup, S., 1989, Recapitulation, epigenesis and heterochrony, *Geobios Mém. Spéc.*, **12**: 269–281.

Lowe, P.R., 1926, On the callosities of the ostrich (and other Palæognathæ) in connection with the inheritance of acquired characters, *Ibis*, **68**: 667–679.

Lowe, P.R., 1928a, Studies and observations bearing on the phylogeny of the ostrich and its allies, *Proc. Zool. Soc. London*, **1928**: 185–247.

Lowe, P.R., 1928b, A description of *Atlantisia rogersi*, the diminutive and flightless rail of Inaccessible Island (southern Atlantic), with some notes on flightless rails, *Ibis*, **70**: 99–131.

Lowe, P.R., 1933, On the primitive characters of the penguins, and their bearing on the phylogeny of birds, *Proc. Zool. Soc. London*, **1933**: 483–541.

Lowe, P.R., 1934, On the evidence for the existence of two species of steamer-duck (*Tachyeres*), and primary and secondary flightlessness in birds, *Ibis*, **76**: 467–495.

Lowe, P.R., 1935, On the relationship of the Struthiones to the dinosaurs and to the rest of the avian class, with special reference to the position of *Archæopteryx*, *Ibis*, **77**: 398–432.

Lowe, P.R., 1942, Some additional anatomical factors bearing on the phylogeny of the Struthiones, *Proc. Zool. Soc. London*, **1942**: 1–20.

Lowe, P.R., 1944, An analysis of the characters of *Archæopteryx* and *Archæornis*. Were they reptiles or birds? *Ibis*, **86**: 517–543.

Lutz, H., 1942, Beitrag zur Stammesgeschichte der Ratiten: Vergleich zwischen Emu-Embryo und entsprechendem Carinatenstadium, *Rev. Suisse Zool.*, **49**: 299–399.

Marsh, O.C., 1880, *Odontornithes: a monograph on the extinct toothed birds of North America*, Rep. Geol. Expl. 40th Parallel, Washington, D.C.

Martin, L.D. and Tate, J., Jr., 1976, The skeleton of *Baptornis advenus* from the Cretaceous of Kansas, *Smithsonian Contr. Paleobiol.*, **27**: 35–66.

Mayr, E., 1994, Recapitulation reinterpreted: the somatic program, *Quart. Rev. Biol.*, **69**: 223–232.

Mayr, E. and Amadon, D., 1951, A classification of Recent birds, *Am. Mus. Novit.*, No. **1496**: 1–42.

McDonald, M.A. and Smith, M.H., 1990, Speciation, heterochrony, and genetic variation in Hispaniolan palm-tanagers, *Auk*, **107**: 707–717.

McDonald, M.A. and Smith, M.H., 1994, Behavioral and morphological correlates of heterochrony in Hispaniolan palm-tanagers, *Condor*, **96**: 433–446.

McDowell, S., 1948, The bony palate of birds. Part I, the Palaeognathae, *Auk*, **65**: 520–549.

McGowan, C., 1982, The wing musculature of the brown kiwi *Apteryx australis mantelli* and its bearing on ratite affinities, *J. Zool. London*, **197**: 173–219.

McGowan, C., 1984, Evolutionary relationships of ratites and carinates: evidence from ontogeny of the tarsus, *Nature*, **307**: 733–735.

McGowan, C., 1986, The wing musculature of the weka (*Gallirallus australis*), a flightless rail endemic to New Zealand, *J. Zool. London*, **210**: 305–346.

McKinney, M.L. and McNamara, K.J., 1991, *Heterochrony: the evolution of ontogeny*, Plenum Press, New York.

McNab, B.K., 1994, Energy conservation and the evolution of flightlessness in birds, *Am. Nat.*, **144**: 628–642.

Merrem, B., 1813, Tentamen systematis naturalis avium, *Abh. Königel. Akad. Wiss. Berlin (Physikal.)*, **1812–13**: 237–259.

Montagna, W., 1945, A re-investigation of the development of the wing of the fowl, *J. Morphol.*, **76**: 87–113.

Müller, H.J., 1963, Die Morphologie und Entwicklung des Craniums von *Rhea americana* Linné. II. Viszeralskelett, Mittelohr, und Osteocranium, *Z. Wiss. Zool.*, **168**: 35–118.

Nauck, E.T., 1930, Beiträge zur Kenntnis des Skeletts der paarigen Gliedaßen der Wirbeltiere. VII. Der Coracoscapularwinkel am Vogelschultergürtel, *Gegenbaurs Morphol. Jahrb.*, **64**: 541–557.

Nelson, G.J., 1978, Ontogeny, phylogeny, paleontology, and the biogenetic law, *Syst. Zool.*, **27**: 324–345.

Newton, A., 1896, *A dictionary of birds*, 4 Volumes, Adam and Charles Black, London.

Olson, S.L., 1973a, A classification of the Rallidae, *Wilson Bull.*, **85**: 381–416.

Olson, S.L., 1973b, Evolution of the rails of the South Atlantic islands (Aves: Rallidae), *Smithsonian Contr. Zool.*, **152**: 1–53.

Olson, S.L., 1977, A great auk, *Pinguinis* [sic], from the Pliocene of North Carolina (Aves: Alcidae), *Proc. Biol. Soc. Washington*, **90**: 690–697.

Olson, S.L., 1982, A critique of Cracraft's classification of birds, *Auk*, **99**: 733–739.

Olson, S.L., 1983, Lessons from a flightless ibis, *Nat. Hist.*, **92**: 40.

Olson, S.L., 1985, The fossil record of birds. In D.S. Farner and J.R. King (eds), *Avian biology*, Volume 8, Academic Press, New York: 79–238.

Owen, R., 1842, Monograph on *Apteryx australis*, including its myology, *Proc. Zool. Soc. London*, **1842**: 22–41.

Parker, T.J., 1882, On the skeleton of *Notornis mantelli*, *Trans. N.Z. Inst.*, **14**: 245–258.

Parker, T.J., 1892, Observations on the anatomy and development of the *Apteryx*, *Philos. Trans. Royal Soc. London* (Ser. B), **182**: 125–134.

Parker, W.K., 1888a, On the remnants or vestiges of amphibian and reptilian structures found in the skulls of birds, both Carinatæ and Ratitæ, *Proc. Philos. Soc. London*, **43**: 397–402.

Parker, W.K., 1888b, On the structure and development of the wing in the common fowl, *Philos. Trans. Royal Soc. London* (Ser. B), **179**: 385–598.

Parkes, K.C. and Clark, G.A., Jr., 1966, An additional character linking ratites and tinamous, and an interpretation of their monophyly, *Condor*, **68**: 459–471.

Patterson, C., 1983, How does phylogeny differ from ontogeny? In B.C. Goodwin, N. Holder and C.C. Wylie (eds), *Development and evolution*, Cambridge University Press, Cambridge: 1–31.

Poplin, F. and Mourer-Chauviré, C., 1985, *Sylviornis neocaledoniae* (Aves, Galliformes, Megapodidae), oiseau géant éteint de l'Ile des Pins (Nouvelle-Calédonie), *Geobios*, **18**: 73–97.

Prager, E.M., Wilson, A.C., Osuga, D.T. and Feeney, R.E., 1976, Evolution of flightless land birds on southern continents: transferrin comparison shows monophyletic origin of ratites, *J. Mol. Evol.*, **8**: 283–294.

Pycraft, W.P., 1900, On the morphology and phylogeny of the Palaeognathæ (Ratitæ and Crypturi) and Neognathæ (Carinatæ), *Trans. Zool. Soc. London*, **15**: 149–290.

Pycraft, W.P., 1907, On some points in the anatomy of the emperor and Adélie penguins, *Nat. Antarct. 1901–04 Exped. Rep. (Vert. 3)*, *Zool.*, **2**: 1–36.

Raikow, R.J., 1985, Locomotor system. In A.S. King and J. McLellund (eds), *Form and function in birds*, Volume 3, Academic Press, London: 57–147.

Rieppel, O., 1990, Ontogeny—a way forward for systematics, a way backward for phylogeny, *Biol. J. Linnean Soc.*, **39**: 177–191.

Sibley, C.G. and Ahlquist, J.E., 1990, *Phylogeny and classification of birds: a study in molecular evolution*, Yale University Press, New Haven.

Simpson, G.G., 1946, Fossil penguins, *Bull. Amer. Mus. Natur. Hist.*, **87**: 1–99.

Stapel, S.O., Leunissen, J.A.M., Versteeg, M., Wattel, J. and de Jong, W., 1984, Ratites as oldest offshoot of avian stem—evidence from α-crystallin A sequences, *Nature*, **311**: 257–259.

Steiner, H., 1922, Die ontogenetische und phylogenetische Entwicklung des Vogelflügel-skelettes, *Acta Zool.*, **25**: 307–360.

Storer, R.W., 1971, Classification of birds. In D.S. Farner and J.R. King (eds), *Avian biology*, Volume 1, Academic Press, New York: 1–18.

Strickland, H.E. and Melville, A.G., 1848, *The dodo and its kindred: or the history, affinities, and osteology of the dodo, solitaire, and other extinct birds of the islands Mauritius, Rodriguez, and Bourbon*, Reeve, Benham, and Reeve, London.

Sullivan, G.E., 1962, Anatomy and embryology of the wing musculature of the domestic fowl (*Gallus*), *Aust. J. Zool.*, **10**: 458–518.

Thulborn, R.A., 1985, Birds as neotenous dinosaurs, *Rec. N.Z. Geol. Surv.*, **9**: 90–92.

Tissot, B.N., 1988, Multivariate analysis. In M.L. McKinney (ed.), *Heterochrony in evolution: a multidisciplinary approach*, Plenum Press, New York: 35–51.

Tucker, B.W., 1938, Some observations on Dr. Lowe's theory of the relationship of the Struthiones to the dinosaurs and to other birds. In F.C.R. Jourdain (ed.), *Proceedings of the Eighth International Ornithological Congress*, University Press, Oxford: 222–224.

Vleck, D., Vleck, C.M. and Hoyt, D.F., 1980, Metabolism of avian embryos: ontogeny of oxygen consumption in the rhea and emu, *Physiol. Zool.*, **53**: 125–135.

Webb, M., 1957, The ontogeny of the cranial bones, cranial peripheral and cranial parasympathetic nerves, together with a study of the visceral muscles of *Struthio, Acta Anat.*, **38**: 81–203.

Wiglesworth, J., 1900, Inaugural address on flightless birds, *Trans. Liverpool Biol. Soc.*, **14**: 1–33.

Wiman, C. and Hessland, I., 1942, On the garefowl, *Alca impennis* L., and the sternum of birds, *Nova Acta Reg. Soc. Sci. Upsaliensis* (Ser. 4), **13**: 1–28.

Wray, R.S., 1887, On some points in the morphology of the wings of birds, *Proc. Zool. Soc. London*, **1887**: 343–357.

Part Three

HETEROCHRONY AND ECOLOGY

Chapter 10

CAUSES AND CONSEQUENCES OF HETEROCHRONY IN EARLY ECHINODERM DEVELOPMENT

Gregory A. Wray

INTRODUCTION

Heterochronic changes in adult morphology form a pervasive and important component of evolutionary change (de Beer, 1940; Gould, 1977; McKinney and McNamara, 1991). Much less is known about heterochronies in early life history phases, such as embryos and larvae (Raff and Wray, 1989). It has long been known, and is becoming increasingly apparent, that heterochronies in early development are common in many groups of animals (Garstang, 1929; de Beer, 1940; Berrill, 1931; Jägersten, 1972; Strathmann, 1978; Wray and McClay, 1988; Wray and Raff, 1991a; Jeffery and Swalla, 1992). Heterochronies in early development have received less attention partly because some definitions of heterochrony are explicitly based on the timing of sexual maturation (Gould, 1977), which is an inherently poor yardstick for evolutionary changes in the timing of embryogenesis, and partly because some of the best evidence for heterochrony comes from the fossil record, which is notoriously poor in embryonic and larval remains. However, neontological data can provide evidence of heterochronic change, and during the past several years considerable progress has been made towards characterizing heterochronies in early development, as well as understanding their developmental bases and their ecological and evolutionary consequences.

This chapter examines heterochronies in the early development of echinoderms, a phylum of marine invertebrates that includes such familiar animals as starfishes and sea urchins. These heterochronies are interesting for several reasons, but three are particularly relevant. First, many of them show a strong phylogenetic association with life history strategy, rather than with adult morphology. This suggests that they are either caused by or have important ecological or

Evolutionary Change and Heterochrony. Edited by Kenneth J. McNamara © 1995 The Editor and Contributors.
Published in 1995 by John Wiley & Sons Ltd.

evolutionary consequences that are quite independent of producing a particular adult morphology (Wray and Raff, 1991a; Wray and Bely, 1994). Second, several phylogenetically independent, but functionally parallel, changes in life history strategy have evolved in echinoderms. These "replicate" evolutionary events provide a very important opportunity to test specific hypotheses regarding ecological and evolutionary consequences of these heterochronies, as well as their developmental bases (Wray and Bely, 1994). And third, it is becoming possible to dissect apart these heterochronies in terms of developmental mechanisms. The large and rapidly growing comparative developmental data base for echinoderms includes comparisons of gene expression, cell interactions, and morphogenesis during early development among several species (e.g., Wray and McClay, 1989; Wray and Raff, 1990; Henry et al., 1992).

THE EVOLUTION OF COMPLEX LIFE CYCLES

Among animals, a complex life cycle refers to indirect development via an intermediate life history phase that contains structures not present in the adult (Figure 10.1) (Jägersten, 1972; McEdward and Janies, 1993). In many species, this intermediate life history phase is a free-living larva while in others it develops within a brood pouch, or egg case, or attached to a placenta. In each case, specialized structures for the acquisition of nutrition are present transiently. Often, but not always, additional transient structures are present that provide defence from predation, aid in locomotion or dispersal, or facilitate the transition to an adult life style.

These transient structures are often interpreted as adaptations to the environment of pre-adult life (Garstang, 1929; Jägersten, 1972; Emlet, 1991; Wray, 1992a). Such adaptations are expected, since early life history phases often experience an environment that is biotically and abiotically very distinct from that of adult life (Ebenman, 1992; Wray, 1995). Support for hypotheses of pre-adult adaptation come from evidence that some transient structures do in fact increase performance or survivorship of larvae (e.g., Emlet, 1991; Morgan, 1995); that many are not developmentally required to build adult morphology (Wray and Raff, 1991b); and that many have evolved independently on separate occasions, suggesting functional utility (Emlet, 1991; Wray, 1992a; Hadfield et al., 1995).

Complex life cycles are present in most animal phyla, where they are often widely distributed phylogenetically. The majority of animals, including most insects, amphibians, and marine invertebrates, have complex life cycles with free-living larvae (Jägersten, 1972; Strathmann, 1978). Indeed, completely direct development, without any vestige of larval or nutritive traits, is exceptional among animals (McEdward and Janies, 1993). Complex life cycles are numerically dominant within most animal phyla, with vertebrates and chaetognaths being among the few exceptions. Complex life cycles can persist over macroevolutionary time scales, as evidenced by the presence of homologous larvae in phyla and classes that diverged as long ago as the Cambrian (Strathmann, 1978; Wray, 1995).

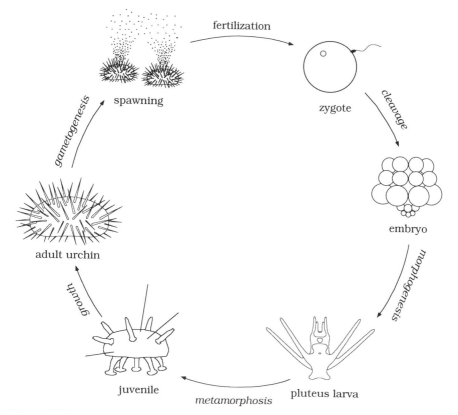

Figure 10.1 *A complex life cycle.* The vast majority of metazoans have complex life cycles, with an intermediate life history phase that undergoes metamorphosis to produce an adult. In most echinoderms, the intermediate life history phase is a free-living larva that swims in the plankton and feeds on unicellular algae; in contrast, the adult is benthic, much larger, and feeds in a very different manner. The biotic and abiotic challenges faced by larvae and adults are often very different and have produced numerous adaptations for one life history phase without affecting the other.

Complex life cycles are also remarkably diverse. Within most large clades of animals there exist several ecologically distinct life history strategies. Even among tetrapods, with their relatively direct development, there is considerable diversity in the source of nutrition during embryogenesis, rates of growth, and the timing of hatching or birth. Echinoderms exhibit a large, but not unusually large, diversity of developmental modes. Among starfishes (Asteroidea), for example, a variety of habitats, food sources, and modes of development are present (Figure 10.2) (Chia and Walker, 1991; McEdward and Janies, 1993). The diversity shown in Figure 10.2 does not include various asexual modes of reproduction also known to be present within the group (Chia and Walker, 1991). A comparable

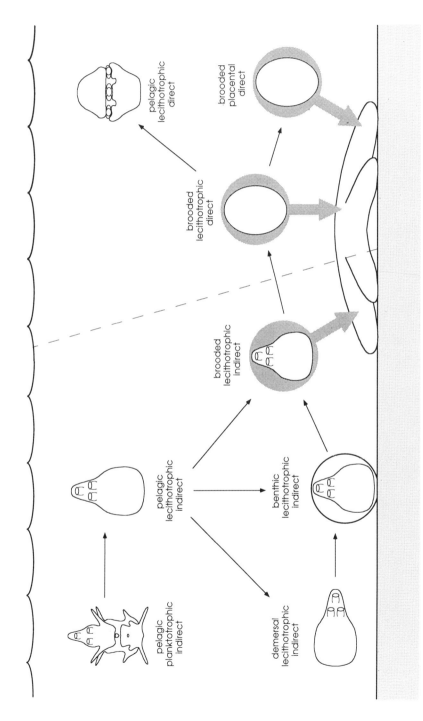

Figure 10.2 *Diversity of echinoderm developmental strategies.* A wide diversity of developmental strategies has evolved within each of the five extant echinoderm classes. In general, it is convenient to organize this diversity according to three criteria: habitat (pelagic, demersal, brooded), source of nutrition (planktotrophy, lecithotrophy, direct maternal transfer), and directness (indirect, direct). This figure illustrates some of the diversity known from the Asteroidea (seastars). A comparable diversity exists in the other classes, with the possible exception of the Crinoidea (sea lilies), which is smaller and not as well characterized. Not included in this

diversity of developmental modes is present in three of the other echinoderm classes: sea urchins and sand dollars (Echinoidea) (Emlet *et al.*, 1987; Raff, 1987), brittle stars (Ophiuroidea) (Hendler, 1991), and sea cucumbers (Holothuroidea) (Smiley, 1991). The remaining extant order, the sea lilies (Crinoidea), may be somewhat less diverse in terms of life history strategy (Holland, 1991), but this may be a result of smaller clade size and less complete comparative information.

This chapter first examines evolutionary changes within the plesiomorphic (ancestral) life history strategy in echinoderms, which is indirect development by means of a planktotrophic (feeding) larva, and then examines the evolution of various derived developmental modes, including lecithotrophic (non-feeding) and brooded development. Throughout, the emphasis is on the various heterochronies associated with evolutionary changes in developmental mode, their developmental bases, and their ecological and evolutionary consequences.

FORM AND FUNCTION IN FEEDING LARVAE

Diversity in form and function among feeding larvae

The presence of a planktotrophic (feeding) larva is almost certainly ancestral for echinoderms (Jägersten, 1972; Strathmann, 1978, 1993). This is evident from the extensive suite of morphological and functional traits shared by the feeding larvae of echinoderms, and from the considerable diversity associated with other developmental modes. The comparative data strongly suggest a common origin prior to the divergence of the extant echinoderm classes (Neilsen, 1987; Wray, 1992a; Strathmann, 1993), which occurred between about 550 and 450 million years ago (Smith, 1988).

Despite these shared traits, the feeding larvae of echinoderms have diversified both morphologically and functionally to a considerable extent (e.g., Mortensen, 1921; Fell, 1948; Strathmann, 1974; Wray, 1992a) (Figure 10.3). The evolutionary history of this diversification is sufficiently well understood to allow some general conclusions to be drawn.

The feeding larvae of each class share a suite of derived traits that suffice to distinguish them readily from those of the other classes (Figure 10.3). Indeed, feeding echinoderm larvae are traditionally given distinct names that reflect these differences: bipinnaria and brachiolaria (Asteroidea), auricularia (Holothuroidea), echinopluteus (Echinoidea) and ophiopluteus (Ophiuroidea).

Within each class feeding larvae have undergone further diversification, although of a generally smaller magnitude. While a core of features is preserved, morphological changes are neither rare nor limited to subtle modifications. Thus, all feeding ophioplutei have an internal skeleton, composed of two bilaterally symmetrical spicules, and this skeleton supports at least eight arms, of which the same two are typically longer than the others (Mortensen, 1921). Yet ophioplutei are also quite diverse in terms of arm shape, the proportions between arm lengths,

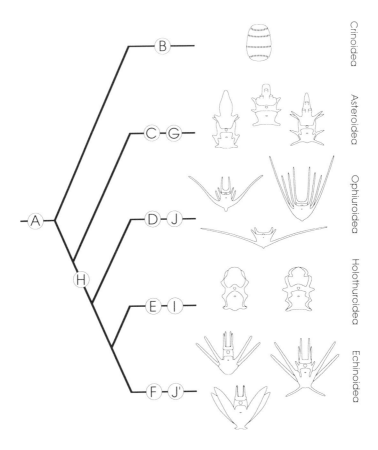

Figure 10.3   *Diversity of feeding larvae of echinoderms.* The planktotrophic (feeding) larvae of echinoderms, and those of enteropneust hemichordates, share a number of morphological and functional traits that are therefore likely to be plesiomorphic for the phylum (A). There are extant planktotrophic larvae in four echinoderm classes, while in the fifth, the Crinoidea, only non-feeding larvae and brooded development are known (B). The morphology of feeding larvae has diversified considerably within each of the four classes that contain them, but all feeding larvae within a class can readily be assigned to class on the basis of several synapomorphies (C–F). Within the Asteroidea, brachiolar arms are present (G). Within two classes, the Ophiuroidea and Echinoidea, feeding larvae contain endoskeletons. These skeletons may represent a single origin (H), followed by a loss in the stem lineage of the Holothuroidea (I), or may represent independent evolutionary origins (J, J').

spination patterns on arms, the occasional presence of an unpaired additional spicule, and the shape of ectoderm overlying the arms (Mortensen, 1921). Intraclass diversity of feeding larvae is even more pronounced in echinoids (Wray, 1992a), although this may be as much an artifact of more extensive data as it is a real difference.

Particularly interesting is the phenomenon of uncoupled morphological change across life history phases. Intervals of morphological change in larvae do not always correspond with periods during which adult morphology is changing, nor is the reverse always true. For example, larvae have diversified considerably and adults less so during the radiation of camarodont echinoids, while adults have diversified substantially but larvae very little during the radiation of clypeasteroid echinoids (Wray, 1992a). Uncoupled evolution between life history phases is often very dramatic during switches from feeding to non-feeding life history strategies (Wray, 1995), a phenomenon considered in more detail later.

Heterochrony and diversity among feeding larvae

A substantial proportion of morphological diversity in feeding larvae may be the product of heterochrony. First, the proportions of particular body parts often differ among related larvae (Figure 10.3). Overall body shape (length : width) and arm length both vary extensively within the four classes that contain feeding larvae. For example, the arms of most echinoplutei are less than 0.5 mm long, but those of some species in the echinoid order Diadematacea exceed 5 mm (Mortensen, 1921). Much of this diversity may be due to differences in relative growth rates, although there are few data that address this issue directly. Second, body size can vary, even among closely related species. Most echinoderm larvae are less than 0.5 mm long, for instance, but the bipinnaria larvae of the asteroid *Luidia sarsi* approach 35 mm in length (Domanski, 1984). In some cases, differences in larval size are transient: the initially smaller larvae of *Strongylocentrotus purpuratus* eventually "catch up" to the larger larvae of *S. droebachiensis* (Sinervo and McEdward, 1988).

Third, some morphological novelties may be the product of changes in the timing of developmental processes. One possible case is the larval skeleton, a derived feature with a significant impact on larval form and function in echinoderms (Pennington and Strathmann, 1990; Emlet, 1991; Hart, 1991). Larval skeletons are present in all the feeding larvae within two echinoderm classes, the Echinoidea and Ophiuroidea. Differences in skeletal morphology account for much of the larval diversity in these classes (Mortensen, 1921; Wray, 1992a). A second possible case involves the brachiolar arms of asteroid larvae. These structures are also functionally significant, providing a firm attachment to the substratum during settlement. Both larval skeletons and brachiolar arms are probably the products of dramatic heterochronies involving the co-option of previously existing adult structures into larvae, as discussed below.

In addition, several echinoderms are known to exhibit an ecophenotypic plasticity based on larval food levels that has been described as "heterochronic plasticity" (Strathmann *et al.*, 1992). In these cases, well fed larvae divert energy towards producing an adult rudiment more rapidly, thereby shortening the time to metamorphosis, while starved larvae build larger feeding structures at the expense of later metamorphosis (Boidron-Metairon, 1988; George, 1990;

Strathmann *et al.*, 1992). The accelerated metamorphosis in well fed larvae involves in part an earlier onset of morphogenetic events involved in building the adult rudiment, as well as a faster growth of the rudiment once it appears. Echinoderms, like most marine invertebrates, experience significant mortality as larvae, on the order of 10–20 % per day (Rumrill and Chia, 1984; Morgan, 1995). It thus seems likely that such heterochronic plasticity, which dynamically allocates energy resources in a way that minimizes time spent in the plankton, has a significant adaptive value (Strathmann *et al.*, 1992). Since the plankton is a patchy environment from the perspective of larval food sources, this kind of adaptive phenotypic plasticity may be more widespread.

Developmental bases for changes in larval morphology

Changes in size, shape and growth rate of larvae
Development of the larval skeleton has been studied extensively in echinoids (reviewed by Ettensohn and Ingersoll, 1992). This work offers some clues as to how modifications in development might have produced changes in the morphology of larval skeletons. Biomineralization begins at two, bilaterally symmetrical sites, where tetrahedral spicules are soon evident. Skeletogenic cells elongate the four branches at very different rates and in different shapes to produce a complex final morphology (Gordon, 1926; Ettensohn and Malinda, 1993). If the ectoderm overlying elongating arm rods is ablated, the skeletogenic cells will cease depositing biomineral matrix, suggesting that a critical inductive interaction regulating skeletal growth takes place at the arm tips (Ettensohn and Malinda, 1993). If skeletogenic cells are grown *in vitro*, they synthesize larger than normal spicules, again suggesting an inductive interaction (Armstrong *et al.*, 1993). Differences among species in arm length, and in the presence of fewer or more arms, both of which are prominent features of the evolutionary history of echinoid larval morphology (Wray, 1992a), may have their basis in alterations in the timing and location of this inductive interaction. The skeletogenic cells appear to synthesize other aspects of larval skeletal morphology without inductive cues, however. If skeletogenic cells are transplanted to host embryos of another species with a distinct skeletal morphology, they will synthesize arm rods of the donor, rather than host, embryo (von Ubisch, 1939; Armstrong and McClay, 1994). This suggests that at least some aspects of skeletal morphology are intrinsic to the skeletogenic cells themselves.

The time at which skeletal growth begins and the initial rate of skeletal growth both vary among echinoid species. In members of the subclass Cidaroidea, skeletogenic cells begin to differentiate, undergo their migration to the sites of skeletal synthesis, and first deposit biomineral matrix much later than during the development of species in the subclass Euechinoidea (Wray and McClay, 1988). In addition, there are initially about three or four times as many skeletogenic cells in euechinoid larvae as in cidaroid larvae (Wray and McClay, 1988). It is likely that this latter difference directly affects rates of skeletal growth: experimental manipulations of skeletogenic cell number suggest that rates of skeletal synthesis, but not final size, vary in proportion to the number of skeletogenic cells present (Ettensohn and Malinda, 1993). The later onset and slower rates of skeleton synthesis are probably both partially responsible for the significantly later appearance of arms in cidaroid versus euechinoid larvae.

The initial shape and growth rate of the larval skeleton depends upon body size, at least in some cases. As mentioned earlier, larvae of *Strongylocentrotus purpuratus* and *S. droebachiensis* eventually converge on same final size and shape, but grow at different rates (Sinervo and McEdward, 1988). That this is primarily a property of their different initial sizes can be shown by artificially reducing the size of *S. droebachiensis* embryos to those of *S. purpuratus*. The skeletons produced by larvae reared from these dwarf embryos are morphometrically identical to those that develop from unmanipulated *S. purpuratus* embryos (Sinervo and McEdward, 1988). Growth rates are presumably affected because initially larger larvae feed faster (Hart, 1991). Smaller eggs produce larvae that are thus doubly compromised: they have farther to grow, and are less efficient feeders. Differences in egg size are common among closely related echinoderms (Emlet *et al.*, 1987), and can evolve very quickly (Wray, 1992b). It thus seems likely that egg size evolution has had an important impact on larval growth trajectories throughout the phylum.

Heterochronic co-options of adult structures into larvae
Larval skeletons and brachiolar arms may have been heterochronically co-opted into echinoderm larvae from previously existing adult structures. The biomineral matrix of echinoderm larval and adult skeletons is composed of very similar material: the inorganic component is calcite in both cases, precipitated as a single crystal axis (Okazaki, 1960), and at least some of the protein components of the biomineral matrix are also the same in larvae and adults (Wilt *et al.*, 1985; Drager *et al.*, 1989). In addition, the larval skeleton is temporarily continuous with a few adult skeletal elements in some echinoids (Gordon, 1926; Emlet, 1989). Similarly, the brachiolar arms of asteroid larvae are morphologically and functionally very similar to adult tube-feet. In both cases, a mesodermally derived hydrostatic system controls movement, and the structures are used for attachment and locomotion on the benthos.

It seems likely that the larval skeleton and brachiolar arms originally existed in somewhat different forms in adults and were later co-opted into larvae. Adult skeletons are present in all known adult echinoderms, including fossils that date back at least to the Cambrian (Smith, 1988), but larval skeletons are restricted to two clades, the class Echinoidea and the Ophiuroidea. Similarly, tube-feet are present in adults of all extant echinoderm classes, but brachiolar arms are present only in some asteroid orders (Blake, 1987). Synthesis of a skeletal material or of tube-feet at a very different time during development would constitute rather dramatic heterochronies in complex developmental programmes. In the case of tube-feet, the co-option into larvae probably happened only once, given the restricted phylogenetic distribution and morphological similarity of brachiolar complexes in asteroids. It is not as clear how many times larval skeletons have evolved within echinoderms. The morphological uniformity of larvae within ophiuroids and echinoids (Mortensen, 1921; Fell, 1948; Wray, 1992a) suggests no more than two origins; however, there are sufficient morphological and development differences between the two clades to raise the possibility of separate origins.

Ecological and macroevolutionary consequences

Many of the differences in larval morphology discussed above appear to improve feeding rates, increase swimming speed, provide protection from predation, or aid in settlement and metamorphosis. A plausible interpretation is that these represent adaptations for the distinct environment that feeding larvae inhabit.

Feeding rates
Echinoderm larvae from different classes have significantly different maximum clearance rates, and achieve maximum rates at different times during development (Strathmann, 1974; Hart, 1991). Several aspects of larval morphology seem to have important consequences for feeding rates, and some of these are probably based on heterochronies. One of the primary factors influencing maximum clearance rate is the total length of ciliated band (Strathmann, 1974; Hart, 1991). It is possible to increase ciliated band length by increasing overall body size, by adding convolutions to the band, and by extending the band along arms. For those larvae with skeletons, the presence of internal support for arms can greatly increase the length that can be achieved, as can the evolution of additional arms. Each of these methods of increasing ciliated band length has evolved on several occasions in echinoderms, consistent with the hypothesis that they represent larval adaptations (Wray, 1992a).

Besides longer ciliated bands, at least three other morphological changes may increase feeding rates. Although none of these three changes seem likely to result from heterochronies, they may all have been the heterochronic consequence of decreasing the time to metamorphosis, which in turn may lower mortality rates significantly. First, increasing the number of cilia per unit length of band may increase particle capture rates, simply by raising the probability that an algal cell in the neighbourhood of a larva will be contacted by a cilium. Since all ectodermal cells are monociliated in echinoderms (Neilsen, 1987), the only ways to pack more cilia into a unit length of band are to either increase the density of cells or increase the width of the band. Both kinds of apomorphies are evident in echinoderm larvae (Wray, 1992a). Second, flatter arms may increase feeding rates, not by enhancing the number of particles captured per unit time, but rather by increasing the efficiency with which they are transported to the oral field prior to ingestion (Hart, 1991). Third, beginning larval life at a larger size is doubly beneficial, as pointed out earlier: there is not only less growing to do before metamorphosis, but the initial ciliated band length, and therefore feeding rate, is higher (Hart, 1991). Egg size, which to a first approximation determines initial larval size, also affects many other life history parameters, however, and may be under balancing selection (Wray, 1995).

Swimming

A larval skeleton may provide benefits for swimming. In an interesting experiment, Pennington and Strathmann (1990) "deboned" echinoplutei by dissolving the calcite of the biomineral matrix. They found that dead echinoplutei passively orient if they contain a skeleton, but not if their skeletons are removed. This may provide an energetic saving in the turbulent waters where echinoplutei graze. A second possible advantage of a larval skeleton also involves an energy conservation argument. Emlet (1983) showed, using larger-than-life models in a fluid whose viscosity approximated the Reynold's number experienced by living larvae, that arms significantly increase drag. Since feeding larvae are negatively buoyant, it is possible that the added drag of arms allow echinoplutei and ophioplutei to remain suspended in the water column with less expenditure of energy (Pennington and Strathmann, 1990). Since drag increases with arm length, heterochronies in skeletal growth could therefore have an impact on the energy budget of larvae.

Defence

There are basically two ways that feeding larvae can reduce their risk of predation: defence and avoidance. Many invertebrate larvae bear structures that may be defensive, such as setae in polychaetes and shells in molluscs. It has been suggested that a larval skeleton may deter certain predators of echinoderm larvae (Emlet, 1983), and some empirical evidence supports this hypothesis (Pennington *et al.*, 1986). An alternative approach is to reduce the risk of predation by minimizing the time to metamorphosis. The increases in feeding rate and heterochronic plasticity discussed earlier may well be beneficial in this regard, as both reduce the time spent in the plankton.

Settlement

A critical phase in many complex life cycles is the transition from larval to adult modes of life. For most marine invertebrates, this means a transition from a planktic (swimming) to a benthic (bottom dwelling) existence. Swimming speed and attachment structures may be critical parameters during this transition, since most marine invertebrate larvae are tiny and their preferred settlement sites are often turbulent. Within echinoderms, variations in the distribution of cilia that appear to increase swimming speed (Emlet, 1991) have evolved on multiple occasions (Wray, 1992a), but these changes are not obviously heterochronic in nature. The origin of brachiolar arms, which may be the result of a dramatic heterochrony as argued above, may well represent an adaptation for settlement.

Tube-feet are important sensory and locomotory structures in adult asteroids, and brachiolar arms may play a role in identifying appropriate settlement sites or in attachment. Larvae of the asteroid order Paxillosida are unique in lacking brachiolar arms, which may be functionally related to the fact that the adults of this order inhabit soft sediments, whereas adults of other asteroid groups live on hard substrates.

LIFE HISTORY STRATEGY AND FURTHER CHANGES IN LARVAL FORM

Diversity of life history strategies in echinoderms

Like most phyla, the Echinodermata contains species representing a wide variety of life history strategies (Figure 10.2) (Strathmann, 1978; Wray and Raff, 1991a; McEdward and Janies, 1993). Boundaries between strategies are not always distinct, but categories based on three criteria have proven useful for understanding this diversity, particularly from an ecological perspective: source of nutrition prior to metamorphosis, the location of early development, and how directly the adult body plan is built (McEdward and Janies, 1993). Further diversity in echinoderm life history strategies include hermaphroditism and various modes of asexual reproduction (Chia and Walker, 1991), but these modes are not considered in this review.

Planktotrophic, planktic, indirect development is the ancestral mode of development for echinoderms (Jägersten, 1972; Strathmann, 1978; Wray, 1992a). This mode remains numerically dominant within the phylum as a whole, and occurs in species from a wide diversity of habitats as adults (Emlet et al., 1987; Raff, 1987; Hendler, 1991; Smiley, 1991). However, various derived developmental strategies have evolved in echinoderms on numerous occasions, differing from the ancestral condition in one or more of the three criteria listed above (Emlet et al., 1987; Wray and Raff, 1991a; McEdward and Janies, 1993). With regard to the source of nutrition, premetamorphic development may be supported by obligate planktotropy (larval feeding is required to reach metamorphosis), obligate lecithotrophy (inability to feed, with a reliance on yolk reserves), facultative planktotrophy (yolk reserves are sufficient, but can be supplemented by feeding), and placentation (direct nutrient transfer from the mother via a specialized structure). The location of early development is similarly diverse: larvae may be planktic (swim high in the water column) or demersal (swim near the benthos); most or all of pre-metamorphic development may be completed in benthic egg masses that are abandoned or guarded; or embryos may be brooded externally, internally, or released. In the vast majority of known cases, development is indirect (transient structures are formed and later lost during metamorphosis), but in a very few cases, it approaches a direct trajectory to the adult.

Not all permutations of these three life history parameters are known to exist in echinoderms (McEdward and Janies, 1993). For example, all known planktotrophs have markedly indirect development; that is, their larvae are morphologically very differerent from adults and undergo a radical metamorphosis. For obvious reasons, completing premetamorphic development within egg masses or being internally brooded precludes planktotrophy, but it is not so clear why few, if any, dermersal planktotrophs exist. Most of the known species approaching fully direct development, such as the echinoid *Abatus cordatus* (Schatt, 1985), are brooded, and only one known case, the asteroid *Pteraster tesselatus* (McEdward, 1992), is planktic.

Derived life history strategies have evolved on many occasions within echinoderms. For example, larval feeding has been lost on at least 20 separate occasions in echinoids alone, and brooding has evolved in at least half of these lineages (Emlet, 1989; my unpublished tallies). Such estimates rely on counting the number of cases where sister taxa contain the ancestral and derived life history patterns; this in turn requires accurate phylogenies (see Strathmann, 1978, for a discussion). Lack of phylogenetic resolution hampers similar estimates for asteroids, ophiuroids, and holothuroids, but it is clear that multiple shifts in life history strategy have evolved within each of these groups as well (McEdward and Janies, 1993; Hendler, 1991; Smiley, 1991). Since there are many extant echinoderms whose developmental modes are not known, and since echinoderms have an evolutionary history that spans over half a billion years, it seems very likely that hundreds of switches to derived developmental modes have happened within the phylum.

These switches in life history strategy provide valuable opportunities to study the causes and consequences of evolutionary modifications in early development, including heterochronies. As the following sections discuss, independent cases of derived life history strategies in echinoderms share several informative developmental and ecological correlates (Wray and Raff, 1991a; Wray and Bely, 1994). Some of these correlates involve changes in developmental timing, and are relevant to this review. Indeed, switches in life history modes provide particularly good cases for "replicate" studies of the consequences of phenotypic transformations, precisely because these heterochronies have understandable ecological or evolutionary consequences.

Heterochronic changes that accompany life history transformations

There are several overt morphological changes associated with derived life history modes in echinoderms that are heterochronic in nature. The kinds of changes discussed below have evolved repeatedly in most or all cases for which the relevant data are available, suggesting that they are either developmentally necessary to achieve the derived developmental mode, or are an adaptive response to the new life style that is driven ecologically.

Larger eggs
The eggs of echinoderms with lecithotrophic developmental modes are invariably larger than those of their closest relatives with planktotrophic larvae. The size difference is substantial, typically involving one to three orders of magnitude in egg volume. In echinoids, the most thoroughly studied group, eggs of planktotrophs are in the range of 70–250 μm in diameter, while those of lecithotrophs are 350–2000 μm in diameter (Emlet *et al.*, 1987; Raff, 1987). The correlation is remarkable: the eggs of facultative planktotrophs fall between these size ranges, while brooding, the most derived mode of lecithotrophic development, tends to involve the largest eggs of all (Wray and Raff, 1991a). Similar correlations exist in the Asteroidea, Ophiuroidea and Holothuroidea (Emlet *et al.*, 1987; Hendler, 1991; Smiley, 1991). Even exceptions to this trend are intelligible: species with the highly derived developmental mode of placental nutrition have egg sizes that are relatively small (in the range of planktotrophs), because maternal nutrition transferred during development substitutes for sources stored in the egg. These smaller egg sizes are presumably the result of secondary reductions from lecithotrophic ancestors with large eggs.

The large eggs of lecithotrophs are the result of peramorphic oogenesis: either oogenesis is prolonged (hypermorphosis) or happens faster (acceleration), or both. Although few details are known, acceleration must at least be involved, given that oogenesis, which takes months in echinoderms that produce small eggs, cannot be prolonged several orders of magnitude in species that reproduce annually to produce eggs that are several orders of magnitude larger. There are also differences in the proportions of various egg constituents between planktotrophs and lecithotrophs that may be the result of differences in biosynthetic rates or durations. For example, in the genus *Heliocidaris*, the yolk protein vitellogenin is much less abundant in the eggs of the lecithotrophic than the planktotrophic species (Scott *et al.*, 1990).

Large egg size may be a necessary condition for the evolution of non-feeding development in echinoderms, given that maternal nutrient reserves alone must fuel development through metamorphosis in lecithotrophs. There is no direct experimental evidence to support this hypothesis, although tests are certainly feasible (McEdward, 1992). For example, artificially increasing the size of the eggs of an obligate planktroph to 350 μm in diameter should result in facultative planktotrophy, while decreasing the size of a facultative planktotroph's eggs below 200 μm should make feeding obligate. Despite the lack of experimental evidence, the phylogenetic correlation between egg size and life history mode provides strong indirect evidence for the importance of increases in egg size. The correlation is highly significant within echinoids by Maddison's (1990) concentrated changes test (Wray and Bely, 1994), but the other classes have not been formally examined.

More cells
A second derived feature common to most lecithotrophic life cycles in echinoderms is higher cell numbers at most phases of embryogenesis and larval life. For

example, within the genus *Heliocidaris*, a gastrula has ~1000 cells in *H. tuberculata* (planktotroph) and ~9000–13 000 in *H. erythrogramma* (lecithotroph), and by early larval life the discrepancy is even more pronounced (Parks *et al.*, 1988). The size of postmetamorphic juveniles is roughly similar between related species with planktotrophic and lecithotrophic development (Emlet *et al.*, 1987), suggesting similar overall cell numbers just after metamorphosis in either developmental mode. Thus, the evolution of lecithotrophy seems to involve an early period (embryogenesis) of hypermorphic cell proliferation relative to the ancestral state to produce an early larva with more cells. This may be followed by a truncated period of cell proliferation in later larval life to produce a final cell count that is not vastly different, given the much shorter larval periods of lecithotrophs. Careful counts of cell numbers in related species with alternative developmental modes could test these hypotheses about changes in cell proliferation that accompany these transformations.

Shorter time to metamorphosis
Species with lecithotrophic larvae often take substantially less time to reach metamorphosis than their closest relatives with planktotrophic larvae (Wray and Raff, 1991a). It might be expected that this difference exists simply because lecithotrophic larvae do not need to grow (and indeed, cannot), whereas planktotrophic must grow, and are limited in this process by the rate at which they capture food. However, planktotrophs and lecithotrophs arrive at metamorphosis by very different routes (Wray and Raff, 1991a; Byrne, 1991; Janies and McEdward, 1993), suggesting that the situation is actually more complex.

From a strictly morphological perspective, the abbreviated pre-metamorphic development of species with lecithotrophic larvae seems to arise in two distinct ways. First, some larval structures are deleted, reduced, or altered, particularly those no longer needed in the absence of feeding. For example, most lecithotrophic larvae never form a larval mouth, nor do their guts undergo morphological differentiation (e.g., Williams and Anderson, 1975; Byrne, 1991; Amemiya and Emlet, 1992); the arms of plutei, which aid in capturing food (see above), are either reduced in number or lost entirely in echinoids (e.g., Okazaki, 1975; Williams and Anderson, 1975) and ophiuroids (e.g., Mladenov, 1979; Hendler, 1982); and the ciliated band is either lost entirely or repositioned in such a way as to increase swimming speed (Emlet, 1991; Wray, 1995). Many of these are cases of non-terminal deletion. The parallel nature of these transformations has resulted in convergently simplified lecithotrophic larvae (Figure 10.4) called "schmoos" (Wray, 1995).

The other class of morphological change that probably contributes to a shorter premetamorphic period is the earlier appearance of adult structures. In nearly all species for which data are available, asymmetric growth of the coeloms and morphogenesis of the water vascular system, both early steps in metamorphosis to a radial adult body plan, begin much sooner relative to other developmental events than in the most closely related species with planktotrophy, and

A

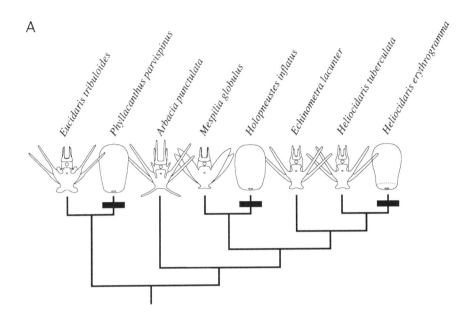

B

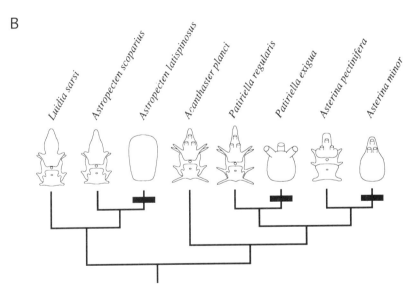

Figure 10.4   *Parallel origins of schmoo larvae in echinoderms.* The transformation from planktotrophic to lecithotrophic larval development has happened many times within echinoderms. In nearly all such cases, several parallel apomorphies have evolved, including a general simplification of larval morphology, a repositioning or loss of the ciliated bands, reduction or loss of the larval skeleton when present, and a shorter planktic phase. These convergently similar larvae are known as schmoos. Shown here is a sampling of morphological diversity in larvae from the Echinoidea (sea urchins) (A) and the Asteroidea (seastars) (B).

pedicellariae and tube-feet, which are exclusively adult structures, often appear much earlier as well (e.g., Williams and Anderson, 1975; Amemiya and Emlet, 1992; McEdward, 1992).

Developmental bases for switches in life history strategy

There is little doubt that changes in the timing of developmental processes are an important part of how transformations in life history strategy evolve. However, recent research into the nature of these changes has revealed a surprising complexity of modification in underlying developmental mechanisms. In the cases that have been examined, these modifications include many of the earliest and most crucial processes of early development. Many, but not all, of the modifications can be reasonably interpreted as adaptations that shorten the time to metamorphosis.

Formation of embryonic axes
In asteroids and echinoids with planktotrophic larvae, the dorsoventral axis of the embryo is specified well after fertilization (Hörstadius, 1973; Kominami, 1988; Henry *et al.*, 1992). In *Heliocidaris erythrogramma* and *Holopneustes inflatus*, two echinoids representing independent evolutionary derivations of lecithotrophy, the dorsoventral axis is specified significantly earlier in development (Wray and Raff, 1989; Henry *et al.*, 1992), and in *H. erythrogramma*, the dorsoventral axis is also committed (fixed) much earlier (Henry and Raff, 1990). It is possible that the earlier conclusion of these events, which are the initial steps in axis formation, is a part of the generally accelerated premetamorphic development of lecithotrophs, a hypothesis that could be tested with additional comparative data.

Cell fate specification
The accelerated formation of the dorsoventral axis in *H. erythrogramma* is almost immediately reflected in differences in cell fates along this axis (Wray and Raff, 1989, 1990). In the planktotrophs that have been examined, in contrast, early cell fates have a radially symmetrical distribution (Hörstadius, 1973; Cameron *et al.*, 1986), reflecting the later specification of the dorsoventral axis. Other derived features of cell fate specification in *H. erythrogramma* include changes in the timing and order of particular specification events. For example, ectodermal specification begins earlier and skeletogenic and gut cell specifications occur later. These timing differences change the proportional volume of the embryo allocated to particular cell types, and alter the ratios of various differentiated cell types in the larvae. Some of these changes make functional sense with regard to the derived life history mode (Wray and Raff, 1989, 1990). For example, gut, which is non-functional in lecithotrophic larvae, is specified later and as a smaller proportion of cells, whereas adult ectoderm, which is needed for the accelerated

development of the adult rudiment, is specified sooner and in much greater volume. Whether parallel changes in cell fate specification have evolved in other cases where larvae have become lecithotrophic remains unknown.

Gastrulation
Another important developmental process that has been modified in association with the evolution of lecithotrophy is gastrulation. The blastopore is unusually large in early gastrulae of several independent cases, including the ophiuroid *Amphioplus abditus* (Hendler, 1977), the asteroid *Pteraster tesselatus* (McEdward, 1992), and the echinoid *Abatus cordatus* (Schatt, 1985). The blastopore closes soon after gastrulation is completed in many cases, including the three just cited, as well as the echinoids *Heliocidaris erythrogramma* (Williams and Anderson, 1975) and *Phyllacanthus parvispinus* (Parks *et al.*, 1989). This is not the case in species with planktotrophic larvae, where the blastopore serves a function in feeding, becoming the larval anus. Cell movements during gastrulation have only been examined in one species with lecithotrophic development, *H. erythrogramma* (Wray and Raff, 1991b). Several dramatic modifications in cell movements have evolved in this case, including the imposition of a pronounced dorsoventral asymmetry in cell movements and a different mechanism of archenteron extension. The asymmetric cell movements may be functionally related to the asymmetric cell fate specification described in the previous paragraph (Wray and Bely, 1994).

Skeletogenesis
As mentioned earlier, the skeletons characteristic of the planktotrophic larvae of ophiuroids and echinoids are invariably lost or reduced in species with obligate lecithotrophy. This reduction or loss may represent the neutral loss of a structure no longer under positive selection for a formerly crucial function (feeding), or may be actively selected as a means of accelerating the time to metamorphosis by dispensing with developmentally costly structures that are no longer needed. Whichever the case, in three independent cases, the reduction or loss is mirrored by a dramatic change in the timing of gene expression associated with skeletogenesis (Wray and Bely, 1994) (Figure 10. 5). Synthesis of the larval skeleton of echinoids involves the expression of several genes whose products are incorporated into the spicule sheath or biomineral matrix (Harkey *et al.*, 1988; Sucov *et al.*, 1987; Wilt *et al.*, 1985; Livingston *et al.*, 1991). Although the timing of expression of some of these genes varies slightly among species with planktotrophic larvae (Wray and McClay, 1989), there is a large delay in expression in each of the three independent cases of lecithotrophic larvae that have been examined: *Heliocidaris erythrogramma* (Parks *et al.*, 1988), *Phyllacanthus parvispinus* (Parks *et al.*, 1989), and *Asthenosoma ijimai* (Amemiya and Emlet, 1992) (Figure 10.5A). This phylogenetic association is highly significant by the concentrated changes test (Wray and Bely, 1994) (Figure 10.5B), suggesting that the heterochrony in gene expression is either involved in causing the transformation to lecithotrophy, or, more likely, is a strongly selected response to it.

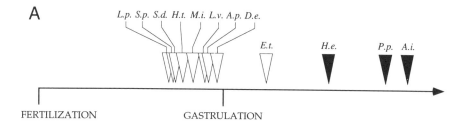

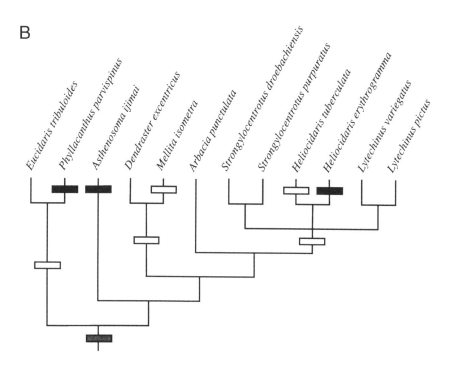

Figure 10.5 *Heterochronies in gene expression associated with lecithotrophy.* The timing of expression of *msp130*, a gene involved in skeletogenesis, correlates with developmental mode in echinoids. A, Developmental timelines, showing differences in the onset of *msp130* expression in planktotrophs and lecithotrophs. In species with planktotrophic larvae (white arrowheads), expression begins relatively early, just before the synthesis of the larval skeleton. In species with lecithotrophic larvae (black arrowheads), the onset of expression is substantially delayed. B, Phylogenetic distribution of species with different times of expression. Expression data are available for twelve species, including three independent cases of lecithotrophic larvae. By the concentrated changes test (Maddison, 1990), the correlation between developmental mode and timing of *msp130* expression is highly significant ($P < 0.0002$), suggesting a functional role.

Coelomogenesis and rudiment formation
Changes in later morphogenesis have also evolved in echinoderm species with lecithotrophic larvae. The coelomic pouches arise just after gastulation in echinoderms with planktotrophic development, but do not grow or change form until the adult rudiment begins to develop in preparation for metamorphosis. In several species with lecithotrophic development, morphogenesis of the coeloms is modified (Gemmill, 1912; Parks *et al.*, 1989; Wray and Raff, 1989; Amemiya and Emlet, 1992; Janies and McEdward, 1993). In most of these cases, the coeloms do not arise earlier, but they form at a larger size, and almost immediately begin to undergo further morphogenetic movements to form the water vascular system. Particularly dramatic changes in coelomogenesis have evolved in *Pteraster tesselatus*, which lacks any trace of the ancestral bilateral pouches (Janies and McEdward, 1993). Formation of the adult rudiment itself is also modified in at least some species with derived developmental modes. In echinoid species with planktotrophic larvae, for example, the vestibule that houses the rudiment is initially a tiny structure, and requires several days or weeks of growth before it achieves its final size and adult structures can develop fully. In contrast, in *H. erythrogramma* the vestibule invaginates at full size (Wray and Raff, 1989), allowing the immediate elaboration of adult structures. The overall timing of rudiment formation, like *msp130* expression, has changed independently on several occasions, suggesting a functional role (Figure 10.6).

Larval cell number
In contrast to the foregoing changes, many of which seem to be the result of selection, the greater number of cells in lecithotrophic larvae may well be a passive consequence of another change: the increase in egg size that invariably accompanies the origin of lecithotrophy in echinoderms. In some echinoids, and in many other embryos, the rapid early cell divisions of cleavage end when a critical nucleus: cytoplasm volume ratio is achieved. (During early development cytoplasmic volume decreases with each mitosis since there is no net growth, but nuclear volume is regenerated and remains constant.) An increase in egg volume may be sufficient to explain the greater number of cells in lecithotrophic larvae, simply as a prolongation of cleavage required to achieve the same nucleus: cytoplasm ratio (Parks *et al.*, 1988).

Ecological and macroevolutionary consequences

Short-term consequences
Changes in developmental mode almost certainly have important ecological and population-level consequences (Emlet *et al.*, 1987; Roughgarden, 1989; Strathmann, 1993). Many of these involve classic trade-offs of the type familiar to students of life history theory (Stearns, 1992; Roff, 1992). For example, in all known cases, the loss of larval feeding has been accompanied by substantially larger egg sizes

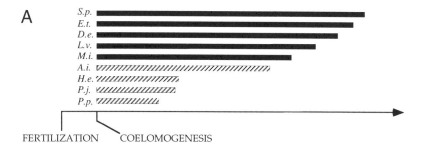

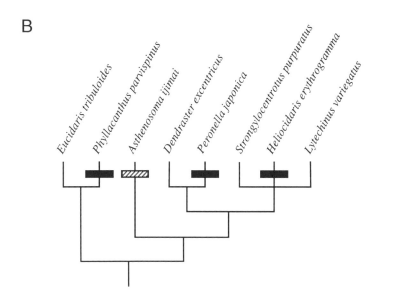

Figure 10.6   *Heterochronies in morphogenesis associated with lecithotrophy.* The duration of morphogenesis of the adult body rudiment correlates with developmental mode in echinoids. A, Developmental timelines, showing differences in the duration of morphogenesis of the adult rudiment. This process is defined as beginning with coelom formation and concluding with metamorphosis. Species with planktotrophic larvae (black bars) take longer to complete this process than species with lecithotrophic larvae (grey bars). B, Phylogenetic distribution of species with different durations of rudiment morphogenesis. Three species with lecithotrophic larvae have substantially shorter rudiment development times, but the fourth (grey bar) has only a marginally shorter duration. If only the three "big" differences are counted as significant, the concentrated changes test suggests that the correlation is significant ($P < 0.01$); if all four lecithotrophs are counted as shorter than all five planktotrophs, the correlation is much stronger ($P < 0.00002$). In either case, the concentrated changes test suggests a functional role for the shortened duration of rudiment morphogenesis.

and smaller clutch sizes in echinoderms. Thus, lowered fecundity and greater maternal investment per propagule are both strongly correlated with the change in developmental mode (Emlet *et al.*, 1987; Raff, 1987). The switch from planktic development to benthic egg masses or brooding generally involves a further increase in egg size and drop in brood size (Wray and Raff, 1991a).

The age distribution of mortality in early life history phases is probably also quite different among life history strategies. Few field data are available to support this contention, in large measure because such data are extraordinarily difficult to obtain for marine organisms. The reason for believing it is that larvae suffer mortality that is, to a first approximation, a direct consequence of the time they spend in the plankton (Rumrill and Chia, 1984; Morgan, 1995). Since mortality rates are very high in the plankton, even modest changes in the duration of the planktic period may shift the age distribution of mortality appreciably. Dispersal is also affected by developmental mode in echinoderms. Both theoretical and empirical data support the conclusions that dispersal is somewhat diminished by the shorter planktic period of lecithotrophic larvae, and severely limited by development in egg masses and brooding (Palmer and Strathmann, 1981; McMillan *et al.*, 1992).

Together, the differences just outlined will clearly have consequences for both local and species-wide population structure. Because egg sizes and brood sizes span orders of magnitude among echinoderm species, these consequences are probably not subtle.

Long-term consequences
Besides these short term, microevolutionary consequences, there may also be long term, macroevolutionary consequences of developmental mode (Jablonski and Lutz, 1983; Jablonski, 1986). The lower dispersal of non-planktotrophic developmental modes may result in elevated speciation rates, due to reduced gene flow and the greater likelihood of geographically isolated populations. Decreased dispersal may also result in elevated extinction rates, due to the generally smaller geographic ranges of species with lecithotrophic larvae or that brood. There is some empirical evidence to support these predictions from the fossil record of molluscs (Jablonski, 1986; Hansen, 1983) and bryozoans (Taylor, 1988). It is possible to distinguish planktotrophic and lecithotrophic developmental modes from fossils in some echinoid orders (Emlet, 1989), and comparable studies are now in progress in echinoderms (R. B. Emlet, pers. comm.). It is suggestive that some echinoderm clades composed of species that brood appear to be substantially larger than related clades. Examples include the echinoid genera *Phyllacanthus* and *Abatus*, and the asteroid genus *Pteraster*. In the absence of diversity measures over geological timescales, however, sister clade size comparisons can only provide indirect evidence of higher speciation rates.

Planktotrophy is apparently absent from the extant members of the class Crinoidea, the sub-class Phrynophiuroidea (Ophiuroidea) and the orders Echinothurioida (Echinoidea) and Velatida (Asteroidea). Several families and sizable genera of echinoderms are also likely to be entirely composed of species with derived developmental modes. In no cases have all constituent species been examined, and it remains possible that the clades defined by the derived developmental modes are somewhat smaller. Nonetheless, it seems likely that at least some switches in developmental strategy in echinoderms have been followed by diversification to produce sizable clades. This suggests that derived modes of development are not a barrier to evolutionary success; whether they can be beneficial under certain circumstances is not yet clear, and merits further study.

ACKNOWLEDGMENTS

Special thanks to Larry McEdward, Rudy Raff and Richard Strathmann for many stimulating discussions about heterochrony in early development. Ken McNamara provided helpful comments on an early draft of this chapter.

REFERENCES

Amemiya, S. and Emlet, R.B., 1992, The development and larval form of an echinothurioid echinoid, *Asthenosoma ijimai*, revisited, *Biol. Bull.*, **182**: 15–30.
Armstrong, N., Hardin, J. and McClay, D.R., 1993, Cell–cell interactions regulate skeleton formation in the sea urchin embryo, *Development*, **119**: 833–840.
Armstrong, N. and McClay, D.R., 1994, Skeletal pattern is specified autonomously by the primary mesenchyme cells in sea urchin embryos, *Dev. Biol.*, **162**: 329–338.
Berrill, N.J., 1931, Studies in tunicate development, II: abbreviation of development in the Molgulidae, *Phil. Trans. Roy. Soc. Lond.* **B219**: 281–346.
Blake, D.B., 1987, A classification and phylogeny of post-Paleozoic sea stars (Asteroidea: Echinodermata), *J. Nat. Hist.*, **21**: 481–528.
Boidron-Metairon, I.F., 1988, Morphological plasticity in laboratory-reared echinoplutei of *Dendraster excentricus* (Eschscholtz) and *Lytechinus pictus* (Lamarck) in response to food conditions, *J. Exp. Mar. Biol. Ecol.*, **119**: 31–41.
Byrne, M., 1991, Developmental diversity in the starfish genus *Patiriella* (Asteroidea: Asterinindae). In T. Yanigasawa, I. Yasumasu, C. Oguro, N.E. Suzuki and T. Motokawa (eds), *Biology of Echinodermata*, Balkema, Rotterdam: 499–508.
Cameron, R.A., Hough-Evans, B., Britten, R.J. and Davidson, E.H., 1986, Lineage and fate of each blastomere of the eight-cell sea urchin embryo, *Genes and Development*, **1**: 75–84.
Chia, F.-S. and Walker, C.W., 1991, Echinodermata, Asteroidea. In A.C. Giese and J.S. Pearse (eds), *Reproduction of marine invertebrates*, vol. 6, Boxwood Press, Palo Alto.
de Beer, G., 1940, *Embryos and ancestors*, Clarendon Press, London.
Domanski, P.A., 1984, Giant larvae: prolonged planktonic larval phase in the asteroid *Luidia sarsi*, *Mar. Biol.*, **80**: 189–195.
Drager, B.J., Harkey, M.A., Iwata, M. and Whiteley, A.H., 1989, The expression of embryonic primary mesenchyme genes of the sea urchin, *Strongylocentrotus purpuratus*, in the adult skeletogenic tissues of this and other echinoderms, *Dev. Biol.*, **133**: 14–23.
Ebenman, B., 1992, Evolution in organisms that change their niches during the life cycle, *Am. Nat.*, **139**: 990–1021.

Emlet, R.B., 1983, Locomotion, drag, and the rigid skeleton of larval echinoderms, *Biol. Bull.*, **164**: 433–445.

Emlet, R.B., 1989, Apical skeletons of sea urchins (Echinodermata: Echinoidea): two methods for inferring mode of larval development, *Paleobiology*, **15**: 223–254.

Emlet, R.B., 1991, Functional constraints on the evolution of larval forms of marine invertebrates: experimental and comparative evidence, *Am. Zool.*, **31**: 707–725.

Emlet, R.B., McEdward, L.R. and Strathmann, R.R., 1987, Echinoderm larval ecology viewed from the egg. In M. Jangoux and J.M. Lawrence (eds), *Echinoderm studies*, Balkema, Amsterdam, **2**: 55–136.

Ettensohn, C.A. and Ingersoll, E., 1992, Morphogenesis of the sea urchin embryo. In E.F. Rossomando and S. Alexander (eds), *Morphogenesis: an analysis of the development of biological form*, Marcel Dekker, New York: 189–262.

Ettensohn, C.A. and Malinda, K.M., 1993, Size regulation and morphogenesis: a cellular analysis of skeletogenesis in the sea urchin embryo, *Development*, **119**: 155–167.

Fell, H.B., 1948, Echinoderm embryology and the origin of chordates, *Biol. Rev.*, **23**: 81–107.

Garstang, W., 1929, The origin and evolution of larval forms, *Br. Ass. Adv. Sci.*, **1929**: 77–98.

Gemmill, J.F., 1912, The development of the starfish *Solaster endeca* Forbes, *Trans. Zool. Soc. Lond.*, **20**: 1–71.

George, S.J., 1990, Population and seasonal differences in egg quality of *Arbicia lixula* (Echinodermata: Echinoidea), *Int. J. Repro. Dev.*, **17**: 111–121.

Gordon, I., 1926, The development of the calcareous test of *Echinocardium cordatum*, *Phil. Trans. Roy. Soc. Lond.* **B215**: 255–313.

Gould, S.J., 1977, *Ontogeny and phylogeny*, Belknap Press, Cambridge.

Hadfield, K.A., Swalla, B.J. and Jeffery, W.R., 1995, Multiple origins of anural development in ascidians inferred from rDNA sequences, *J. Mol. Evol.*, **40**: 413–427.

Hansen, T., 1983, Modes of larval development and rates of speciation in early Tertiary neogastropods, *Science*, **220**: 501–502.

Harkey, M.A., Whiteley, H.R. and Whiteley, A.H., 1988, Coordinate accumulation of five transcripts in the primary mesenchyme during skeletogenesis in the sea urchin embryo, *Dev. Biol.*, **125**: 381–395.

Hart, M., 1991, Particle capture and the method of suspension feeding by echinoderm larvae, *Biol. Bull.*, **180**: 12–27.

Hendler, G., 1977, Development of *Amphioplus abditus* (Verrill) (Echinodermata: Ophiuroidea): I. Larval biology, *Biol. Bull.*, **152**: 51–63.

Hendler, G., 1982, An echinoderm vitellaria with a bilateral larval skeleton: evidence for the evolution of ophiuroid vitellariae from ophioplutei, *Biol. Bull.*, **163**: 431–437.

Hendler, G., 1991, Echinodermata: Ophiuroidea. In A.C. Giese and J.S. Pearse (eds), *Reproduction of marine invertebrates*, vol. 6, Boxwood Press, Palo Alto.

Henry, J.J. and Raff, R.A., 1990, Evolutionary change in the process of dorsoventral axis determination in the direct developing sea urchin, *Heliocidaris erythrogramma*, *Dev. Biol.*, **141**: 55–69.

Henry, J.J., Klueg, K.M. and Raff, R.A., 1992, Evolutionary dissociation between cleavage, cell lineage and embryonic axes in sea urchin embryos, *Development*, **114**: 931–938.

Holland, N., 1991, Echinodermata: Crinoidea. In A.C. Giese and J.S. Pearse (eds), *Reproduction of marine invertebrates*, vol. 6, Boxwood Press, Palo Alto.

Hörstadius, S., 1973, *Experimental embryology of echinoderms*, Oxford University Press, Oxford.

Jablonski, D., 1986, Larval ecology and macroevolution in marine invertebrates, *Bull. Mar. Sci.*, **39**: 565–587.

Jablonski, D. and Lutz, R.A., 1983, Larval ecology of marine benthic invertebrates: paleobiological implications, *Biol. Rev.*, **58**: 21–89.

Jägersten, G., 1972, *Evolution of the metazoan life cycle*, Academic Press, London.

Janies, D.A. and McEdward, L.R., 1993, Highly derived coelomic and water-vascular morphogenesis in a starfish with pelagic direct development, *Biol. Bull.*, **185**: 56–76.

Jeffery, W.R. and Swalla, B.J., 1992, Evolution of alternative modes of development in ascidians, *BioEssays*, **14**: 219–226.

Kominami, T., 1988, Determination of dorso-ventral axis in early embryos of the sea urchin, *Hemicentrotus pulcherrimus*, *Dev. Biol.*, **127**: 187–196.

Livingston, B.T., Shaw, R., Bailey, A. and Wilt, F., 1991, Characterization of a cDNA encoding a protein involved in formation of the skeleton during development of the sea urchin *Lytechinus pictus*. *Dev. Biol.*, **148**: 473–480.

Maddison, W.P., 1990, A method for testing the correlated evolution of two binary characters: are gains or losses concentrated on certain branches of a phylogenetic tree?, *Evolution*, **44**: 539–557.

McEdward, L.R, 1992, Morphology and development of a unique type of pelagic larva in the starfish *Pteraster tesselatus* (Echinodermata: Asteroidea), *Biol. Bull.*, **182**: 177–187.

McEdward, L.R. and Janies, D.A., 1993, Life cycle evolution in asteroids: what is a larva?, *Biol. Bull.*, **184**: 255–268.

McKinney, M.L. and McNamara, K.J., 1991, *Heterochrony: the evolution of ontogeny*, Plenum Press, New York.

McMillan, W.O., Raff, R.A. and Palumbi, S.R., 1992, Population genetic consequences of developmental evolution in sea urchins (genus *Heliocidaris*), *Evolution*, **46**: 1299–1312.

Mladenov, P.V., 1979, Unusual lecithotrophic development of the Caribbean brittle star *Ophiothrix oerstedi*, *Mar. Biol.*, **55**: 55–62.

Morgan, S., 1995, Life and death in the plankton: larval mortality adaptation. In L. McEdward (ed.), *Larval ecology of marine invertebrates*, in press, CRC Press, Boca Raton FL.

Mortensen, T.H., 1921, *Studies of the development and larval forms of echinoderms*, GEC Gad, Copenhagen.

Neilsen, C., 1987, Structure and function of metazoan ciliary bands and their phylogenetic significance, *Acta Zool.*, **68**: 205–262.

Okazaki, K., 1960, Skeleton formation of sea urchin larvae. II. Organic matrix of the spicule, *Embryologica*, **5**: 283–320.

Okazaki, K., 1975, Normal development to metamorphosis. In G. Czihak (ed.), *The sea urchin embryo*, Springer-Verlag, Berlin: 177–232.

Palmer, A.R. and Strathmann, R.R., 1981, Scale of dispersal in varying environments and its implications for life histories of marine invertebrates, *Oecologia*, **48**: 308–318.

Parks, A.L., Parr, B.A., Chin, J., Leaf, D.S. and Raff, R.A., 1988, Molecular analysis of heterochronic changes in the evolution of direct developing sea urchins, *J. Evol. Biol.*, **1**: 27–44.

Parks, A.L., Bisgrove, B.W., Wray, G.A. and Raff, R.A., 1989, Direct development in the sea urchin *Phyllacanthus parvispinus* (Cidaroidea): phylogenetic history and functional modification, *Biol. Bull.*, **177**: 96–109.

Pennington, J.T., Rumrill, S.S. and Chia, F.-S., 1986, Stage-specific predation upon embryos and larvae of the Pacific sand dollar, *Dendraster excentricus*, by 11 species of common zooplankton predators, *Bull. Mar. Sci.*, **39**: 234–240.

Pennington, J.T. and Strathmann, R.R., 1990, Consequences of the calcite skeleton of planktonic echinoderm larvae for orientation, swimming, and shape, *Biol. Bull.* **179**: 121–133.

Raff, R.A., 1987, Constraint, flexibility, and phylogenetic history in the evolution of direct development in sea urchins, *Dev. Biol.*, **119**: 6–19.

Raff, R.A. and Wray, G.A., 1989, Heterochrony: developmental mechanisms and evolutionary results, *J. Evol. Biol.*, **2**: 409–434.

Roff, D.A., 1992, *The evolution of life histories: theory and analysis*, Chapman and Hall, New York.

Roughgarden, J., 1989, The evolution of marine life cycles. In M.W. Feldman (ed.), *Mathematical evolutionary theory*, Princeton University Press, Princeton: 270–300.

Rumrill, S. and Chia, F., 1984, Differential mortality during the embryonic and larval lives of northeast Pacific echinoids. In B.F. Keegan and B.D.S. O'Connor (eds), *Echinodermata: proceedings of the fifth international echinoderm conference*, Balkema, Amsterdam: 333–338.

Schatt, P., 1985, *Developpement et croissance embryonnaire de l'oursin incubant* Abatus cordatus *(Echinoidea:Spatangoidea)*. Ph.D. thesis, Université Pierre et Marie Curie, Paris.

Scott, L.B., Lennarz, W.J., Raff, R.A. and Wray, G.A., 1990, The "lecithotrophic" sea urchin *Heliocidaris erythrogramma* lacks typical yolk platelets and yolk glycoproteins, *Dev. Biol.*, **138**: 188–193.

Sinervo, B. and McEdward, L.R., 1988, Developmental consequences of an evolutionary change in egg size: an experimental test, *Evolution*, **42**: 885–899.

Smiley, S., 1991, Echinodermata, Holothuroidea. In A.C. Giese and J.S. Pearse (eds), *Reproduction of marine invertebrates*, vol. 6, Boxwood Press, Palo Alto.

Smith, A.B., 1988, Fossil evidence for the relationship of extant echinoderm classes and their times of divergence. In C.R.C. Paul and A.B. Smith (eds), *Echinoderm phylogeny and evolutionary biology*, Clarendon Press, Oxford: 85–97.

Stearns, S.C., 1992, *The evolution of life histories*, Oxford University Press, Oxford.

Strathmann, R.R., 1974, Introduction to function and adaptation in echinoderm larvae. *Thal. Jugoslavica*, **10**: 321–339.

Strathmann, R.R., 1978, The evolution and loss of feeding larval stages of marine invertebrates, *Evolution*, **32**: 894–906.

Strathmann, R.R., 1993, Hypotheses on the origins of marine larvae, *Ann. Rev. Ecol. Syst.*, **24**: 89–117.

Strathmann, R.R., Fenaux, L. and Strathmann, M.F., 1992, Heterochronic developmental plasticity in larval sea urchins and its implications for evolution of nonfeeding larvae, *Evolution*, **46**: 972–986.

Sucov, H.M., Benson, S., Robinson, J.J., Britten, R.J., Wilt, F. and Davidson, E.H., 1987, A lineage-specific gene encoding a major matrix protein of the sea urchin embryo spicule II. Structure of the gene and derived sequence of the protein, *Dev. Biol.*, **120**: 507–519.

Taylor, P.D., 1988, Major radiation of cheilostome bryozoans: triggered by the evolution of a new larval type?, *Hist. Biol.*, **1**: 45–64.

von Ubisch, L., 1939, Keimblattchimärenforschung an Seeigellarven, *Biol. Rev.* **14**: 88.

Williams, D.H.C. and Anderson, D.T., 1975, The reproductive system, embryonic development, larval development and metamorphosis of the sea urchin *Heliocidaris erythrogramma* (Val.) (Echinoidea: Echinometridae), *Aust. J. Zool.*, **23**: 371–403.

Wilt, F.H., Benson, S. and Uzman, J.A., 1985, The origin of the micromeres and formation of the skeletal spicules in developing sea urchin embryos. In R.H. Sawyer and R.M. Showman (eds), *The cellular and molecular biology of development*, University of South Carolina, Columbia: 297–310.

Wray, G.A., 1992a, Rates of evolution in developmental processes, *Am. Zool.*, **32**: 123–134.

Wray, G.A., 1992b, The evolution of larval morphology during the post-Paleozoic radiation of echinoids, *Paleobiology*, **18**: 258–287.

Wray, G.A., 1995, Evolution of larvae and developmental modes. In L. R. McEdward (ed.), *Ecology of marine invertebrate larvae*, CRC Press, Boca Raton, FL, in press.

Wray, G.A. and Bely, A.E., 1994, The evolution of early development in echinoderms is driven by several distinct factors, *Development*, 1994 supplement: 97–105.

Wray, G.A. and McClay, D.R., 1988, The origin of spicule-forming cells in a "primitive" sea urchin (*Eucidaris tribuloides*) which appears to lack primary mesenchyme cells, *Development*, **103**: 305–315.

Wray, G.A. and McClay, D.R., 1989, Molecular heterochronies and heterotopies in early echinoid development, *Evolution*, **43**: 803–813.

Wray, G.A. and Raff, R.A., 1989, Evolutionary modification of cell lineage in the direct-developing sea urchin *Heliocidaris erythragramma*, *Dev. Biol.*, **132**, 458–570.

Wray, G.A. and Raff, R.A., 1990, Novel origins of lineage founder cells in the direct-developing sea urchin *Heliocidaris erythrogramma*, *Dev. Biol.*, **141**: 41–54.

Wray, G.A. and Raff, R.A., 1991a, Rapid evolution of gastrulation mechanisms in a sea urchin with lecithotrophic larvae, *Evolution*, **45**: 1741–1750.

Wray, G.A. and Raff, R.A., 1991b, The evolution of developmental strategy in marine invertebrates, *Trends Ecol. Evol.*, **6**: 45–50.

Chapter 11

EFFECT OF PAEDOMORPHOSIS IN EYE REDUCTION ON PATTERNS OF EVOLUTION AND EXTINCTION IN TRILOBITES

Raimund Feist

INTRODUCTION

Trilobites were primitively oculated arthropods, possessing compound eyes. This was an early character state which preceded eventual eye loss in some later forms, this being a derived character (Fortey and Whittington, 1989). Most trilobite larvae possessed eyes or eye-like structures (Speyer and Chatterton, 1989). In early larval stages (protaspis) the eye was located at the anterior or anterolateral dorsal margin and migrated, during the meraspid stages, inward and backward together with associated structures such as eye ridges and facial sutures (Whittington, 1957). Normally, the adult trilobite eye was situated near to the glabella in a backward position on the adaxial genal field. Consequently, the retention into the adult stage of ancestral larval and juvenile features such as reduced size or even absence of the eye, anteriorly positioned eye lobes, eye ridges, broad fixigenae with reduced or obsolete palpebral lobes and distally or submarginally situated facial sutures, indicates that paedomorphic processes were operating during ontogenetic development.

The evolution of the visual system in trilobites is known to have been affected by heterochronic processes. The schizochroal eye in phacopids, where large lenses with individual corneal coverings are separated from each other by cuticular material, derived from an earlier more primitive holochroal type with closely packed tiny lenses covered by a single corneal membrane. An example of paedomorphosis is clearly indicated where early larval eyes of holochroal-eyed trilobites take on a schizochroal configuration (Clarkson, 1975).

In both kinds of visual system, regression and loss of the eye occurred periodically in many independant taxa during the 300 million years of trilobite

Evolutionary Change and Heterochrony. Edited by Kenneth J. McNamara © 1995 The Editor and Contributors.
Published in 1995 by John Wiley & Sons Ltd.

evolutionary history. However, cases of stratigraphically controlled phyletic sequences of species where a paedomorphic trend in eye reduction can be followed continuously, are exceptional. Such examples occur, for instance, in the mid-Ordovician *Ormatops* (Fortey and Owens, 1990) and in the mid-late Silurian *Denckmannites* (Schrank, 1973). Further examples are reported here from late Devonian phacopid and proetid lineages which all emphasise the prominent role of paedomorphic processes among intrinsic responses to environmental constraints. This chapter analyses the interrelationship between eye-loss trends and ecological control, especially during stressful periods of worldwide biotic crises such as occurred, in particular, at the end of the Frasnian (Kellwasser crisis) and Famennian (Hangenberg crisis) stages of the Devonian (Feist, 1991). The examples are taken from sequences of subtropical, outer shelf, bottom-level communities of the trilobite-rich, cephalopod limestone realm. This environment existed largely beyond neritic influences and was thus mostly homogeneous in time and space, cosmopolitan in distribution and well correlated by fine scaled conodont biostratigraphy. It provides perhaps the best opportunity to show evidence of the adaptive role of heterochronic trends among trilobites.

EXTRINSIC CONSTRAINTS AND ADAPTATION

The Late Devonian was a period of global sea-level highstand which, beginning with the mid-Givetian Taghanic onlap, resulted from a series of positive transgression–regression cycles (Johnson *et al.*, 1985). Intermittent short-term eustatic perturbations characterised by transgressive–regressive couplets, led to the temporary rise of oceanic oxygen-depleted bottom waters which, in repeatedly invading outer shelf habitats, were the direct causes of mass extinctions among obligate bottom-level biotas including trilobites (Wilde and Berry, 1984; Becker, 1993).

Already decreasing in diversity earlier in the Devonian (Feist, 1991), trilobites suffered strong selective pressure when environmental conditions changed globally in one direction by repeated eustatic deepening. Only a few lineages of hitherto exclusively neritic trilobites adapted to deeper shelf environments. Adaptive evolution as a biological response to deepening beyond the penetration of light primarily involved the visual organ. Thus, one of the predominant morphological changes was eye reduction that affected different taxonomic entities evolving contemporaneously along the same environmental gradient of increasing water depth. It is striking that the percentage increases of blind forms or those with reduced sight coincide with periods of pronounced oceanic deepening (Figure 11.1). This supports the view that eye reduction is adaptive and blindness is a homoplasous character that may occur repeatedly in different lineages of the same family (Campbell, 1975).

Blind Late Devonian trilobites occur in oxygenated muddy sediments of the outer shelf cephalopod realm. They probably had a shallow burrowing mode of

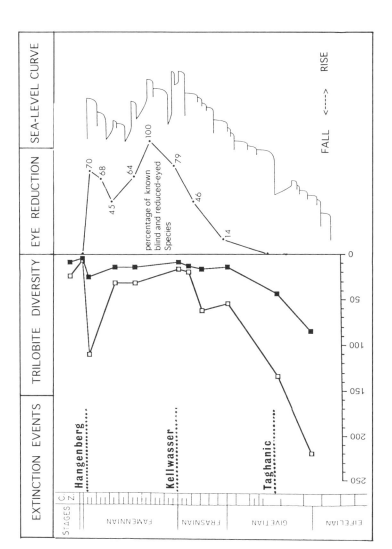

Figure 11.1 Comparison of the Late Devonian global sea-level curve (modified from Johnson et al., 1985) with the fluctuation in the percentage of reduced-eyed (including blind) trilobites. Trends in deepening are closely accompanied by an increase in eye reduction. Maximum sea-level highstand coincides with spread of oceanic anoxia and reduction of species diversity culminating in mass extinctions. □, species; ■, genus; C.Z., conodont zones; Hangenberg, Kellwasser, Taghanic: major trilobite extinction events associated with global eustatic perturbations.

life as indicated by the gross morphology of their smooth exoskeleton (Campbell, 1975; Feist and Clarkson, 1989). Adaptation to a specialised infaunal life as deposit feeders in the uppermost layers of the fine-grained sediment was preceded by a period of adaptive evolution leading to eye loss in a globally deepening environment. After adopting an endobenthic mode of life these trilobites were independent of water depth and would occasionally co-occur with sighted epifaunal taxa when the photic zone extended to the sea floor during periods of sea-level lowering (for example, the association of blind *Trimerocephalus* with oculated cyrtosymbolines in the early Famennian). However, the continuation of endobenthic taxa was rapidly halted when the eustatically driven rise of oceanic anoxia led to the poisoning of the outer shelf sea-bottom.

The adopted biological adaptive strategy under control of extrinsic constraints such as strong selective pressure towards reduced sight under lower light conditions is the reduction, by paedomorphic processes, in the morphological development of the visual complex. This will be demonstrated with a few examples.

LATE DEVONIAN CASE HISTORIES OF EYE REDUCTION

The last representatives of the declining scutelluids, odontopleurids, harpids and dalmanitids were apparently all normally oculated and no case of blindness has been reported. By contrast, it has been known for a long time that eye loss was a common phenomenon among both Frasnian and Famennian proetids and phacopids (Richter and Richter, 1926; Delo, 1937; Erben, 1961; Alberti, 1970). Up to now, out of 115 proetacean species with a known visual area, 42 were blind and another 26 exhibit a reduced eye morphology. Out of 45 phacopids, 19 were blind. There are many examples of isolated blind or taxa with reduced eyes for which phyletic lineages leading to normally sighted ancestors have not been established. The mid-Frasnian *Clavibole*, apparently the last representative of the Dechenellinae, and the early Famennian phacopidellines *Dienstina* and *Ductina* with reduced and obsolete vision respectively, constitute such examples. In addition to these, a newly discovered, but so far undescribed, species of *Otarion* from the mid-Frasnian of southern France would represent the first case of blindness in aulacopleurids (Figure 11.2). However, phyletic series of species as well as intraspecific evolution with unidirectional changes of the visual complex leading to blindness, have been demonstrated in drevermanniines (Richter and Richter, 1926; Feist, 1991; Feist and Schindler, 1994), cyrtosymbolines (Yuan, 1988), tropidocoryphines and pteropariines (Feist, 1976; Feist and Clarkson, 1989), as well as in different lineages of phacopines (Richter and Richter, 1926, 1955; Maksimova, 1955; Osmólska, 1963; Chlupáč, 1977; Feist, 1991; Feist and Schindler, 1994).

In these trilobites, the visual complex constituting the eye consists of the palpebral lobe in connection to the eye ridges (where present) and the surrounding visual surface carrying the ocular lenses. The palpebral lobe and visual surface are separated from one another by the facial suture along which the

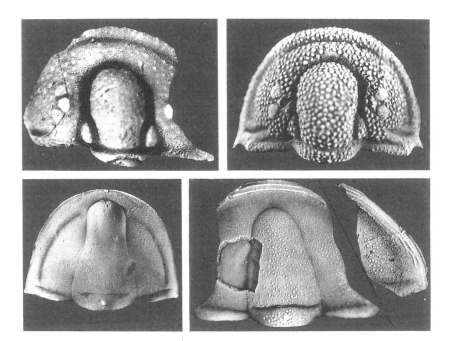

Figure 11.2   Eye reduction in Late Devonian proetids. Top, Aulacopleurinae, left: *Otarion stigmatophthalmus* Richter, internal mould of cephalon with attached left librigena displaying small globular eye lobe between two cushion-like eye platforms, × 17.5, Late Frasnian, Sessacker, Germany (Senckenberg Museum, coll. Lippert Sl/13); right: *Otarion* n.sp., external mould of cephalon with completely vanished eye lobe between well developed cushion-like eye platforms and minute, elevated palpebral lobe at the fixigenal margin, × 11.5, Late Frasnian, Coumiac, France (Univ. Montpellier, USTM-RF 113). Bottom, left, Cyrtosymbolinae: *Helioproetus subcarintiacus* (Richter, 1913), external mould of cephalon with partially fused librigenae, × 6.9, Late Famennian, Apricke, Germany (Univ. Montpellier, USTM-RF 114); right, Drevermanniinae: *Palpebralia nodannulata* (Richter, 1913), cranidium with partially exfoliated left fixigena showing discontinuous ocular ridge, × 8; detached librigena, external mould, × 14.2, both associated in latest Frasnian, Sessacker, Germany, (Senckenberg Museum, SMF 55833 and 55835).

genal field is split during ecdysis into the fixigenal part including the palpebral lobe, which remains with the cranidium, and the librigenal part remaining with the visual surface (Figure 11.3). The facial suture which defines the moulting behaviour is thus intimately linked with the visual complex and its course would obligatorily change when paedomorphic processes lead to eye reduction.

Modes of eye reduction in proetids

Chatterton (1971) published the ontogenetic growth series of the late Early Devonian *Proetus talenti* from which the early larval development of Devonian

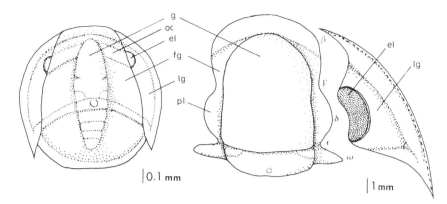

Figure 11.3   Morphological features of early larval and adult configurations of *Proetus talenti* Chatterton (after Chatterton, 1971); α–ω = turning points of the facial suture, pl = palpebral lobe, fg = fixigena, lg = librigena, g = glabella, el = eye lobe, or = ocular ridge.

proetids can be depicted. The protaspis exhibits opisthoparian facial sutures that are rather distant from the glabella and only slightly divergent from the anterior to the posterolateral margin. Small eye lobes are developed on the librigenae in the anterior third of the suture and faint eye ridges run from them to the front of the glabella. Palpebral lobes cannot be distinguished, but developed later during the meraspid stage when the growing eye migrated backward along the facial suture (Figure 11.3).

In all normally oculated adult proetids, well defined palpebral lobes developed between the sutural incurvation points γ and ε (Figure 11.3). As Erben (1961) has shown, eye reduction was accompanied by the regression of the palpebral lobe (*Piriproetus* mode of eye reduction). Generally, the palpebral lobe was initially shortened from behind and regressed. It then underwent several successive stages of change in shape until achieving complete obsolescence. This development occurred in harmony with a change of the sutural organisation that became more and more stretched during the continuous outward migration of the incurvation points γ and ε. In the terminal phase, the suture became rectilinear or concave towards the exterior, when no turning points existed between β and ω (Figure 11.3). This resulted in a considerable enlargement of the fixigenae. The visual surface diminished in size and prominence whilst migrating forward, becoming more and more flattened until flush with the genal field. By this stage, the lenses had generally disappeared, but a very small number of them may sometimes have remained. In *Pteroparia oculata* these residual lenses were relatively larger and less densely packed. This would mirror the schizochroal configuration observed in early larval stages of normally oculated proetids (Clarkson and Zhang, 1991). Conversely, in some rare cases, such as in *Piriproetus*, a rudimentary palpebral lobe was still developed in blind species (Alberti, 1970). These discrepancies probably resulted from dissociation in the timing of morphological development that affected different elements of the visual complex.

The *Piriproetus* mode of eye reduction is considered to be a succession of morphological stages within a long ranging general trend observed in many Devonian to Lower Carboniferous proetids. However, short-term intra- and interspecific unidirectional changes in this mode can be observed in several Late Devonian lineages (Figure 11.4).

Drevermanniinae

The late Frasnian *Palpebralia* is one of those rare proetids in which ocular ridges were still present in the adult stage. As seen among the oldest representatives, these ridges run continuously from the palpebral lobe to the glabella as in early larvae of *Proetus talenti*. Early morphotypes of *Palpebralia palpebralis* exhibit a large kidney-shaped visual surface which tended to diminish in younger morphotypes. This became flush with the genal field in descendant species, whereas the palpebral lobe concomitantly regressed by outward migration of the facial suture (Figure 11.4B). The eye ridge was still present on the adaxial part of

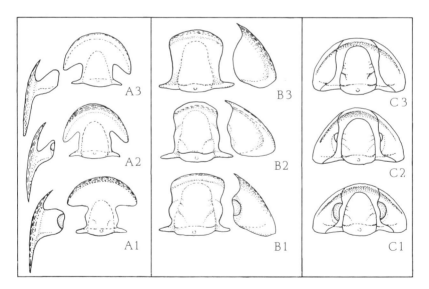

Figure 11.4 Examples of eye reduction leading to blindness in short-term evolutionary lineages (less than 3 million years) of Late Devonian proetids. A, Pteropariinae, Frasnian stage: A1 *Pterocoryphe languedociana* Feist, A2 *Pteroparia oculata* Feist and Clarkson, A3 *P. coumiacensis* Feist (after Feist, 1991). B, Drevermanniinae, Frasnian stage: B1 *Palpebralia palpebralis globoculata* Feist, B2 *P. palpebralis palpebralis* (Richter and Richter), B3 *P. brecciae* (Richter) (after Feist, 1991). C, Cyrtosymbolinae, Famennian stage: C1 *Chaunoproetus (Chaunoproetoides) orientalis similis* Yuan, C2 *C.(Chaunoproetoides) ignorans* Yuan, C3 *C. (Chaunoproetus) palensis* (Richter) (after Yuan, 1988).

the enlarged fixigenae but did not reach the suture (for example in *P. nodannulata*, Figure 11.2). When the final stage with a concave sutural organisation was reached in *P. brecciae*, the eye ridges had completely vanished.

Cyrtosymbolinae
An analogous case of paedomorphic eye reduction is seen in a phyletic series of species of the Late Famennian *Chaunoproetus* lineage (Yuan, 1988). In this biostratigraphically unbroken range of five successive species a number of developmental trends are particularly well documented (Figure 11.4C). These include: the progressive shortening and regression of the palpebral lobe, the concomitant forward migration and regression until obsolescence of the visual field and the increasing regressing and outward displacement of the suture. Similarly, the blind *Helioproetus subcarintiacus* with comparably broad fixigenae and concave sutures (Figure 11.2) is claimed to constitute the final stage of eye loss in a phyletic series of species that originated from oculated cyrtosymboline ancestors (Yuan, 1988). In both cases the final stage of morphological development in the adults bears striking similarities to the shape of the protaspis stage in the ancestral *Proetus talenti*.

Tropidocoryphidae
Silurian to mid-Devonian tropidocoryphids are normally oculated proetaceans with strongly diverging anterior sutures. However, in the Late Givetian, a slight reduction of both the palpebral lobe and the visual surface is perceptible in *Longicoryphe*. In the descendant represented by the Lower Frasnian *Erbenicoryphe* (the last representative of the main lineage) merely a slight anterior curvature of the otherwise straight palpebral suture indicates the position of the former palpebral lobe and all that remains of the eye is a smooth swollen surface lacking lenses (Feist and Clarkson, 1989). However, the fixigenae are only enlarged in the anterior part of the palpebral area. The main tropidocoryphine rootstock gave rise, at the beginning of the Late Devonian, to the pteropariines, characterised by the progressive backward swing of the anterior facial sutures and the consecutive enlargement of the anterolateral genal fields (Figure 11.4A). Whereas in the earlier *Pterocoryphe* palpebral lobes and extended visual surfaces are normally developed, the earliest *Pteroparia* species are characterised by the straightening of the palpebral suture and the regression and forward migration of the still prominent visual surface.

However, this stage of eye reduction was not restricted to early forms of *Pteroparia* such as *P. oculata* but is maintained without major change in the latest Frasnian *P. ziegleri* (Feist and Schindler, 1994). Further development in eye reduction is seen in the descendants such as blind, late mid- to end-Frasnian representatives of *Pteroparia* including *P. columbella*, *P. coumiacensis* and *P. aekensis*, where the visual surface became flush with the genal field.

The straightening of the palpebral suture was accomplished, in contrast to the cases seen in proetids, by the outward migration of the suture incurvation at turning point γ alone, whereas ε remained stable and very near to the glabella. Apparently, eye reduction in tropidocoryphids did not occur in the typical proetid *Piriproetus* mode, which may emphasise the peculiarity of this proetacean family.

Modes of eye reduction in phacopids

Environmentally controlled small-eyed and blind phacopids occurred sporadically from the Late Silurian, but the overwhelming majority of Silurian to Middle Devonian taxa are normally oculated. Their prominent schizochroal eyes occupy most of the genal fields and are provided with kidney-shaped visual surfaces carrying more than a hundred lenses. By contrast, eye reduction leading to blindness is a general trend in Late Devonian lineages. Indeed, with the exception of the latest Famennian *Omegops*, which is restricted to shallow platform habitats, there are no taxa with large eyes known within the considerable time span of some 20 million years between the mid-Givetian worldwide eustatic deepening and the end-Famennian extinction of the group.

Migration, reduction and subsequent disappearance of the visual complex and associated dorsal facial sutures that occurred in this family during the Famennian constitute an outstanding example of an evolutionary trend (Fortey and Owens, 1990). The final stage, represented by blind forms with marginal or submarginal sutures, would have resulted from paedomorphosis (Hupé, 1953). However, the undoubted overall paedomorphic tendency does not constitute a single evolutionary lineage as is often represented (Hupé, 1953; Harrington, 1959) but in fact contains developmental stages of taxa of various phylogenetic sources and ages. Instead, it obviously results from repeated trends in different lineages which were derived from persisting phacopine or phacopidelline root stocks. It is within these lineages that species-to-species changes are to be sought (Figure 11.5).

Early ontogenetic stages of phacopid eye development are poorly known and the only published descriptions of early meraspid cranidia belong to the late Early Devonian *Phacops spedeni* (Chatterton, 1971) and to the Late Famennian *Phacops accipitrinus* (Alberti, 1972). In both specimens elevated palpebral lobes are situated at the anterior margin of the fixigenae whereas the corresponding adults exhibit large eyes reaching the posterior margin of the ankylosed genal field. Consequently, the ancestral eye underwent major changes in size and shape during ontogeny.

In Late Devonian representatives of *Phacops*, the eye was progressively reduced in both sagittal length and dorsoventral height. The visual surface, while maintaining its typical kidney-shaped outline, had fewer vertical rows with a restricted number of lenses in comparison with the mid-Devonian ancestors. Paedomorphic features such as forward migration of the eye by shortening from behind and increase of the postocular field, along with a diminishing number of lenses in the posterior part, already occurred in early Frasnian taxa. However, species with a variable, but still significant, number of rows (15) and lenses (50)

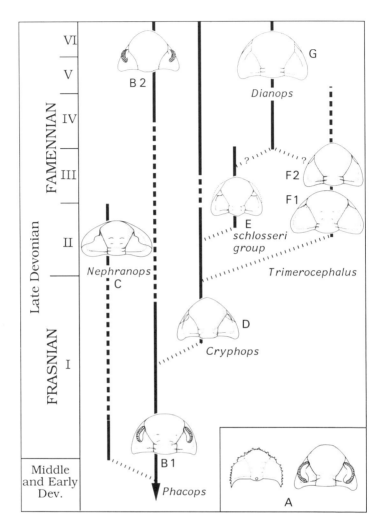

Figure 11.5   Reduction and loss of the eye in Late Devonian phacopine lineages.
A, early larval and adult configuration of *Phacops spedeni* Chatterton, Emsian
(after Chatterton, 1971); B, *Phacops* branch, B1 *Phacops* n. sp. a, B2 *P. granulatus*
(Münster); C, *Nephranops incisus* (Roemer); D, *Cryphops acuticeps* (Kayser); E,
"Cryphops" schlosseri Richter and Richter; F, *Trimerocephalus* branch, F1
Trimerocephalus dianopsoides Osmólska, F2 *T. polonicus* Osmólska; G, *Dianops
limbatus* (Richter). Subdivision of the Late Devonian series is after recognised
ammonoid stages.

persisted until the end-Devonian extinction. By contrast, no case of eye loss is
known in this main lineage. The *Nephranops* branch was derived from the main
rootstock in the Frasnian. In this lineage, the shape of the *Phacops* eye was
maintained but was reduced in height. The unchanged reniform visual surface is

devoid of any lenses; only a single specimen has been found that exhibits two residual lenses (Richter and Richter, 1926). The existence of such an intermediate stage of lens reduction is emphasised by the subsequent discovery of a second case with two remaining lenses. On the other hand, a slightly younger occurrence of the oculated *Nephranops miserrimus* might point to polyphyly and differences in the rate of eye reduction as this taxon could not have been derived from blind ancestors. Basically, the configuration of the visual complex seen in *Phacops* is unchanged. The instantaneous loss of ocular lenses might be due to postdisplacement by extreme retardation of dissociated morphological development of this feature.

Cryphops
By the mid-Frasnian, the main phacopine rootstock gave rise to the *Cryphops* lineage. There is a gradual transition in size and position of the eyes between *Phacops* species such as *P. cryphoides* and early morphotypes of the *Cryphops acuticeps* group. The latter still exhibits the ancestral reniform outline of the visual field, though with a reduced number of lenses. At that stage it is difficult to separate both groups by features of the visual complex alone.

Representatives of the *Cryphops* lineage are better characterised by the anteriorly pointed, protruding glabella, the special configuration of the anteroventral part of the cephalon and the short, lens-shaped pygidium. However, when the number of lenses decreases to less than 13 in the latest Frasnian (typical morphotypes of *C. acuticeps*) the visual surface adopts an elliptical outline (*cryptophthalmus* pattern, Richter and Richter, 1926) (Figure 11.6). This pattern has been observed in many older phacopids with reduced eyes such as *Eocryphops*, *Struveaspis*, *Plagiolaria* and *Reedops*. On the other hand, Lütke (1968) observed the occurrence of both the reniform and the *cryptophthalmus* pattern within the same population of *Nephranops miserrimus*. Apparently, a decrease in lens number and the elliptical shape of the visual surface are combined features that result from delayed morphological development. It is to be expected that the reniform visual surface that characterises the adult eye in *Phacops* was derived from an elliptical one during early larval development.

Several meraspid cranidia of *Cryphops* have been discovered recently and are currently under investigation. They are nearly identical with equivalent early meraspid stages of *Phacops spedeni*. In particular, the functional facial sutures likewise separate the cephalic doublure including minute dorsal shields with lens-bearing visual surfaces. During later ontogeny and up to the adult stage, the ancestral larval configuration (i.e., a small-sized eye in an anterior position) was maintained in all representatives of the *Cryphops* lineage. Dissociated neoteny (McKinney, 1984) is the most likely process responsible for the reduction in the rate of morphological development of the visual organ during the juvenile phase. As the adults are considerably smaller in overall size than the ancestor but develop the same number of thoracic segments (11), precocious maturation may be reached by sequential progenesis when intermoult stages are shortened

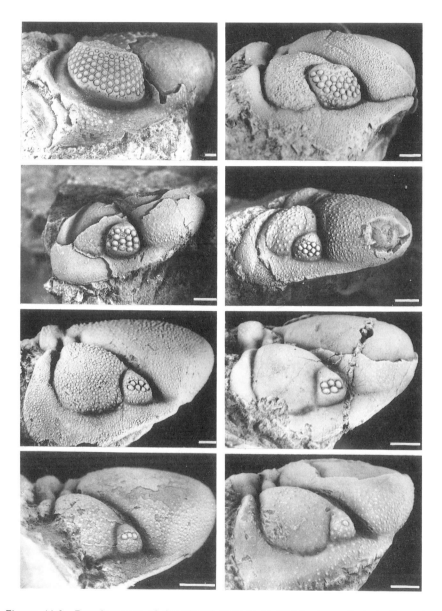

Figure 11.6   Development of the visual surface in Middle to early Late Devonian phacopines (lateral views of holaspid cephala displaying visual surfaces are shown). Top left: *Phacops* n. sp. b. with large, kidney-shaped visual surface displaying 86 lenses in 18 straight rows, Eifelian, Montagne Noire, France (Univ. Montpellier, USTM-RF 115); top right: *Phacops* n. sp. a. with kidney-shaped visual surface carrying 27 lenses in 10 irregular rows, Early Frasnian, Montagne Noire, France (Univ. Montpellier, USTM-RF 116); second row, left: *Cryphops acuticeps* (Kayser), morphotype with kidney-shaped visual surface displaying 18 lenses in 5 straight rows, latest

(McNamara, 1983). Both paedomorphic processes seem to operate simultaneously. We shall see that a third process that concerns the rate of eye lens reduction may be involved concurrently.

The *acuticeps* group is characterised by highly variable numbers of lenses developed in different individuals of the same population. Morphotypes having between 1 and 18 lenses co-occur (Figure 11.6). Of special interest is the distribution of the mean lens number in five continuous populations collected in the last beds immediately below the end-Frasnian mass-extinction in the Frasnian/Famennian boundary stratotype section at Coumiac, southern France (Figure 11.7A). It appears that in an estimated time-span, according to conodont data, of less than two hundred thousand years the mean lens number gradually decreased from an initial 9.59 to 4.42. In each succeeding population an increasing quantity of high lens number morphotypes were replaced by forms with fewer and fewer lenses. However, the former persisted until the final sudden extinction of all morphotypes at the base of the Upper Kellwasser-bed. It is noteworthy that in the last population before extinction, more than two-thirds of the total number of individuals possessed fewer than six lenses. This is below the minimum number observed up to now and diagnostically defined for the genus. Even though holaspids were exclusively taken into account, the possibility cannot be excluded that lens development may have continued as long as the individual continued to moult. However, when lens number is plotted against size (i.e., the sagittal length of the cephalon), it is clear that individuals with fewer lenses do not differ fundamentally in terms of mean size than those with more (Figure 11.7B). Morphotypes with only a single lens co-occur with others having the same size but exhibiting 3, 7 or 12 lenses. This may indicate that an intra-specific gradual decrease of the number of lenses took place which resulted from paedomorphic processes such as either postdisplacement (McNamara, 1986), when onset of lens development was delayed during early ontogenetic stages, or neoteny, whereby the rate of lens development would have been reduced.

The derivation of *Trimerocephalus* from *Cryphops* in early Famennian times may have been accompanied by slow morphological development of the eye which reduced it to a minute protuberance without lenses and situated in the

Frasnian, Coumiac, France (Univ. Montpellier, USTM-RF 117); second row, right: *Cryphops acuticeps* (Kayser), morphotype with kidney-shaped visual surface displaying 12 lenses, late Mid Frasnian, Montagne Noire, France (Univ. Montpellier, USTM-RF 118); third row, left: *Cryphops acuticeps* (Kayser), morphotype with elliptical visual surface displaying 7 lenses, latest Frasnian, Coumiac, France (Univ. Montpellier, USTM-RF 104); third row, right: *Cryphops acuticeps* (Kayser), morphotype with elliptical visual surface displaying 5 lenses, latest Frasnian, Coumiac, France (Univ. Montpellier, USTM-RF 102); bottom left: *Cryphops acuticeps* (Kayser), morphotype with 3 lenses, latest Frasnian, Coumiac, France (Univ. Montpellier, USTM-RF 119); bottom right: *Cryphops acuticeps* (Kayser), morphotype with 1 lens, , latest Frasnian, Coumiac, France (Univ. Montpellier, USTM-RF 120). Scale: 1 mm.

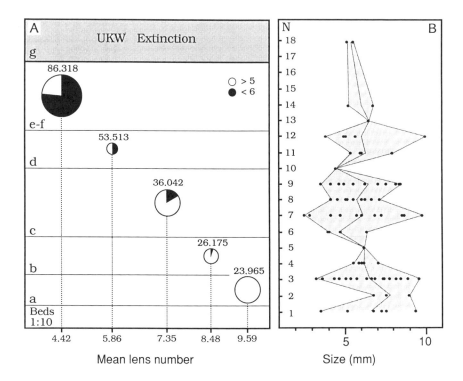

Figure 11.7   Reduction of ocular lenses in latest Frasnian *Cryphops acuticeps* from the Frasnian/Famennian stratotype section at Coumiac, France. A, Series of succeeding populations in the last beds (a to e–f) prior to the Upper Kellwasser (UKW) extinction level (g); size of populations (number of individuals considered): bed a = 34, bed b = 25, bed c = 34, bed d = 14, bed e–f = 55. A steady percentage increase of individuals with less than six lenses is obvious from older to younger beds, whereas the coefficient of variance increases at the same time reaching a maximum (86.318) just below the extinction event. B, Graph showing individual lens numbers plotted against cephalic length from all specimens studied where the visual surface is preserved. Mean size of different morphotypes does not vary significantly.

anterior genal angle (Richter and Richter, 1926; Osmólska, 1963). It disappeared completely in the adults of descendant species.

   Dianops was blind and devoid of any relictual eye protuberance even in larval stages. It is characterised by its functional submarginal sutures. This taxon may have originated in the mid-Famennian *Cryphops schlosseri* group according to Richter and Richter (1955) rather than in *Trimerocephalus* (Maksimova, 1955; Osmólska, 1963; Chlupáč, 1977). The latter authors believe that *Dianops* was descended directly from *Trimerocephalus* after further outward migration of the facial sutures. However, the known ancestral ontogenetic trajectories, especially in *Phacops* and *Cryphops*, reveal the close linkage of the facial suture to the eye lobe which separates the distal visual surface from the adaxial palpebral lobe. In *Cryphops schlosseri*, both the eye lobe and the suture are situated outside the genal field on the enlarged anterolateral border of the cephalon. This would thus represent the most paedomorphic taxon of the *Cryphops* lineage. A further step in this unidirectional evolution is seen in *Dianops* when the final, most primitive, phacopid larval stage prior to eye appearance was maintained on account of dissociated neoteny.

ANKYLOSIS AND BLINDNESS

Cases of partial ankylosis of the dorsal facial sutures have been reported from late Devonian blind proetids such as *Helioproetus, Typhloproetus, Chaunoproetus* and *Palpebralia* (Richter and Richter, 1926; Erben, 1961). This might indicate adaptation to conditions where consolidation of the dorsal cephalon by repeated postecdysial recalcification of the sutures would be advantageous. On the other hand, in these trilobites, where the course of the suture lines had become extremely simple after disappearance of the eye protuberances, the ecdysial gap might easily have been reclosed after ecdysis (Henningsmoen, 1975). The presence of equal-sized specimens with completely detached librigenae would point to the latter interpretation.

By contrast, ankylosis of the sutures in phacopids might be controlled by heterochrony. Whereas they were still functional in early holaspids of the Silurian *Acernaspis* (Ramsköld, 1988) they became ankylosed in Devonian representatives of *Phacops* such as *P. zinkeni* (Jahnke, 1969) well before the end of the larval period. In these trilobites the neck joint opened to facilitate exuviation when the whole dorsal shield of the cephalon had to be removed during ecdysis. This special moulting behaviour, which differs from that of most other trilobites, has been described in detail by Speyer (1985). The so-called Salterian mode of moulting, where the cephalic shield is inverted and reversed, has been observed in all small-eyed or blind Late Devonian descendants of *Phacops*. In these trilobites, the reappearance of functional sutures in holaspids is of special interest. In *Dianops*, in particular, functional sutures were retained into the adult stage (Richter and Richter, 1955), which implies that onset of ankylosis and subsequent change in moulting behaviour had been retarded. Osmólska (1963) has previously shown, in a continuous growth series of *Trimerocephalus dianopsoides*, that ankylosis occurred at the end of the meraspis stage and was thus later than in the phacopine ancestor. In a recently discovered growth series of another apparently younger *Trimerocephalus*, ankylosis began at an even later stage, already within the early holaspid period. A similar trend can be observed in the latest, reduced-eyed representatives of the *Cryphops* lineage. Within the *schlosseri* group, in particular, no specimen with fused dorsal sutures is known so far. As a result, ancestral larval moulting behaviour was maintained into the postlarval stage on account of the delayed onset of ankylosis. It is striking that eye reduction, distal position of facial sutures and retarded onset of ankylosis occurred concomitantly and may all represent the effects of paedomorphosis.

RATES OF EYE REDUCTION AND EXTRINSIC CONSTRAINTS

From the examples given above it would appear that eye reduction and the associated changes of the sutural pattern were unidirectional and irreversible. No case has been reported from trilobite evolutionary history where blind ancestors gave rise to oculated descendants. Eye loss can be interpreted as an effect of paedomorphic heterochrony where the onset of eye development is extremely delayed to such an extent that it never occurs. This feature then becomes genetically fixed within the population. Consequently, as there is no instance of eye development during the entire ontogenetic trajectory, a recapitulated reappearance of the visual organ in descendant populations would scarcely be possible. Blind forms were best adapted to deep water environments and/or for infaunal modes of life. All descendants would therefore have stayed blind so long as the environmental conditions remained unchanged. These highly specialised blind descendants of the lineage would have become extinct when further deepening occasionally led to dispersal of oceanic anoxia or, on the contrary, a sharp global regression elevated the off-shore habitat into the photic zone. They were thus constrained from both above and below.

Though trends in eye reduction occur repeatedly within the same long-lived clade in different periods of time, reduction and loss of the visual organ was apparently accomplished relatively rapidly within phyletic series of species with short time spans. In this regard, it is scarcely probable that pre-adaptively reduced-eyed Early Devonian *Piriproetus* species would have given rise, as Erben (1961) claimed, to blind representatives of *Palpebralia* which existed some 20 million years later. On the one hand, blind forms of *Piriproetus* are recognised in the Early Devonian (Alberti, 1970), on the other, sighted ancestral species of *Palpebralia* occur shortly before the end-Frasnian blind forms (Feist and Schindler, 1994).

Isolated occurrences of blind proetids have often been phylogenetically linked to other blind, and therefore most similar, species, even though they are separated from them by several million years of interval without intermediate documentation. Such repeatedly occurring and misleading similarities should not be used as evidence of relationships (Whittington, 1992). According to Charig (1990) the reduction and loss of the same characters shared by two separate groups cannot be compared in order to ascertain an immediate common ancestry. For instance, the late Famennian *Bapingaspis* is considered to be derived from the latest Frasnian *Palpebralia brecciae* and the late Tournaisian *Drevermannia pruvosti* supposedly derived from late Famennian drevermanniines (Yuan, 1988; Hahn *et al.*, 1994). In both cases major extinction events, such as both the Kellwasser and Hangenberg crises which fatally affected obligate outer shelf bottom biotas and especially endobenthic blind trilobites, have not been taken into consideration. Evolutionary processes such as heterochronic developments in phyletic lineages under the control of environmental constraints as well as global extinction events, should be taken into account in considering systematic concepts.

We are left with the question of how heterochronic processes were initiated by extrinsic constraints. McNamara (1983) argued that heterochronic variation in trilobites might result from changes to the regulation of ecdysone secretion which determines the moulting process and thus controls maturation. According to Aiken (1980) changes in developmental regulation may be controlled by environmental factors such as light, temperature, water chemistry and substrate conditions. Parson (1987) demonstrated that phenotypic variability, mutation and recombination rates would augment, in particular, under conditions of increased physical stresses. These may affect the regulation of hormonal levels which occasionally leads to genetically fixed phenotypic expression of neoteny (Matsuda, 1982). Thus, it would be expected, in stressful environments, that the increase of selection pressure favours the loss of unused structures by regressive evolution, as often observed in obligate cave inhabitants (Howarth, 1993). Accordingly, the paedomorphic reduction in mean lens-number observed in succeeding populations of *Cryphops acuticeps* might result from environmentally controlled alterations to regulatory gene systems (McNamara, 1983).

Environmental factors on the outer shelf sea-bottom, especially water depth, temperature and chemistry, must have drastically changed on a global scale during the end of the Frasnian and immediately before the Kellwasser Extinction Event, one of the most severe biotic crises of the Phanerozoic (McGhee, 1989). This led to a highly stressful environment in which specific diversity diminished by increased selection pressure (Sanders, 1968; Feist, 1991). For instance, among phacopids, one single species (*Cryphops acuticeps*) is reported worldwide from this unstable environment. However, stress-induced low species diversity would have coincided with a higher degree of intraspecific morphological variability (Bretsky and Lorenz, 1970). This phenomenon can clearly be demonstrated in *C. acuticeps* populations where the coefficient of variance significantly increases towards the level of final extinction of the group (Figure 11.7). This might signify that hormonal mechanisms responsible for the onset and timing of eye lens development were profoundly perturbed by extrinsic constraints.

CONCLUSIONS

Unidirectional evolution in eye reduction of Late Devonian trilobites occurred contemporaneously in independent lineages and may have been linked to periods of global eustatic deepening. It enabled the trilobites to adapt to level-bottom conditions beyond the penetration of light.

Adaptive eye reduction is the result of paedomorphic processes when ancestral early larval configurations of the visual organ are maintained into the adult stage of descendant species. These processes probably resulted from environmentally controlled alteration of hormonal mechanisms which are responsible for growth and morphological development.

Paedomorphic eye reduction is irreversible. Thus, oculated forms cannot have been derived from blind ancestors. Furthermore, blind taxa must have originated from oculated ancestors. Such evolution was extremely short, in fact nearly instantaneous. Once the eye was lost, descendants remained blind until the extinction of the lineage, initiated by stressful, short-term and profound environmental changes such as occur during periods of global eustatic fluctuations.

ACKNOWLEDGMENTS

I am very grateful to Kenneth J. McNamara (Western Australian Museum), Euan N.K. Clarkson (University of Edinburgh), Dieter Korn (University of Tübingen), Nicolas P. Rowe and Serge Legendre (University of Montpellier) for reading an earlier version of the manuscript and for valuable comments, discussion and advice. This work was supported by the Centre National de la Recherche Scientifique, URA 327, contribution No. ISE 94/079.

REFERENCES

Aiken, D.E., 1980, Moulting and growth. In J.S. Cobb and B.F. Phillips (eds), *The biology and management of lobsters*, vol. 1, Academic Press, New York: 91–163.

Alberti, G.K.B., 1970, Zur Augenreduktion bei devonischen Trilobiten, *Paläont. Z.*, **30**: 145–160.

Alberti, H., 1972, Ontogenie des Trilobiten *Phacops accipitrinus*, *N. Jb. Geol. Pal. Abh.*, **141**: 1–3

Becker, R.T., 1993, Anoxia, eustatic changes, and Upper Devonian to lowermost Carboniferous global ammonoid diversity, *Syst. Ass. Spec. vol.*, **47**: 115–164.

Bretsky, P.W. and Lorenz, D.M., 1970, Adaptive response to environmental stability: a unifying concept in paleoecology, *North Am. Paleont. Convention, Chicago*, 1969, **Proc.** E: 522–550.

Campbell, K.S.W., 1975, The functional anatomy of phacopid trilobites: musculature and eyes, *J. Proc. R. Soc. N.S.W.*, **108**: 168–188.

Charig, A.J., 1990, Evolutionary Systematics. In D.E.G. Briggs and P.R. Crowther (eds), *Palaeobiology: a synthesis*, Blackwell Scientific Publications, Oxford: 434–437.

Chatterton, B.D.E., 1971, Taxonomy and ontogeny of Siluro-Devonian trilobites from near Yass, New South Wales, *Palaeontographica* (A), **137**: 1–108.

Chlupáč, I., 1977, The phacopid trilobites of the Silurian and Devonian of Czechoslovakia, *Rozpr. Ústřed. ústav. geol.*, **43**: 1–164.

Clarkson, E.N.K., 1975, The evolution of the eye in trilobite, *Fossils and Strata*, **4**: 7–31.

Clarkson, E.N.K. and Zhang Xi-guang, 1991, Ontogeny of the Carboniferous trilobite *Paladin eichwaldi shunnerensis* (King 1914), *Trans. R. Soc. Edin., Earth Sci.*, **82**: 277–295.

Delo, D.M., 1937, Secondary blinding among the phacopid trilobites and its significance, *Am. Mid. Nat.*, **18**: 1096–1102.

Erben, H.K., 1961, Blinding and extinction of certain Proetidae (Tril.), *J. Pal. Soc. India*, **2** (1958): 82–104.

Feist, R., 1976, Systématique, phyogénie et biostratigraphie de quelques Tropidocoryphinae du Dévonien Français, *Géobios*, **9**: 47–80.

Feist, R., 1991, The late Devonian trilobite crises, *Hist. Biol.*, **5**: 197–214.

Feist, R. and Clarkson, E.N.K., 1989, Environmentally controlled phyletic evolution, blindness and extinction in Late Devonian tropidocoryphine trilobites, *Lethaia*, **22**: 359–373.

Feist, R. and Schindler, E., 1994, Trilobites during the Frasnian Kellwasser Crisis in European Late Devonian cephalopod limestones, *Cour. Forsch. Inst. Senckenberg*, **169**: 195–223.

Fortey, R.A. and Owens, R.M., 1990, Trilobites. In K.J. McNamara (ed.), *Evolutionary trends*, Belhaven, London: 121–142.

Fortey, R.A. and Whittington, H.B., 1989, The Trilobita as a natural group, *Hist. Biol.*, **2**: 125–138.

Hahn, G., Hahn, R. and Brauckmann, C., 1994, Trilobiten mit *Drevermannia*-Habitus im Unter-Karbon, *Cour. Forsch. Inst. Senckenberg*, **169**: 155–193.

Harrington, H.K., 1959, General description of Trilobita. In R.C. Moore (ed.), *Treatise on Invertebrate Paleontology, part O, Arthropoda 1*, University of Kansas Press and Geol. Soc. Am., Lawrence, Kansas: O38–O117.

Henningsmoen, G., 1975, Moulting in trilobites, *Fossils and Strata*, **4**: 179–200.

Howarth, F.G., 1993, High-stress subterranean habitats and evolutionary change in cave-inhabiting arthropods, *Am. Nat.*, **142** (suppl.): S65–S77.

Hupé, P., 1953, Classe des trilobites. In J. Piveteau (ed.), *Traité de Paléontologie*, vol. 3, Masson et Cie., Paris: 44–246.

Jahnke, H., 1969, *Phacops zinkeni* F.A. Roemer 1843—ein Beispiel für eine ontogenetische Entwicklung bei Phacopiden (Trilobitae, Unterdevon), *N. Jb. Geol. Pal. Abh.*, **133**: 309–324.

Johnson, J.G., Klapper, G. and Sandberg, C.A., 1985, Devonian eustatic fluctuations in Euramerica, *Geol. Soc. Amer. Bull.*, **96**: 567–587.

Lütke, F., 1968, Trilobiten aus dem Oberdevon des Südwest-Harzes—Stratigraphie, Biotop und Systematik, *Senck. leth.*, **49**: 119–191.

Maksimova, S.A., 1955, Trilobity srednego i verchnego devona Urala i severnych Mugodschar, *Trudy vsesojusn. nautschno-issled. geol Inst. (VSEGEI)*, **3**: 1–263.

Matsuda, R., 1982, The evolutionary process in talitrid amphipods and salamanders in changing environments, with a discussion of "genetic assimilation" and some other evolutionary concepts, *Can. J. Zool.*, **60**: 733–749.

McGhee, G.R., Jr, 1989, The Frasnian–Famennian extinction event. In S.K. Donovan (ed.), *Mass extinctions: processes and evidence*, Belhaven, London: 133–151.

McKinney, M.L., 1984, Allometry and heterochrony in an Eocene echinoid lineage: morphological change as a by-product of size selection, *Paleobiology*, **10**: 407–419.

McNamara, K.J., 1983, Progenesis in trilobites. In D.E.G. Briggs and P.D. Lane (eds), *Trilobites and other Arthropods: Papers in Honour of H.B. Whittington. F.R.S. Special Papers in Palaeontology*, **31**: 59–68.

McNamara, K.J., 1986, A guide to the nomenclature of heterochrony, *J. Paleont.*, **60**: 4–13.

Osmólska, H., 1963, On some Famennian Phacopinae (Trilobita) from the Holy Cross Mountains, Poland, *Acta palaeont. pol.*, **8**: 495–519.

Parson, P.A., 1987, Evolutionary rates under environmental stress, *Evol. Biol.*, **21**: 311–347.

Ramsköld, L., 1988, Heterochrony in Silurian phacopid trilobites as suggested by the ontogeny of *Acernaspis*, *Lethaia*, **21**: 307–318.

Richter, R., 1913, Beiträge zur Kenntnis Devonischer Trilobiten II. Oberdevonische. *Abh. Senck. Nat. Gesell.* **30**: 245–423.

Richter, R. and Richter, E., 1926, Die Trilobiten des Oberdevons, Beiträge zur Kenntnis

Devonischer Trilobiten IV, *Abh. preuss. geol. L.-A.*, n.F., **99**: 1–314.

Richter, R. and Richter, E., 1955, Oberdevonische Trilobiten, Nachträge. Trilobiten aus der *Prolobites-Stufe* III. 2. Phylogenie der oberdevonischen Phacopidae, *Senck. leth.*, **36**: 49–72.

Sanders, H.L., 1968, Marine benthic diversity: a comparative study, *Am. Natur.*, **108**: 243–282.

Schrank, E., 1973, *Denckmannites caecus* n.sp., ein blinder Phacopide aus dem höchsten Thüringer Silur, *Z. geol. Wiss.*, **1**: 347–351.

Speyer, S.E., 1985, Moulting in phacopid trilobites, *Trans. R. Soc. Edin. (Earth Sci.)*, **76**: 239–253.

Speyer, S.E. and Chatterton, B.D.E., 1989, Trilobite larvae and larval ecology, *Hist. Biol.*, **3**: 27–60.

Wilde, P. and Berry, W.B.N., 1984, Destabilization of the oceanic density structure and its significance to marine "extinction" events, *Palaeogeogr. Palaeoclim. Palaeoecol.*, **48**: 143–162.

Whittington, H.B., 1957, The ontogeny of trilobites, *Biol. Rev.*, **32**: 421–469.

Whittington, H.B., 1992, *Trilobites*, The Boydell Press, Woodbridge.

Yuan, Jinliang, 1988, Proetiden aus dem Jüngeren Oberdevon von Süd-China, *Palaeontographica* A, **201**: 1–102.

Chapter 12

IMPACT OF ENVIRONMENTAL PERTURBATIONS ON HETEROCHRONIC DEVELOPMENT IN PALAEOZOIC AMMONOIDS

Dieter Korn

INTRODUCTION

Heterochrony as a fundamental morphogenetic process has, during the last decade, been detected in many lineages of ammonoid evolution, and the phenomenon of paedomorphosis is particularly well known from Mesozoic forms (Landman, 1988; Dommergues, 1990). Landman and Geyssant (1993) have observed that in the evolution of Mesozoic ammonoids peramorphosis is less important than paedomorphosis, and that within the paedomorphs the ratio between the processes of progenesis and neoteny may vary in different ecological groups of ammonites. Despite many publications describing heterochronic development of ammonites (summarised by Landman and Geyssant, 1993), only a little is known about the relationships between heterochronic development and environmental perturbations in ammonoids in general.

Among Palaeozoic ammonoids, small paedomorphs are especially common in deposits of the Late Famennian (Korn, 1992; 1995) and Middle Permian (Zhao and Zheng, 1977; Frest *et al.*, 1981; Glenister and Furnish, 1988). In both epochs, in addition to the possession of minute shells, these dwarfed forms are characterised by the development of different spectacular conch and sculpture morphologies, making these forms unique components in the ammonoid faunas. Interestingly, many Late Devonian and Middle Permian paedomorphic ammonoids belong to the extremely long-ranging goniatite superfamily Prionocerataceae. This originated in the Middle Famennian (Korn, 1994) and became extinct in the latest Permian Changxingian Stage (Frest *et al.*, 1981). The paedomorphic

Evolutionary Change and Heterochrony. Edited by Kenneth J. McNamara © 1995 The Editor and Contributors.
Published in 1995 by John Wiley & Sons Ltd.

ammonoids of this superfamily predominantly occur in sediments that have been deposited in a comparatively shallow environment.

Extrinsic factors, such as the influence of sea-level changes and their effects on the environment, have been recognised as important influences in the evolutionary history of Palaeozoic ammonoids (Saunders and Swan, 1984; Swan, 1988; House, 1989; Korn, 1992; Becker, 1993), but there is currently much debate about whether eustatic transgressions or regressions led to extinction events and adaptive radiations, respectively. Analysing patterns of heterochrony in Palaeozoic ammonoids may shed further light on this question. Difficulties in the direct correlation of environmental perturbations with the evolution of Palaeozoic ammonoids arise since there exist only a few comprehensive investigations concerning the microfacies and biofacies development of the ammonoid-bearing sections. Furthermore, interpretation of external and internal shell characters of ammonoids does not provide a full picture of the lifestyle of these animals. Thus, suggestions concerning the direct influences of environmental changes on ammonoid evolution are still rather speculative.

HABITATS OF PALAEOZOIC AMMONOIDS

In contrast to Mesozoic ammonoids, only a few studies examining the ecology of Palaeozoic ammonoids have been carried out. Therefore, our knowledge of the habitat—whether nektic, nekto-benthic, or benthic—of most of the Palaeozoic ammonoids is comparatively limited. On account of the shell geometries of Palaeozoic ammonoids it can, however, be assumed that there is some correspondence with the schemes presented by Westermann (1990) for Jurassic and Cretaceous forms, and by Wang and Westermann (1993) for Triassic ammonoids. Apart from the Mesozoic heteromorphs, many conch and sculpture morphologies were already realised in the Palaeozoic, where distribution patterns of ammonoids show similar occurrences of smoothly sculptured, compressed and involute shells in deeper facies, and coarsely ornamented ammonoids with many different shell morphologies in shallower facies.

As Westermann (1990) has pointed out for Mesozoic ammonoids, the similar shell shapes of different ammonoids may reflect similar behaviour and swimming capability, but does not necessarily imply life in analogous habitats. Therefore, an interpretation of the ammonoid habitat using only this external structure must remain equivocal. In Palaeozoic ammonoids, buoyancy, hydrodynamic efficiency and sculpture can be interpreted in terms of functional constraints, but for the majority of forms these are not well understood (Swan and Saunders, 1987). Interpretation of Palaeozoic ammonoid habitats is here based mainly on analyses of functional morphology and palaeoenvironmental data of autochthonous assemblages from Late Devonian rocks of Central Europe and Northern Africa.

Examination of sections through the Wocklum Limestone in Germany and its lateral equivalents allows a sedimentological comparison of basin sections with those on top of submarine elevations ("Tiefschwellen"), caused by Middle and early Late Devonian reef buildups. The distribution of different ammonoid types in the Late Devonian *Wocklumeria* Stufe shows a striking pattern (Figure 12.1). Adult shells of non-paedomorphic ammonoids (genera *Discoclymenia, Sporadoceras, Cymaclymenia, Cyrtoclymenia*, etc.) are especially abundant in the basinal facies and are far less common in the Schwellen environment, where the juveniles of these species occur much more abundantly. One can perceive a significant separation of adult and juvenile shells of these ammonoids, suggesting that during their lifetime these ammonoids migrated from shallower into deeper environments.

The Schwellen facies is dominated by paedomorphs (genera *Balvia, Linguaclymenia, Parawocklumeria*, etc.), associated with juveniles of the non-paedomorphic ammonoids. A similar distribution pattern has been documented by Landman (1989) for Cretaceous scaphitids. Since a separation by water energy can be ruled out (postmortem drifting is unlikely because of the constant separation of the various types), differences in the occurrence of the ammonoid sizes can be correlated with different ecological conditions. Both types of ammonoids, however, are typical for pelagic off-shore facies. They have been collected only rarely in sedimentary rocks with increasing terrestrial influence (except for *Cymaclymenia evoluta*, which is typical for terrigenous sediments), and they appear to be absent in oolitic formations. Generally, the content of ammonoids increases with a decrease of clay intercalations. The ammonoid shells are deposited following a very short transport distance, as indicated by their preservation with the body chamber. However, they are often densely packed due to submarine slumping events (Clausen *et al.*, 1990). Turbidite-like beds are extremely rare. From the sedimentary features it can be concluded that the fossil content of the limestone beds reflects the shell-bearing fauna on the sea bottom and in the water column above it.

DEVONIAN AMMONOID HETEROCHRONY

Material, location, stratigraphy

The correlation of the ammonoid faunal development with eustatic fluctuations is based on bed-by-bed collecting of approximately 5000 specimens from sections in the northern Rhenish Massif, Germany. These sections show a complete succession of the fossiliferous, 3 m thick Wocklum Limestone (corresponding with the lower four zones of the latest Devonian *Wocklumeria* Stufe), which is predominantly composed of thin-bedded nodular limestones alternating with more or less prominent clay intercalations (Figure 12.2). The sections contain a continuous succession of ammonoid faunas, without any unfossiliferous partitions within the Wocklum Limestone. Using ammonoids, the *Wocklumeria* Stufe can be divided into five zones, some of which can be subdivided even further.

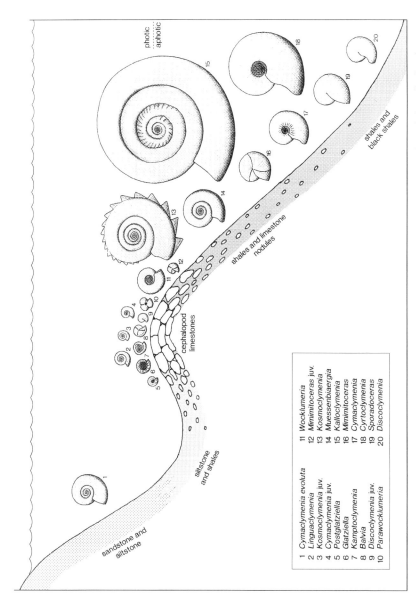

Figure 12.1 Stylised panorama of latest Devonian ammonoid habitats. Note that the majority of the diminutive ammonoids (paedomorphs) as well as juveniles of non-paedomorphs) occur in elevated areas, whereas the larger adults of non-paedomorphs are typical for deeper facies. All ammonoids presented at equal scale.

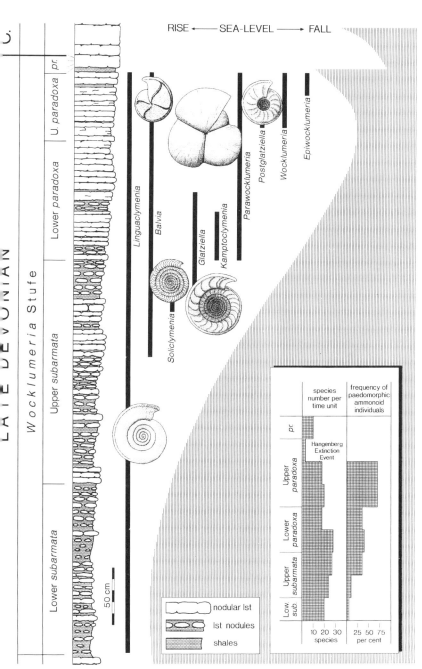

Figure 12.2 Columnar section of the latest Devonian Wocklum Limestone in the Muessenberg area of the Rhenish Massif, with ammonoid stratigraphy, distribution of paedomorphic genera and correlation with proposed sea-level changes. Note the continuous increase of paedomorphic genera paralleling the proposed regressive trend. Also shown in the box at right bottom is the diversity of species and frequency of paedomorphs in the latest Devonian *Wocklumeria* Stufe. C. = Carboniferous, pr. = prorsum zone.

In the Late Devonian *Wocklumeria* Stufe, two different ammonoid types occur:

1. *Non-paedomorphic ammonoids.* Long-ranging genera such as *Cyrtoclymenia, Cymaclymenia, Kosmoclymenia, Mimimitoceras* and *Sporadoceras*. They retain a shell size of at least 50 mm up to the giant *Kalloclymenia* which exceeds 300 mm in diameter. With only a few exceptions, these ammonoids lack strong sculpture. The adult shells display no distinct features, such as ventrolateral grooves or ventrolateral projections of the aperture.
2. *Paedomorphic ammonoids.* This morphological group is composed of ammonoids which reach diameters of only 30 mm, sometimes even much less. Almost all the paedomorphic ammonoids show a somewhat peculiar morphology. Most of them bear ventrolateral grooves, many are ribbed or coarsely ornamented, and some of them, which belong to independent lineages, are triangularly coiled. In many species, terminal growth can be observed, being shown by ventrolateral projections of the aperture. These genera and species are short-ranging, except for *Linguaclymenia similis* which ranges throughout the lower four ammonoid zones of the *Wocklumeria* Stufe.

Heterochronic processes

Various heterochronic processes can be observed in the evolution of latest Devonian ammonoids.

Progenesis

Several independent evolutionary lineages are characterised by a rapid size reduction, resulting in diminutive forms. These dwarfs generally resemble the ancestral juveniles morphologically and are not scaled-down versions of their normal-sized ancestors. Interestingly, there are no intermediates known which link ancestral non-paedomorphs with descendant paedomorphs. Size reduction from *Mimimitoceras* to *Balvia* was 4 : 1 (Figure 12.3), and from *Kosmoclymenia* to *Linguaclymenia* 5 : 1 (Korn, 1995), probably resulting from a drastic shortening of ontogeny by precocious maturation. All the earliest known forms of progenetic genera display a very simple shell morphology, but formed the basis for adaptive radiations leading to new phenotypes with various shell and ornament peculiarities, such as triangular coiling (*Kamptoclymenia, Soliclymenia*), apertural projections (*Linguaclymenia, Glatziella*), ventrolateral grooves (*Linguaclymenia, Balvia, Glatziella, Kamptoclymenia*) and different sculptures with ribs (*Glatziella*).

Neoteny

Shell development in the supposed lineage *Kamptoclymenia endogona* → *Parawocklumeria paradoxa* (Figure 12.4) shows a constant retention of juvenile characters to the adults (Schindewolf, 1937; Korn, 1995).

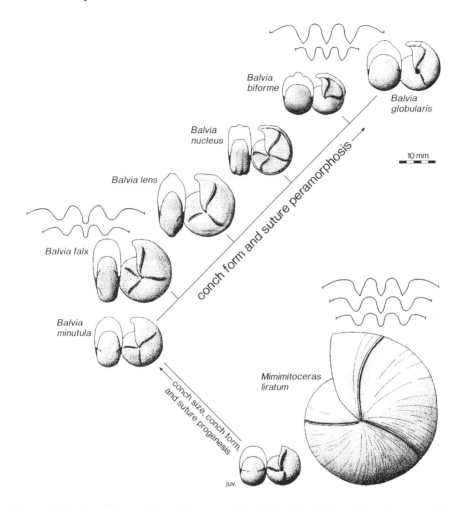

Figure 12.3   Possible paedo- and peramorphoclades in Late Devonian prionoceratid ammonoids. Note the two-phased evolutionary pattern: 1. Rapid size decrease from *Mimimitoceras* to *Balvia* resulting in progenetic dwarfs displaying the preadult ancestral morphology. 2. Conch form and suture development caused by a combination of hypermorphosis and acceleration, producing different conch and ornament features and pointed lobes. All ammonoids presented at equal scale, suture lines not to scale.

Hypermorphosis/acceleration

Sutural development of the investigated lineages documents the transformation of broadly rounded lobes into pointed lobes. This occurs in the lineages *Parawocklumeria* → *Wocklumeria* and *Glatziella* → *Postglatziella* as well as within the genus *Balvia* (Korn, 1995). Besides this, sculpture characteristics are developed during phylogeny, with the most advanced forms typically at the end

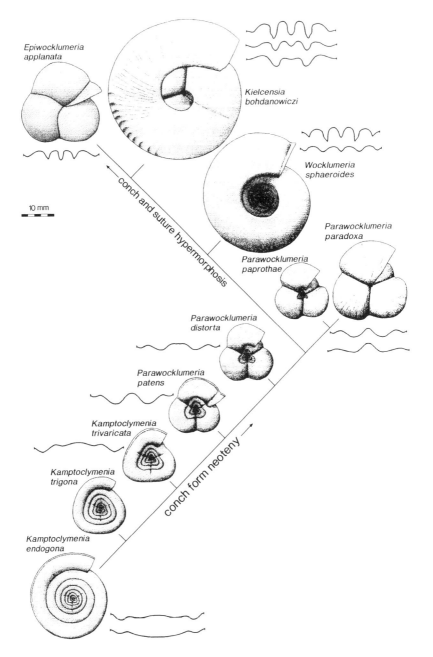

Epiwocklumeria
applanata

Kielcensia
bohdanowiczi

Wocklumeria
sphaeroides

10 mm

conch and suture hypermorphosis

Parawocklumeria
paradoxa

Parawocklumeria
paprothae

Parawocklumeria
distorta

Parawocklumeria
patens

Kamptoclymenia
trivaricata

conch form neoteny

Kamptoclymenia
trigona

Kamptoclymenia
endogona

of the lineages. It is obvious that these processes generally resemble the evolutionary pattern observed in peramorphic ammonoids.

Environmental perturbations and ammonoid paedomorphosis

During the early *Wocklumeria* Stufe, several evolutionary trends within the ammonoid phylogeny occurred, and these parallel the observed regression (Figure 12.2):

1. The development of different conch morphologies increases continuously, with its peak in the lower part of the early *paradoxa* Zone. Particularly striking is the triangular coiling of some genera, the development of ventrolateral grooves, and terminal pertures with ventrolateral projections.
2. Consequently, the number of species increases constantly—from about 20 to approximately 30.
3. The ratio of paedomorphs to non-paedomorphs increases both in species and in individual number.
4. The paedomorphs are usually short-ranging and they change their morphology rapidly during phylogeny. In contrast, the non-paedomorphs are especially long-ranging with only minor modifications of shell form and ornament. Their evolution shows only few innovations.

To understand possible reasons for the latest Devonian ammonoid paedomorphosis and the subsequent proliferation of separate ammonoid groups, it is necessary to determine changes that occurred in the extrinsic regimes. Environmental fluctuations, such as temperature, salinity, oxygen content, nutrient supply, or water chemistry perturbations, are difficult to trace because of diagenetic alteration of the rocks. However, it is possible to document sea-level changes by means of microfacies and biofacies analyses. Smaller-scale trangressions and regressions are barely visible in the basin sections, but affected the more sensitive Schwellen sedimentation much more, resulting in compound changes of clay intercalations or biogene composition and density.

All the examined sections show a fluctuation of the carbonate microfacies within the Wocklum Limestone. At the base of this lithostratigraphic unit,

---

Figure 12.4 Possible paedo- and peramorphoclades in the Late Devonian wocklumeriid ammonoids. Note the two-phase evolutionary trends: 1. Conch form neoteny expressed by continuous transformation of three-segmented or triangular inner whorls to adults in the lineage from *Kamptoclymenia* to *Parawocklumeria*, but without remarkable sutural development. 2. Conch and suture hypermorphosis shown by conch size increase and formation of pointed lobes during ontogeny in the lineage from *Parawocklumeria* to *Wocklumeria*, *Kielensia*, and *Epiwocklumeria*. All ammonoids presented at equal scale; suture lines not to scale.

mudstones occur with prominent intercalations of clay. Towards the top of the Wocklum Limestone, the clay content decreases markedly, with the limestones turning to bioclastic wackestones and sometimes crinoidal packstones. From bottom to near the top of the Wocklum Limestone, an almost continuous increase of biogenic content is recognisable. These data demonstrate a more or less continuous reduction of the water depth.

Further evidence for a regressive trend within the Wocklum Limestone is also provided by the distribution of blind and multiocular trilobites (Hahn *in* Luppold *et al.*, 1984). Frequently, the Wocklum Limestone contains blind trilobites, indicating a habitat depth below the photic zone (see Chapter 11). Trilobites with big eyes—indicating the penetration of light down to the bottom—are scarce and only occur in the higher part of the Wocklum Limestone, together with a more diverse fauna including pelmatozoans, abundant rugose corals and foraminifers.

The adaptive radiation of several independent ammonoid lineages started in the lower part of the *Wocklumeria* Stufe and is closely related to the change in sedimentary regimes. It suggests a close correlation between the rapid ammonoid evolution rate and environmental changes due to a regressive trend, involving instability in the former uniform habitat. The shallowing resulted in a diversification of the pelagic habitat, and is likely to have led to allopatric speciation. This in turn led to the adaptive radiation observed in both goniatites and clymeniids.

These unstable conditions may have been the direct cause of rapid maturation (Mancini, 1978). It is unclear whether the Late Devonian micromorphs exploited a vacant niche, or were simply supported by selection on smaller body size. It is possible that the paedomorphosis in the ammonoids may have been induced by temperature changes. As Zuev *et al.* (1979) have shown for Recent North Atlantic squid, body size can decrease with warmer temperatures. This resembles the observation that the latest Devonian paedomorphs occur in a shallower, and probably warmer, environment than the non-paedomorphic ammonoids.

The regression in the lower two-thirds of the Wocklum Limestone is reversed into a transgression which culminates in the Hangenberg Black Shale. This is considered to be a result of an onlap of anoxic conditions over the shelf margin. Wang *et al.* (1993) have discussed the reasons for a large-scale extinction event at this horizon and did not preclude the possibility of an extraterrestrial impact. The unit only sporadically contains an ammonoid fauna, which is poor in species number (Korn *et al.*, 1994). None of the paedomorphic ammonoid species survived the Hangenberg Event. This is probably because of the close connection between these ammonoids and the benthos that suffered the ascending euxinic conditions covering the shelf margins. The combination of the regression with a subsequent transgression, however, may have been responsible for the mass extinction (Hallam, 1992).

PERMIAN AMMONOID HETEROCHRONY

Material, location, stratigraphy

The prionoceratid ammonoids of the family Shouchangoceratidae described by Zhao and Zheng (1977) are among the most bizarre of all Palaeozoic ammonoids. They display different spectacular ornament types which are rare or unique in Palaeozoic ammonoids (Figure 12.5). These forms are the main components of an endemic fauna that is found especially in the Zhejiang (Dingjiashan Formation), Jiangxi (Hutang Formation), and Hunan (Maokou Formation) provinces in South China. Based on the co-occurring ammonoid genera, the stratigraphic position of the sediments bearing these ammonoids is regarded as being Roadian to Wordian (Frest *et al.*, 1981). All these occurrences belong to the "restricted sea ecological pattern" of the Chinese authors, which is widespread in south China (Zhou *et al.*, 1989). The ammonoid fauna of this ecological type is dominated by small and coarsely ornamented ammonoids, usually with simple suture lines. In contrast to this is the "open sea ecological pattern", containing large and less ornamented ammonoids with derived, multilobed suture lines.

The bizarre prionoceratid ammonoids discussed and illustrated in this chapter came mainly from three localities. The Dingjiashan Formation of Jiande in Zhejiang is a series of about 350 m of dark weathering shale consisting of four members, from bottom to top the Dongwuli Member, the Shimei (stone coal) Member, the Shimentang Member, and the Lixian Coal Series (Zhao and Zheng, 1977). The second and third of these members contain a peculiar ammonoid fauna consisting of some ceratite and goniatite species, the latter group represented by the prionoceratid genera *Shouchangoceras* and *Sangzhites*. The ammonoids occur here in carbonate nodules within the shales. The Hutang Formation of Shangrao in Jiangxi and the Maokou Formation of Sanzhi in Hunan are calcareous correlative representatives of the Dingjiashan Formation. They yielded a similar fauna, but with more representatives of other ammonoid superfamilies.

Ammonoid paedomorphosis

In the Middle Permian, three different evolutionary lineages within the prionoceratid ammonoids are recognisable (Figure 12.5). The origin of these peculiar ammonoids may be the long-ranging genus *Neoaganides* that sporadically occurs in beds from the Pennsylvanian Cherokee Shale up to the Late Permian Dzhulfian Stage (Frest *et al.*, 1981). Species of this genus are characterised by a simple lenticular shell morphology, and usually they are very small (less than 10 mm diameter) with a simple suture of broadly rounded lobes. This is a juvenile character of the prionoceratid ammonoids.

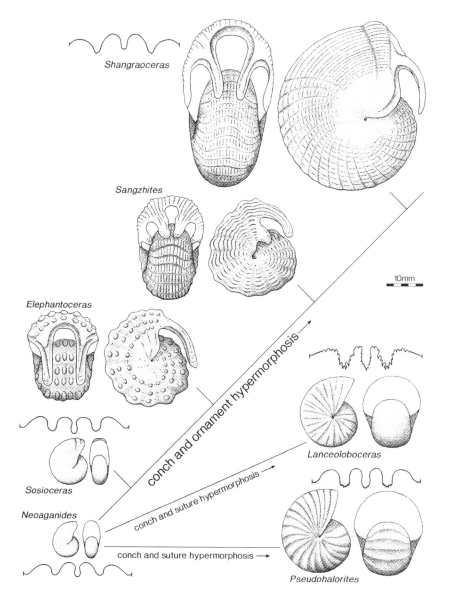

Figure 12.5   Possible peramorphoclades in Middle Permian prionoceratid ammonoids. Note that the proposed tiny ancestor *Neoaganides* is an extremely simple goniatite lacking any conch and ornament peculiarities and with a suture line displaying rounded lobes. Further development in three independant lineages is determined by hypermorphosis, resulting in larger conchs with different sculptures and advanced suture lines. All ammonoids presented at equal scale; suture lines not to scale.

Neoaganides cannot be connected directly to any other ammonoid genus. All the prionoceratid ammonoids that occur between the Late Tournaisian and the Early Pennsylvanian (genera *Imitoceras* and *Irinoceras*) have large shells reaching more than 150 mm diameter and displaying pointed lobes in the adult stage. Thus, *Neoaganides* can be regarded as an extreme progenetic paedomorph, like the Late Devonian genus *Balvia*. The Middle Permian adaptive radiation initiated by *Neoaganides* is typical for hypermorphosis.

The subfamily Shouchangoceratinae consists of forms which display many spectacular sculptures, with a combination of coarse longitudinal and transverse ribs, nodes, spiral lirae and multiple constrictions. The first descendant of *Neoaganides* is possibly the genus *Sosioceras*, known from the Wordian Stage of Sicily. In size, suture and conch form it closely resembles the genus *Neoaganides*, but is characterised by an apertural modification with terminal constrictions (Frest *et al.*, 1981). More advanced genera such as *Elephantoceras*, *Sangzhites*, *Shangraoceras* and *Shouchangoceras* grew larger and developed a striking sculpture and shell morphology. In the most bizarre species, pronounced apertural lappets are formed. Size increase occurs to a maximal diameter of 45 mm (*Shouchangoceras*), but usually remains less than 30 mm. The suture line of these forms remained simple, frequently with rounded lobes in the smaller and pointed lobes in the larger species.

The subfamilies Pseudohaloritinae and Lanceoloboceratinae also include small ammonoids up to 50 mm diameter. Usually they are ribbed, but lack striking ornaments. These forms develop ceratitic (*Pseudohalorites*) or ammonitic (*Lanceoloboceras*) suture lines.

DISCUSSION

Obviously there is some resemblance between the patterns of evolution of the Late Devonian and the Middle Permian ammonoid paedomorphs. Frequently, the development of these forms passes through two phases:

1. An event-like miniaturisation due to progenesis. Examples are the earliest representatives of the latest Devonian genus *Balvia* and the Late Carboniferous to Late Permian genus *Neoaganides*. Shell morphology, suture line and ornament of these ammonoids closely resemble those of juveniles of their normal-sized ancestors.
2. From these dwarf forms, peramorphosis (both hypermorphosis and acceleration) took place with important morphological novelties resulting in radiations with numerous different species. Very often the descendants are characterised by spectacular shell, sculpture and suture developments.

It is difficult to trace satisfying ecological reasons for this two-phase process. However, it could be regarded as a reaction to environmental perturbations which promoted changes in the selection regime. The progenetic event can be regarded as resulting from instability in the conditions of the habitat arising from a regressive eustatic trend that required a faster reproduction rate.this could have been achieved by precocious maturation. The resulting shorter time of development is reflected in the miniaturisation characteristic of these paedomorphs (Gould, 1977; McKinney and McNamara, 1991). In the examples discussed, adaptive radiations were initiated by progenetic miniaturisations, indicating increased specific plasticity which allowed them to occupy a wider range of ecological niches. Most probably, the selective advantage of a smaller body size compensated for the simplification of shell and ornamentation characters. During phylogeny, however, many morphological innovations led to very distinct shell characters, suggesting a slow, but continuous, re-optimisation of shell geometry and sculpture.

Strikingly, the pace of heterochronic development occurs with completely different rates. The drastic size reduction caused by progenesis was achieved at a geologically very rapid rate. However, neotenic or hypermorphotic development was achieved at a slower rate. It can be assumed that response to selective pressure on smaller body size was most effectively achieved with minor genetic change by early offset of ontogenetic development. The result was the evolution of tiny goniatites which differ remarkably from the morphology of their adult ancestors. Later phylogenetic development seems to follow the common evolutionary "rules" that can be observed in other ammonoid lineages. But, at least in the latest Devonian ammonoids, introduction of morphological novelties was achieved at a much faster rate in paedomorphs that in non-paedomorphs (Korn, in prep.).

Further investigations will show if the results presented here can be adopted for explanations of other cases of ammonoid dwarfism. A superficial examination suggests a very similar pattern in different ammonoid lineages in which progenetic development occurs spontaneously. However, this is not necessarily always followed by a new radiation. Especially in the Permian, progenetic ammonoids appear to be occurring at the end of phylogenetic lineages, where they failed to give rise to extensive adaptive radiations, suggesting terminal progenesis (Glenister and Furnish, 1988).

ACKNOWLEDGMENTS

I am indebted to Tim Jones (Tübingen), Kenneth J. McNamara (Perth), and Rainer R. Schoch (Tübingen) for discussion and revision of the article and for many helpful suggestions.

REFERENCES

Becker, R.T., 1993, Anoxia, eustatic changes, and Upper Devonian to lowermost Carboniferous global ammonoid diversity. In M.R. House (ed.), *The Ammonoidea: environment, ecology, and evolutionary change*, Clarendon Press, Oxford: 115–163.

Clausen, C.-D., Korn, D., Luppold, F.W. and Stoppel, D., 1990, Untersuchungen zur Devon/Karbon-Grenze auf dem Muessenberg (nördliches Rheinisches Schiefergebirge), *Bull. Soc. belge Géol.* **98**, 353–369.

Dommergues, J.-L., 1990, Ammonoids. In K.J. McNamara (ed.), *Evolutionary trends*, Belhaven, London: 162–187.

Frest, T.J., Glenister, B.F. and Furnish, W.M., 1981, Pennsylvanian-Permian Cheiloceratacean Ammonoid families Maximitidae and Pseudohaloritidae, *J. Paleont.*, *Mem.* **11**: 1–46.

Glenister, B.F. and Furnish, W.M., 1988, Terminal progenesis in Late Palaeozoic ammonoid families. In J. Kullmann and J. Wiedmann (eds), *Cephalopods, present and past*, Schweizerbart, Stuttgart: 51–66.

Gould, S.J., 1977, *Ontogeny and phylogeny*, Harvard University Press, Cambridge, M.A.

Hallam, A., 1992, Phanerozoic sea-level changes. In D. Bottjer and K. Bambach (eds), *The perspectives in paleobiology and earth history series*, Columbia University Press, New York.

House, M.R, 1989, Ammonoid extinction events, *Phil. Trans. Roy. Soc. London*, B, **325**: 307–326.

Korn, D., 1992, Heterochrony in the evolution of Late Devonian ammonoids, *Acta Palaeont. Polonica*, **37**: 21–36.

Korn, D., 1994, Devonische und karbonische Prionoceraten (Cephalopoda, Ammonoidea) aus dem Rheinischen Schiefergebirge, *Geol. Palaeont. Westf.*, **30**: 1–85.

Korn, D., 1995, Paedomorphosis of ammonoids as a result of sealevel fluctuations in the Late Devonian *Wocklumeria* Stufe, *Lethaia*, in press.

Korn, D, Clausen, C.-D., Belka, Z., Leuteritz, K., Luppold, F.W., Feist, R. and Weyer, D., 1994, Die Devon/Karbon-Grenze bei Drewer (Rheinisches Schiefergebirge), *Geol. Palaeont. Westf.*, **29**: 97–147.

Landman, N.H., 1988, Heterochrony in ammonites. In M.L. McKinney (ed.), *Heterochrony in evolution: a multidisciplinary approach*, Plenum, New York: 159–182.

Landman, N.H., 1989, Iterative progenesis in Upper Cretaceous ammonites, *Paleobiology*, **15**: 95–117.

Landman, N.H. and Geyssant, J.R., 1993, Heterochrony and ecology in Jurassic and Cretaceous ammonites, *Geobios*, **15**: 247–255.

Luppold, F.W., Hahn, G. and Korn, D., 1984, Trilobiten-, Ammonoideen- und Conodonten-Stratigraphie des Devon-Karbon-Grenzprofiles auf dem Muessenberg (Rheinisches Schiefergebirge), *Cour. Forsch.–Inst. Senckenberg*, **67**: 91–121.

Mancini, E.A., 1978, Origin of micromorph faunas in the geologic record, *J. Paleont.*, **52**: 311–322.

McKinney, M.L. and McNamara, K.J., 1991, *Heterochrony: the evolution of ontogeny*, Plenum, New York.

Saunders, W.B. and Swan, A.R.H., 1984, Morphology and morphologic diversity of mid-Carboniferous (Namurian) ammonoids in time and space, *Paleobiology*, **10**: 195–228.

Schinderwolf, O.H., 1937, Zur stratigraphie und palaeontologie der Wocklumer Schichten (Oberdevon). *Abh. Preuss. Geol. Landes.*, new series **178**: 1–132.

Swan, A.R.H., 1988, Heterochronic trends in Namurian ammonoid evolution, *Palaeontology*, **31**: 1033–1051.

Swan, A.R.H. and Saunders, W.B., 1987, Function and shape in late Palaeozoic (mid-Carboniferous) ammonoids, *Paleobiology*, **13**: 297–311.

Wang, Y. and Westermann, G.E.G., 1993, Paleoecology of Triassic ammonoids, *Geobios*, **15**: 373–392.

Wang, K., Attrep, M. and Orth, C.J., 1993, Global iridium anomaly, mass extinction, and redox change at the Devonian–Carboniferous boundary, *Geology*, **21**: 1071–1074.

Westermann, G.E.G., 1990, New developments in ecology of Jurassic–Cretaceous ammonoids. *Atti II Conv. Int. F.E.A. Pergola*, 87, ed. Pallini *et al.*: 459–478.

Zhao Jinke and Zheng Zhuoguong, 1977, Ammonoids of the late Early Permian Period from western Zhejiang and northwestern Jiangxi, *Acta Paleont. Sinica*, 16: 217–251 [in Chinese with English summary].

Zhou Zuren, Glenister, B.F. and Furnish, W.M., 1989, Two-fold or three-fold? Concerning geological time-scale of Permian period, *Acta Paleont. Sinica*, **28**: 269–282 [in Chinese with English summary].

Zuev, G.V., Nigmatulin, C.M. and Nikolsky, V.N., 1979, Growth and life span of *Sthenoteuthis pteropus* in the east-central Atlantic, *Zool. Zh.*, **58**: 1632–1641 [in Russian].

Chapter 13

HETEROCHRONY AND THE EVOLUTION OF ECOGEOGRAPHIC SIZE VARIATION IN MALAGASY SIFAKAS

Matthew J. Ravosa, David M. Meyers and Kenneth E. Glander

> *A plausible argument could be made that evolution is the control of development by ecology. Oddly, neither area has figured importantly in evolutionary theory since Darwin, who contributed much to each.*
>
> (Van Valen, 1973, p. 488)

INTRODUCTION

In recent reviews of the systematics and behavioural ecology of Malagasy primates, both Richard and Dewar (1991) and Tattersall (1992) remark on the importance of ecological factors in lemur evolution. Living Malagasy lemurs range in body weight from 60 g to upwards of 10 000 g. When recently extinct species are included, the range in body size is comparable to that evinced for all other extant primates. As such, a major characteristic of the Malagasy primate radiation is significant body-size diversity coupled with considerable taxonomic diversity.

In this regard, Albrecht *et al.* (1990) recently described a pervasive trend in ecogeographic size variation in adult Malagasy lemurs such that among sister taxa, the largest forms are found in the central domain and progressively smaller-bodied sister taxa are located respectively in the east, west, north and south (see also Godfrey *et al.*, 1990). They suggest that differences in resource productivity among ecogeographic regions are the potential causes of this pattern, with the southern dry forests supposedly containing the poorest food productivity and the central plateau having the highest productivity (at least

Evolutionary Change and Heterochrony. Edited by Kenneth J. McNamara © 1995 The Editor and Contributors. Published in 1995 by John Wiley & Sons Ltd.

until the recent past). We believe that, while correctly documenting the pattern of ecogeographic size variation, Albrecht *et al.* (1990) conflate two issues in linking body size to resource productivity: the relationship between body size and habitat quality, and the relationship between body size and resource seasonality.

Folivorous Malagasy sifakas of the genus *Propithecus* provide a test of models regarding ecogeographic size variation as *P. diadema* (diademed sifaka: 5837 g) is located in the moist eastern rainforest, *P. verreauxi* (western sifaka: 3545 g) is found in the dry northwestern and semiarid southwestern forests, and *P. tattersalli* (golden-crowned sifaka: 3493 g) inhabits the dry northern forest. The purpose of this chapter is to detail patterns of sifaka cranial *and* postcranial ontogeny so as to investigate models of ecogeographic size variation among Malagasy lemurs from an allometric and heterochronic perspective. In a broader sense, our ultimate goal is to demonstrate the importance of ecological *and* life history data for a better understanding of patterns of phyletic size change among mammalian sister taxa (Gould, 1975, 1977; Shea, 1983, 1988; Leigh, 1994).

Samples

In this analysis the sifaka cranial ontogenetic samples consist of 73 crania of the western sifaka, *Propithecus verreauxi* (42 adults, 31 non-adults), and 38 crania of the diademed sifaka, *P. diadema* (22 adults, 16 non-adults). All subspecies are represented in the museum samples for both species. The only cranial specimen of the golden-crowned sifaka, *P. tattersalli*, that of the adult holotype, is also included in bivariate comparisons. All crania were measured by M.J.R.

For between-species comparisons of sifaka craniofacial development, the data are grouped into six dental age classes (1 = infant; 6 = adult). In *P. verreauxi* there are at least five skulls per non-adult dental age classes 1–5. The *P. diadema* sample has at least four specimens per dental age classes 1, 4 and 5, whereas dental age classes 2 and 3 have only one case apiece.

The postcranial ontogenetic sample of *P. diadema* consists of 21 individuals (13 adults, 8 non-adults) trapped and measured in the Ranomafana rainforest of southeast Madagascar by K.E.G. from 1987 to 1989. The ontogenetic sample of *P. tattersalli* consists of 40 individuals (18 adults, 22 non-adults) trapped and measured in the Daraina dry forest of northeast Madagascar by D.M.M. in 1989 and 1990. Adult data for *P. verreauxi* consist of 15 adults trapped and measured in the Ankijabe dry forest of northwest Madagascar by K.E.G. in 1982 and 1984, and seven adults trapped and measured in the Bevola semiarid forest of southwest Madagascar by K.E.G. in 1984. Body weight data on neonate *P. tattersalli* and *P. verreauxi* at the Duke University Primate Center are also included in this study.

For between-species comparisons of sifaka somatic development, the data are grouped into six age classes based on half-year or yearly intervals (as well as data availability): 1 = 0–6 months (infant); 2 = 6 months–1.5 years; 3 = 1.5–2.5

years; 4 = 2.5–4 years; 5 = 4–5 years; 6 = 5+ years (adult). Chronological age for animals in age classes 1–3 was often estimated directly or easily determined due to the highly seasonal nature of sifaka birth patterns. For older animals in age classes 4–6, the chronological age of several adults was estimated from tooth wear (Glander *et al.*, 1992). In *P. tattersalli* there are at least four cases per age classes 1–5. The *P. diadema* sample has at least four cases per age classes 1 and 5, only two cases for age classes 2 and 4, and three cases in age class 3.

Measurements

Craniometric data were recorded with digital calipers accurate to 0.1 mm. A total of 19 linear measures were taken on each museum specimen. Measures from bilateral structures such as ramus height were taken on the right side of the skull. Body weight, postcanine toothrow length and neural volume were obtained only among adults. Neural volume was calculated by filling the braincase with barley at the foramen magnum and then pouring the barley into a graduated cylinder (millilitres).

Postcranial data are from individuals sedated and trapped in the field. Body weights were taken by suspending infants from a 1 kg portable spring scale and adults from a 20 kg portable spring scale, and recorded to within 10 g (Glander *et al.*, 1991, 1992). In addition to body weight, eight linear and three circumferential measures were taken on each individual; these data were recorded in millimetres with a 3 metre metal tape (Glander *et al.*, 1991, 1992). Three linear measures were derived from the original data: body length, arm length and leg length. Upper arm and thigh circumference were not collected for *P. verreauxi*.

Statistical analyses

Within each of the ontogenetic series, least-squares bivariate regression ($p < 0.05$) was applied to log-transformed metric data to describe allometric growth trajectories. In order to reduce the effects of disproportionately large numbers of adults on the slopes of the regression lines, the data are averaged by dental age for cranial analyses and age class for postcranial analyses. Basicranial length is used as the independent variable in all bivariate comparisons of cranial scaling patterns, whereas body weight is used as the independent variable in most postcranial scaling comparisons. Analysis of covariance (ANCOVA, $p < 0.05$) is used to test for differences in patterns of relative growth between regression lines derived for each growth series, namely whether sifakas are ontogenetically scaled.

As ontogenetic series are unavailable for *P. tattersalli* in cranial comparisons and *P. verreauxi* in postcranial comparisons, an alternative method is used to test for ontogenetic scaling among sister taxa. If the relative growth trajectory for one taxon intersects the adult data scatter for another taxon, then it is most parsimonious to infer that these species are ontogenetically scaled. However, if the adult data fall entirely above or below the regression line, then it is inferred that morphological differences between adults of each taxon do not result from the differential extension of common relative growth patterns (Ravosa, 1992; Ravosa *et al.*, 1993).

T-tests ($p < 0.05$) are used to test for species differences in the absolute size of cranial dimensions at common dental ages and postcranial dimensions at common age classes. Given a pervasive pattern of ontogenetic scaling among sifakas, if at common ages non-adult means for sifaka measures are significantly different, then this indicates that species differences in adult form develop via acceleration ("rate hypermorphosis" of Shea, 1983, 1988), with the larger-bodied diademed sifaka growing at a faster rate, but not for a longer time period, than the two smaller-bodied taxa. On the other hand, if species differences are noted only between adult sifakas, then this indicates that morphological differences are attained via hypermorphosis ("time hypermorphosis" of Shea, 1983, 1988), with diademed sifakas growing for a longer duration, but not faster, than western and golden-crowned sifakas. It is of importance to note that phyletic size differentiation can evolve via the heterochronic processes of both acceleration and hypermorphosis.

ONTOGENETIC SCALING IN SIFAKAS

Cranial ontogenetic regression lines for diademed and western sifakas are highly correlated and highly significant. In 15 of 18 between-species comparisons, ANCOVAs of cranial growth trajectories for *P. diadema* and *P. verreauxi* are not significantly different (Table 13.1). Generally, the results indicate that morphological variation in adult skull form between diademed and western sifakas is due to the differential extension of common patterns of relative growth (Figure 13.1). As the single specimen of *P. tattersalli*, an adult male, plots in the data scatter for adult *P. verreauxi* and lies on the regression line for both diademed and western sifakas, ontogenetic scaling of cranial proportions can also be inferred for the golden-crowned sifaka (Table 13.1; Figure 13.1) (Ravosa, 1992). Only three measures show allometric differences among sifakas—upper palate breadth, interorbital breadth and ramus height (Table 13.1).

Postcranial ontogenetic regression lines for diademed and golden-crowned sifakas are highly correlated and highly significant. In 17 of 19 between-taxon comparisons, ANCOVAs of postcranial growth patterns for *P. diadema* and *P. tattersalli* are not significantly different (Table 13.2). When the adult data for *P. verreauxi* are plotted with the two growth series, 12 of 14 comparisons likewise cluster along common scaling trajectories (Figure 13.2) (Ravosa *et al.*, 1993). Among all three species, the two cases that do not indicate ontogenetic scaling are hindfoot length and tail length versus body weight (Table 13.2).

In sum, *Propithecus diadema*, *P. tattersalli* and *P. verreauxi* share common patterns of relative growth in the limbs and trunk, as well as in the skull and jaws

Table 13.1 Between-species ANCOVAs for *P. diadema* and *P. verreauxi* cranial growth series and tests of positional differences for all three sifakas.

Variable	vs Basicranial length[a–e]
Interorbital breadth	*D < T < V
Ramus height	*D < T < V
Upper palate breadth	*D > T > V
Outer biorbital breadth	NS
Palate length	NS
Lower skull length	NS
Anterior face length	NS
Bizygomatic breadth	NS
M2 bite point length	NS
Masseter lever arm length	NS
Temporalis lever arm length	NS
Bigonial breadth	NS
Bicondylar breadth	NS
Bicoronoid breadth	NS
Symphysis height	NS
Symphysis width	NS
M2 mandibular corpus height	NS
M2 mandibular corpus width	NS

[a] ANCOVAs are based on regression lines which are highly significant ($p < 0.05$) and mostly highly correlated ($r > 0.90$).
[b] * = significant Y-intercept difference between sifaka least-squares regression lines ($p < 0.05$) and the position of adult *P. tattersalli* with respect to both scaling lines.
[c] NS = no significant difference between sifaka regression lines and no difference between the position of adult *P. tattersalli* with respect to both regression lines.
[d] D > T > V = *P. diadema* regression line is transposed above adult *P. tattersalli*, which is in turn transposed above the *P. verreauxi* ontogenetic trajectory.
[e] D < T < V = *P. verreauxi* regression line is transposed above adult *P. tattersalli*, which is in turn transposed above the *P. diadema* scaling trajectory.

(Tables 13.1, 13.2; Figures 13.1, 13.2). Differences in adult cranial and postcranial proportions for diademed sifakas, golden-crowned sifakas and western sifakas occur primarily via ontogenetic scaling (Ravosa, 1992; Ravosa *et al.*, 1993). This evolutionary pattern of body-size enlargement via ontogenetic scaling (Gould, 1975) is termed peramorphosis or phyletic gigantism (Gould, 1977). As sifakas share similar relative growth patterns in the cranium and postcranium, it is most parsimonious to assume that selection for rapid body-size change has occurred with little alteration of body proportions apart from those expected due to ontogenetic scaling (Gould, 1975, 1977; Shea, 1983, 1988).

BODY-SIZE DIFFERENTIATION IN SIFAKAS

T-tests between cranial measures for adult *P. diadema* and adult *P. verreauxi* are significant in 20 of 22 comparisons (Table 13.3), with diademed sifakas being

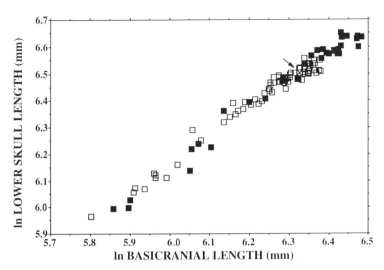

Figure 13.1   A plot of ln lower skull length length (mm) versus ln basicranial length (mm). Note that the growth series for *P. verreauxi* and *P. diadema* are coincident, such that variations in skull form are due to the differential extension of shared allometric patterns. While there is only one available adult cranial specimen of *P. tattersalli*, it plots with the adult data for *P. verreauxi* and therefore all three taxa appear to be ontogenetically scaled. This pattern of body-size differentiation typifies the vast majority of sifaka cranial scaling comparisons. ⊠, *P. tattersalli*; ■, *P. diadema*; □, *P. verreauxi*.

larger in size than western sifakas. This pattern is demonstrated to a lesser extent at dental age class 5 (subadult), where 9 of 19 cases are significantly different. At dental age class 4, only one T-test is significant, although 100% of the *P. diadema* means exceed the mean dimensions for *P. verreauxi*. While similar comparisons are not possible at dental age classes 2 and 3, results indicate that at dental age class 1 (infant), there are no significant differences between cranial measures for *P. diadema* and *P. verreauxi* (Ravosa, 1992).

T-tests for limb, trunk and body weight dimensions between adult *P. diadema* and adult *P. tattersalli* are significant in all 15 cases (Table 13.4). Diademed sifakas are consistently larger in size than golden-crowned sifakas. This pattern is also indicated at age class 5, where 14 of 15 comparisons are significantly different, and at age class 4, where 11 of 15 T-tests are significant. Likewise, this is demonstrated to a lesser extent at age class 3, where nine of 15 comparisons are significantly different. At age class 2, only three significant differences are noted between sifaka postcranial measures. Although not always statistically significant at age class 2, 100% of the means for *P. diadema* measurements exceed the means for *P. tattersalli* measures. While similar analyses are not possible for most measures at age class 1, based on body weights, *P. diadema*, *P. tattersalli* and *P. verreauxi* all have similarly sized neonates ranging from 84 to 155 g (Ravosa *et al.*, 1993).

Allometric differences among sifakas develop mainly via heterochronic

Table 13.2   Between-species ANCOVAs for *P. diadema* and *P. tattersalli* postcranial growth series and tests of positional differences for all three sifakas.

Variable	vs Body weight[a–g]
Tail length	**D < T < V
Hindfoot length	*D > T > V
Body length	NS
Tail–crown length	NS
Chest circumference	NS
Forelimb length	NS
Forefoot length	NS
Arm length	NS
Thumb length	NS
Upper arm circumference	NS
Hindlimb length	NS
Leg length	NS
Big toe length	NS
Thigh circumference	NS
Forelimb length vs hindlimb length	NS
Arm length vs leg length	NS
Upper arm circumference vs thigh circumference	NS
Upper arm circumference vs arm length	NS
Thigh circumference vs leg length	NS

[a] ANCOVAs are based on regression lines which are highly significant ($p < 0.05$) and mostly highly correlated ($r > 0.90$).
[b] ** = significant slope and Y-intercept differences between sifaka least-squares regression lines ($p < 0.05$).
[c] * = significant Y-intercept difference between sifaka least-squares regression lines ($p < 0.05$) and the position of adult *P. tattersalli* with respect to both scaling lines.
[d] NS = no significant difference between sifaka regression lines and no difference between the position of adult *P. verreauxi* with respect to both regression lines.
[e] D > T > V = *P. diadema* regression line is transposed above the *P. tattersalli* ontogenetic trajectory, which is in turn transposed above the adult *P. verreauxi* data.
[f] D < T < V = *P. verreauxi* adult scatter is transposed above the *P. tattersalli* regression line, which is in turn transposed above the *P. diadema* scaling trajectory.
[g] No data are available for analyses of upper arm or thigh circumference in *P. verreauxi*.

changes in growth rate, with *P. diadema* attaining larger adult body size than *P. verreauxi* and *P. tattersalli* mainly by growing at a higher or faster growth rate (Tables 13.3, 13.4). The growth pattern for *P. diadema* follows the prediction for acceleration (Shea, 1983, 1988) in that significant differences in the mean size of cranial and postcranial measures become expressed with greater frequency in later ontogenetic stages (Ravosa, 1992; Ravosa *et al.*, 1993).

This is evident in the patterns of body weight increase for *P. diadema* and *P. tattersalli*. As neonates, both diademed and golden-crowned sifakas are similar in size. At one year of age, juvenile *P. diadema* weigh on average 2950 g, which amounts to 51% of the adult mean of 5837 g. At one year of age, juvenile *P. tattersalli* weigh on average 2384 g, which amounts to 68% of the adult mean of 3493 g. Thus, whereas a lesser percentage of the adult body weight is attained by

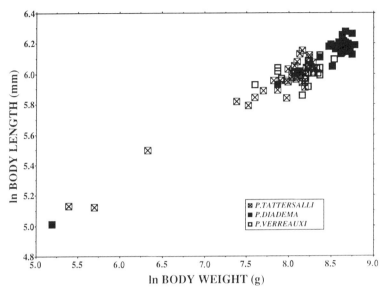

Figure 13.2   A plot of ln body length (mm) versus ln body weight (g). As in most of the sifaka postcranial comparisons, the ontogenetic scaling trajectories for *P. diadema* and *P. tattersalli* are similar, and intersect the adult data for *P. verreauxi*. This further indicates a pervasive pattern of ontogenetic scaling of sifaka body proportions. ⊠, *P. tattersalli*; ■, *P. diadema*; □, *P. verreauxi*.

52 weeks in diademed sifakas, this amounts to an absolutely greater increase in body weight because *P. diadema* weighs more and gains more weight in a year than *P. tattersalli*.

Another reason why acceleration appears to be more important than hypermorphosis in the development of sifaka body-size differentiation is that sexual maturity is reached at approximately the same age of four to five years old in populations of *P. diadema*, *P. tattersalli* and *P. verreauxi* (Richard *et al.*, 1991; Ravosa, 1992; Meyers and Wright, 1993; Ravosa *et al.*, 1993). Moreover, species differences in adult size (dental age and age class 6) are not especially marked when compared with species differences in subadult size (dental age and age class 5). In sum, the larger-bodied *P. diadema* grows at a faster annual rate than smaller-bodied *P. tattersalli* and *P. verreauxi*, but matures at a similar chronological age.

ALLOMETRY, HETEROCHRONY, ECOLOGY AND EVOLUTION IN SIFAKAS

Using ecological data as well as morphological data on the pattern of size differentiation in sifakas, we test two alternative models regarding the evolution

Table 13.3 *P. diadema* and *P. verreauxi* cranial T-tests at each dental age.

Variable	*P.d.* vs *P.v.*[a–d]
Bizygomatic breadth	1,4,**5,6**
Outer biorbital breadth	1,4,**5,6**
Upper palate breadth	1,4,**5,6**
Lower skull length	1,4,**5,6**
Basicranial length	1,4,**5,6**
M2 bite point length	1,4,**5,6**
Symphysis height	1,4,**5,6**
Symphysis width	1,4,**5,6**
M2 Mandibular corpus width	1,4,**5,6**
Interorbital breadth	1,4,**5,6**
Palate length	1,4,**5,6**
Anterior face length	1,4,**5,6**
Masseter lever arm length	1,4,**5,6**
Temporalis lever arm length	1,4,**5,6**
Bigonial breadth	1,4,**5,6**
Bicondylar breadth	1,4,**5,6**
Bicoronoid breadth	1,4,**5,6**
Body weight	**6**
Neural volume	**6**
Postcanine toothrow length	**6**
M2 Mandibular corpus height	1,4,**5,6**
Ramus height	1,4,**5,6**

[a] Dental age classes 1–6: cases where *P. diadema* measures are significantly larger than *P. verreauxi* are noted in **bold** ($p < 0.05$); otherwise T-tests are not significant.
[b] Variables are ranked in ascending order by the dental age at which species differences in **bold** develop.
[c] Not enough *P. diadema* data are available for T-tests at dental age classes 2 and 3.
[d] By dental age class 4, though not always statistically significant, 100% of the means for *P. diadema* measures exceed the means for *P. verreauxi* measures; this pattern of body-size enlargement is maintained into adulthood.

of ecogeographic variation in congeneric Malagasy lemurs: (1) adult body size is set by habitat or forage quality, and (2) adult body size is set by dry-season constraints on food quality and distribution. Ultimately, sifaka postnatal ontogeny is examined so as to discern the relative effects of these selective factors.

The first model is based on a well-documented relationship between body size and forage quality in mammalian herbivores. Larger sister taxa tend to ingest greater percentages of lower-quality forage, which is facilitated by the energetic and digestive benefits of larger body size. In particular, larger mammals have: longer gut passage time due to an absolutely larger digestive tract allowing for greater absorption of nutrients from an otherwise lower-quality diet (Janis, 1976; Chivers and Hladik, 1984; Demment and van Soest, 1985; Sailer *et al.*, 1985; McNaughton and Georgiadis, 1986), energetically more efficient locomotion (Schmidt-Nielsen, 1972, 1984), and relatively lower metabolic demands (Kleiber, 1961; Schmidt-Nielsen, 1972, 1984). Therefore, among close relatives, larger-bodied forms should be associated with poorer-quality habitats.

Table 13.4.   *P. diadema* and *P. tattersalli* postcranial T-tests at each age class.

Variable	*P.d.* vs *P.t.*[a–d]
Hindfoot length	**2,3,4,5,6**
Forefoot length	**2,3,4,5,6**
Hindlimb length	**2,3,4,5,6**
Body weight	**2,3,4,5,6**
Leg length	**2,3,4,5,6**
Forelimb length	**2,3,4,5,6**
Thumb length	**2,3,4,5,6**
Big toe length	**2,3,4,5,6**
Thigh circumference	**2,3,4,5,6**
Body length	2,**3,4,5,6**
Upper arm circumference	2,**3,4,5,6**
Tail-crown length	2,**3,4,5,6**
Arm length	2,**3,4,5,6**
Chest circumference	2,**3,4,5,6**
Tail length	2,3,4,**5,6**

[a] Age classes 2–6: cases where *P. diadema* measures are significantly larger than *P. tattersalli* are noted in **bold** ($p < 0.05$); otherwise T-tests are not significant.
[b] Variables are ranked in ascending order by the age class at which species differences in **bold** develop.
[c] Not enough data spanning the entire range of age class 1 are available for T-tests; based on body weight data for all three species, there are no significant differences among sifakas at age class 1.
[d] By age class 2, though not always statistically significant, 100% of the means for *P. diadema* measures exceed the means for *P. tattersalli* measures; this pattern of body-size enlargement is maintained into adulthood.

The second model is based on a study of several primate communities (Terborgh and van Schaik, 1987), most important of which is information from New World monkeys (Terborgh, 1983). Terborgh and van Schaik (1987) suggest that resource seasonality places energetic constraints on adult body size. They note that in seasonal environments the dry season can be a critical period characterised by relatively smaller patch size and lower patch quality. This resource seasonality in turn imposes strong selective pressures for smaller adult body size, such that smaller-bodied sister taxa should be associated with more seasonal habitats.

To test the first model of ecogeographic size variation in sifakas we examined a general measure of folivore habitat quality. This habitat-quality index, which is the ratio of protein to fibre in samples of mature leaves, has been a useful predictor of folivorous anthropoid biomass (Waterman *et al.*, 1988; Oates *et al.*, 1990) and has been tested recently for Malagasy forests (Ganzhorn, 1992). Primate folivore biomass in Madagascar shows a positive relationship with the protein-to-fibre ratio of mature leaves and most Malagasy folivores select for high-ratio foods (Ganzhorn, 1992). Western forests have higher protein-to-fibre ratios than eastern rainforests and correspondingly higher primate folivore biomass (Ganzhorn, 1992), indicating that western Malagasy forests have a higher-quality habitat for folivores such as sifakas. Interestingly, as the eastern rainforest has a lower carrying capacity for lemurs, rainfall is negatively correlated with lemur folivore biomass (Ganzhorn, 1992). Most importantly, as predicted by the first model, the larger-bodied *P. diadema* is found in the poorer-quality eastern rainforest (Ravosa *et al.*, 1993).

Similar support exists for the second model regarding the effects of resource seasonality as a constraint on adult body size (Terborgh and van Schaik, 1987) in sifakas. Seasonality in Madagascar can be described roughly as a "dry" season with low availability of high-protein immature leaves and insects with perhaps a high abundance of some fruits (Overdorff, 1991) and a "wet" season with high availability of immature leaves and insects (Hladik, 1980; Hladik *et al.*, 1980; Meyers and Wright, 1993). In contrast to the eastern region, the west, north and south have greater seasonal scarcity of rainfall (Griffiths and Ranaivoson, 1972) and presumably greater seasonality in food resources (Meyers and Wright, 1993), fluctuating from very good in the wet season (Ganzhorn, 1992) to very poor in the dry season. For instance, a comparative study of resource tracking shows that the eastern rainforest of *P. diadema* has a more even distribution of food over the course of a year than the more seasonal dry forest of northeastern *P. tattersalli* (Meyers and Wright, 1993) and western *P. verreauxi* by inference (Richard, 1978). As progressively smaller-bodied adult sifakas are located in the east, northwest, northeast and southwest (Albrecht *et al.*, 1990; Ravosa, 1992; Ravosa *et al.*, 1993), this supports the second model suggesting that adult body size is set by the duration and severity of the dry season.

The two models defined previously have more general implications for understanding ecogeographic size variation in Malagasy lemurs. Because all primates derive the majority of protein from either leaves or insects (Hladik *et al.*, 1980; Kay, 1984; Richard, 1985) and many herbivorous insects are subject to the same plant defensive strategies as mammals, e.g., fibre digestibility and secondary defensive compounds (Coley, 1982), one possible explanation for the broad pattern of ecogeographic size variation in Malagasy lemurs is that leaf quality directly influences the gross availability of protein to primary *and*, in turn, secondary consumers. Therefore, factors associated with both models presumably affect *all* Malagasy primates more or less irrespective of dietary preference.

As both models predict the observed patterns of body-size variation in sifakas, it is not possible to distinguish between either scenario based solely on adult data. Moreover, given that phyletic size change via ontogenetic scaling is a relatively simple evolutionary process of modifying development in response to selection for differing adult sizes (Gould, 1975, 1977; Shea, 1983, 1988), both ecological models are consistent with a pervasive pattern of ontogenetic scaling in sifaka cranial (Ravosa, 1992) and postcranial (Ravosa *et al.*, 1993) morphology.

Interestingly, heterochronic or developmental data on growth rate provide a unique perspective with which to address the ecological bases of body-size variation in sifakas. For example, the depth of the seasonal trough in food availability is a major determinant of juvenile starvation or mortality risk

(Janson and van Schaik, 1993). If the first model regarding forage quality and body size is the dominant factor influencing body-size variation, then *P. diadema* should grow at a slower yearly rate than *P. tattersalli* and *P. verreauxi* due to the overall poorer-quality habitat of eastern Madagascar. If the second model regarding resource seasonality and body size is the dominant selective factor, *P. diadema* should grow faster, on average, due to the dampened resource oscillations of the eastern rainforest.

As noted previously, results indicate that eastern *P. diadema* grows at a faster annual rate than *P. verreauxi* (Ravosa, 1992) and *P. tattersalli* (Ravosa *et al.*, 1993). This pattern supports the predictions of the second model regarding the effects of resource seasonality as a limit on adult body size (Terborgh and van Schaik, 1987).

Although a more precise determination of the relative importance of each model is possible only with comparative growth data collected at a finer time scale, such data are available for ringtailed lemurs (*Lemur catta*). Pereira (1993) demonstrates the presence of innate seasonal reductions in *L. catta* growth rate corresponding to the dry season of the southern forest. This seasonal reduction in growth rate illustrates one strategy to reduce juvenile mortality risk for lemurs inhabiting highly seasonal environments, thus highlighting the importance of seasonal reductions in resource availability as a constraint on growth rate (Terborgh and van Schaik, 1987). In fact, smaller adult body size could be a correlate of seasonally reduced growth rates if the period of growth from birth to adulthood is similar among sister taxa (due to constraints on the duration of ontogeny). This pattern apparently characterises sifakas, as heterochronic changes in growth rate are the primary means of body-size differentiation (Ravosa, 1992; Ravosa *et al.*, 1993).

With this in mind, it is beneficial to reconsider the ontogenetic patterns of size differentiation in sifakas. In terms of the second model, both *P. tattersalli* and *P. verreauxi* peak growth rates should be elevated in the higher-quality northeastern and western deciduous forests during the wet season, while growth rates for *P. diadema* should be relatively higher in the austral winter due to the dampened seasonal oscillations of the east. Moreover, it is possible that eastern lemurs might attain a higher annual growth rate than their sister taxa despite a lower peak growth rate in the wet season, if their dry season growth rate is significantly higher than that for northeastern and western congeners. The faster annual growth rates for *P. diadema* than for *P. tattersalli* (Ravosa *et al.*, 1993) and *P. verreauxi* (Ravosa, 1992) appear consistent with this latter interpretation.

IMPLICATIONS FOR ECOGEOGRAPHIC VARIATION IN MALAGASY LEMURS

This study confirms that all three species of *Propithecus* fit the ecogeographic size pattern noted for living and extinct lemurs (Albrecht *et al.*, 1990). Body-size differences among Malagasy primates vary systematically such that extant sister taxa are increasingly larger in the south, north, west and east. Interestingly, sifakas share common patterns of relative growth, which is much as expected if variation in adult size were the target of selection. In addition, allometric differentiation in adult body size develops via heterochronic changes in growth rate, such that the largest taxon (*P. diadema*) grows faster, but not for a longer period, than the two smaller taxa (*P. verreauxi* and *P. tattersalli*).

Both of these ontogenetic patterns have implications for models regarding ecogeographic size variation in Malagasy lemurs. For the largest sifaka, *P. diadema*, a successful evolutionary strategy of coping with the lower-quality forage of the eastern rainforests has been to increase adult body size, thus reducing relative metabolic and dietary demands. The higher annual growth rate of *P. diadema* is facilitated by dampened seasonal oscillations of the east. Conversely, the smaller adult sizes of western *P. verreauxi* and northeastern *P. tattersalli* result from increased resource seasonality, which in turn imposes nutritional and energetic constraints on overall growth rates. Due to the widespread pattern of geographic size variation among Malagasy primates, it is likely that similar ecological *and* developmental factors have played a significant role in the evolution of extant and extinct sister taxa. As such, our current understanding of the evolution of ecogeographic size variation in mammals would indeed profit greatly from further study of the network of interrelationships between ecology and development.

CONCLUSIONS

Body-size differences among Malagasy lemurs vary geographically such that extant sister taxa are increasingly larger in the south, north, west and east. In addition to fitting this ecogeographic size pattern, all three species of Malagasy sifakas (*Propithecus*) share common patterns of relative growth. This allometric differentiation in adult body size develops via heterochronic changes in growth rate, with larger-bodied diademed sifakas growing faster. When the morphometric data are considered vis-a-vis ecological data, two patterns emerge. Apparently for *P. diadema* a successful evolutionary strategy for coping with the lower-quality forage of the eastern rainforests of Madagascar has been to increase adult body size, thus reducing relative metabolic and dietary demands. The higher annual growth rate of *P. diadema* is facilitated by dampened seasonal oscillations of the east. Conversely, smaller adult sizes of western *P. verreauxi* and northeastern *P. tattersalli* result from increased resource seasonality, which in turn imposes nutritional and energetic constraints on overall growth rates. Taken a step further, we argue that similar ecological *and* developmental factors have likely played a prominent role in the evolution of ecogeographic size variation in extant and extinct Malagasy primates.

ACKNOWLEDGMENTS

We thank the Government of the Democratic Republic of Madagascar, specifically the Ministry of Higher Education (MINESUP), the Ministry of Animal Production and Water and Forests (MPAEF), and the Direction of Water and Forests (DEF, SPEF—Antsiranana) for permission to collect morphometric data on sifakas. For access to sifaka cranial collections, we thank M. Rutzmoser (Harvard Museum of Comparative Zoology); G. Musser, W. Fuchs, I. Tattersall, S. Anderson (American Museum of Natural History); B. Patterson, L. Heaney, J. Kerbis, W. Stanley, J. Phelps (Field Museum of Natural History); R. Thorington, L. Gordon (National Museum of Natural History); P. Jenkins, M. Sheridan, M. Sheldrick (Natural History Museum, London); J. Roche, M. Tranier, F. Petter, J. Cuisin (Muséum National d'Histoire Naturelle); C. Smeenk, M. Hoogmoed, D. Reider (Rijksmuseum van Natuurlijke Historie); R. Angermann (Museum fur Naturkunde—Humboldt Universität); B. Latimer, L. Jellema, L. Linden (Cleveland Museum of Natural History); F. Sibley, M. Turner, D. Hodgins (Yale Peabody Museum of Natural History); C. Grigson (Odontological Museum—Royal College of Surgeons); M. Coombs (Pratt Museum of Natural History); and B. Engesser, F. Weidenmayer (Naturhistorisches Museum Basel). Financial support for this research was provided by the National Institutes of Health (DE-05595) to MJR, the National Science Foundation (INT-8602286) to KEG, the National Geographic Society (3980-88) to KEG, the Duke University Research Council to KEG, the Douracouli Foundation to DMM, the World Wildlife Fund (US-4523) to DMM, and Mr and Mrs R. Meyers. B. Demes, J. Ganzhorn, L. Godfrey, K. McNamara, A. Richard, C. van Schaik, and B. Shea are thanked for many helpful comments. The aid of E. Fox, R. Absher, J. Drake, D. Haring, Mr. Bienvenue, B. Manantsoa and D. Garrell is also greatly appreciated. This is Duke University Primate Center publication no. 605.

REFERENCES

Albrecht, G.H., Jenkins, P.D. and Godfrey, L.R., 1990, Ecogeographic size variation among the living and subfossil prosimians of Madagascar, *Am. J. Primatol.*, 22: 1–50.
Chivers, D.J. and Hladik, C.M., 1984, Diet and gut morphology in primates. In D.J. Chivers, B.A. Wood and A. Bilsborough (eds), *Food acquisition and processing in primates*, Plenum Press, New York: 213–230.
Coley, P.D., 1982, Rates of herbivory on different tropical trees. In E.G. Leigh, A.S. Rand and D.M. Windsor (eds), *The ecology of a tropical forest*, Smithsonian, Washington, DC: 123–132.
Demment, M.W. and van Soest, P.J., 1985, A nutritional explanation for body-size patterns of ruminant and non-ruminant herbivores, *Am. Nat.*, 125: 641–647.
Ganzhorn, J.U., 1992, Leaf chemistry and the biomass of folivorous primates in tropical forests: test of a hypothesis, *Oecologia*, 91: 540–547.
Glander, K.E., Fedigan, L.M., Fedigan, L. and Chapman, C., 1991, Field methods for capture and measurement of three monkey species in Costa Rica, *Folia Primatol.*, 57: 70–82.
Glander, K.E., Wright, P.C., Daniels, P.S. and Merenlender, A.M., 1992, Morphometrics and testicle size of rain forest lemur species from southeastern Madagascar, *J. Human Evol.*, 22: 1–17.
Godfrey, L.R., Sutherland, M.R., Petto, A.J. and Boy, D.S., 1990, Size, space and adaptation in some subfossil lemurs from Madagascar, *Am. J. Phys. Anthropol.*, 81: 45–66.
Gould, S.J., 1975, Allometry in primates, with emphasis on scaling and the evolution of the

brain. In F.S. Szalay (ed.), *Approaches to primate paleobiology. Contributions to Primatology*, Volume 5, Karger, Basel: 244–292.

Gould, S.J., 1977, *Ontogeny and phylogeny*, Harvard University Press, Cambridge.

Griffiths, J.F. and Ranaivoson, R., 1972. Madagascar. In J.F. Griffiths (ed.), *World survey of climatology*, Volume 10, Junk, Den Haag: 87–144.

Hladik, A., 1980, The dry forest of the west coast of Madagascar: climate, phenology, and food available for prosimians. In P. Charles-Dominique, H.M. Cooper, A. Hladik, C.M. Hladik, E. Pages, G.F. Pariente, A. Petter-Rousseaux, J.J. Petter and A. Schilling (eds), *Nocturnal Malagasy primates. Ecology, physiology, and behavior*, Academic Press, New York: 3–40.

Hladik, C.M., Charles-Dominique, P. and Petter, J.J., 1980, Feeding strategies of five nocturnal prosimians in the dry forest of the west coast of Madagascar. In P. Charles-Dominique, H.M. Cooper, A. Hladik, C.M. Hladik, E. Pages, G.F. Pariente, A. Petter-Rousseaux, J.J. Petter and A. Schilling (eds), *Nocturnal Malagasy primates. Ecology, physiology, and behavior*, Academic Press, New York: 41–73.

Janis, C., 1976, The evolutionary strategy of the Equidae and the origins of rumen and caecal digestion, *Evolution*, **30**: 757–774.

Janson, C.H. and van Schaik, C.P., 1993, Ecological risk aversion in juvenile primates: slow and steady wins the race. In M.E. Pereira and L.A. Fairbanks (eds), *Juvenile primates: life history, development and behavior*, Oxford University Press, Oxford: 57–74.

Kay, R.F., 1984, On the use of anatomical features to infer foraging behavior in extinct primates. In P.S. Rodman and J.G.H. Cant (eds), *Adaptations for foraging in nonhuman primates. Contributions to an organismal biology of prosimians, monkeys, and apes*, Columbia University Press, New York: 21–53.

Kleiber, N., 1961, *The fire of life*, Wiley, New York.

Leigh, S.R., 1994, Ontogenetic correlates of diet in anthropoid primates, *Am. J. Phys. Anthropol.*, **94**: 499–522.

McNaughton, S.J. and Georgiadis, N.J., 1986, Ecology of African grazing and browsing mammals, *Ann. Rev. Ecol. Syst.*, **17**: 39–65.

Meyers, D.M. and Wright, P.C., 1993, Resource tracking: food availability and *Propithecus* seasonal reproduction. In P.M. Kappeler and J.U. Ganzhorn (eds), *Lemur social systems and their ecological basis*, Plenum Press, New York:179–192.

Oates, J.F., Whitesides, G.H., Davies, A.G., Waterman, P.G., Green, S.M., Dasilva, G.L. and Mole, S., 1990, Determinants of variation in tropical forest primate biomass: new evidence from West Africa, *Ecol.*, **71**: 328–343.

Overdorff, D.J., 1991, Ecological correlates of social structure in two prosimian primates: *Eulemur fulvus rufus* and *Eulemur rubriventer*. Ph.D. Thesis, Duke University.

Pereira, M.E., 1993, Seasonal adjustment of growth rate and adult body weight in ringtailed lemurs. In P.M. Kappeler and J.U. Ganzhorn (eds), *Lemur social systems and their ecological basis*, Plenum Press, New York: 205–221.

Ravosa, M.J., 1992, Allometry and heterochrony in extant and extinct Malagasy primates, *J. Human Evol.*, **23**: 197–217.

Ravosa, M.J., Meyers, D.M. and Glander, K.E., 1993, Relative growth of the limbs and trunk in sifakas: heterochronic, ecological, and functional considerations, *Am. J. Phys. Anthropol.*, **92**: 499–520.

Richard, A.F., 1978, *Behavioral variation: case study of a Malagasy lemur*, Bucknell, Lewisburg.

Richard, A.F., 1985, *Primates in nature*, Freeman, New York.

Richard, A.F. and Dewar, R.E., 1991, Lemur ecology, *Ann. Rev. Ecol. Syst.*, **22**: 145–175.

Richard, A.F., Rakotomanga, P. and Schwartz, M., 1991, Demography of *Propithecus verreauxi* at Beza Mahafaly, Madagascar: sex ratio, survival, and fertility, 1984–1988, *Am. J. Phys. Anthropol.*, **84**: 307–322.

Sailer, L.D., Gaulin, S.J.C., Boster, J.S. and Kurland, J.A., 1985, Measuring the relationship between dietary quality and body size in primates, *Primates*, **26**: 14–27.

Schmidt-Nielsen, K., 1972, Locomotion: energy costs of swimming, flying and running, *Science*, **177**: 222–228.

Schmidt-Nielsen, K., 1984, *Scaling. Why is animal size so important?*, Cambridge University Press, Cambridge.

Shea, B.T., 1983, Allometry and heterochrony in the African apes, *Am. J. Phys. Anthropol.*, **62**: 275–289.

Shea, B.T., 1988, Heterochrony in primates. In M.L. McKinney (ed.), *Heterochrony in evolution: a multidisciplinary approach*, Plenum Press, New York: 237–266.

Tattersall, I., 1992, Systematic versus ecological diversity: the example of the Malagasy primates. In N. Eldredge (ed.), *Systematics, ecology and the biodiversity crisis*, Columbia University Press, New York: 25–39.

Terborgh, J., 1983, *Five new world primates. A study in comparative ethology*, Princeton University Press, Princeton.

Terborgh, J. and van Schaik, C.P., 1987, Convergence vs. nonconvergence in primate communities. In J.H.R. Gee and P.S. Giller (eds), *Organization of communities past and present*, Blackwell, Oxford: 205–226.

Van Valen, L., 1973, Festschrift, *Science*, **180**: 488.

Waterman, P.G., Ross, J.A.M., Bennett, E.L. and Davies, A.G., 1988, A comparison of the floristics and the leaf chemistry of the tree flora in two Malaysian rain forests and the influence of leaf chemistry on populations of colobine monkeys in the Old World, *Biol. J. Linn. Soc.*, **34**: 1–32.

INDEX

Text entries in roman; figure entries in **bold**; table entries in *italic*.